Lattice Point Identities and Shannon-Type Sampling

Monographs and Research Notes in Mathematics

Series Editors:

John A. Burns, Thomas J. Tucker, Miklos Bona, Michael Ruzhansky

About the Series

This series is designed to capture new developments and summarize what is known over the entire field of mathematics, both pure and applied. It will include a broad range of monographs and research notes on current and developing topics that will appeal to academics, graduate students, and practitioners. Interdisciplinary books appealing not only to the mathematical community, but also to engineers, physicists, and computer scientists are encouraged.

This series will maintain the highest editorial standards, publishing well-developed monographs as well as research notes on new topics that are final, but not yet refined into a formal monograph. The notes are meant to be a rapid means of publication for current material where the style of exposition reflects a developing topic.

Analytic Methods for Coagulation-Fragmentation Models, Volume II
Jacek Banasiak, Wilson Lamb, Philippe Laurencot

Neutrices and External Numbers
A Flexible Number System
Dinis Bruno, Imme van den Berg

Extending Structures
Fundamentals and Applications
Ana Agore, Gigel Militaru

Inverse Scattering Problems and their Application to Non-Linear Integrable Equations
Pham Loi Vu

Spectral Methods Using Multivariate Polynomials On The Unit Ball
Kendall Atkinson, David Chien, and Olaf Hansen

Glider Representations
Frederik Caenepeel, Fred Van Oystaeyen

Lattice Point Identities and Shannon-Type Sampling
Willi Freeden, M. Zuhair Nashed

For more information about this series please visit: https://www.crcpress.com/Chapman--HallCRC-Monographs-and-Research-Notes-in-Mathematics/book-series/CRCMONRESNOT

Lattice Point Identities and Shannon-Type Sampling

Willi Freeden
M. Zuhair Nashed

CRC Press
Taylor & Francis Group
Boca Raton London New York

CRC Press is an imprint of the
Taylor & Francis Group, an **informa** business
A CHAPMAN & HALL BOOK

CRC Press
Taylor & Francis Group
6000 Broken Sound Parkway NW, Suite 300
Boca Raton, FL 33487-2742

Printed on acid-free paper

International Standard Book Number-13: 978-0-367-37563-8 (Hardback)

Visit the Taylor & Francis Web site at
http://www.taylorandfrancis.com

and the CRC Press Web site at
http://www.crcpress.com

Contents

Preface

In engineering and science, one is usually not confronted with continuous signals but only with discrete versions obtained by measurements. The classical Shannon theory of sampling aims at the following issue: Given discrete samples at lattice points, how can we capture the essential information about the underlying continuous signal?

The algebraic structure relates Shannon sampling to the theory of lattice points. Therefore it is canonical to characterize the interrelations between lattice points and sampling theory in more detail and, subsequently, to arrange a Shannon-type sampling theory based on corresponding lattice point concepts. In fact, our work demonstrates that Shannon-type sampling exists as a bandlimited and – to some extent – also as a non-bandlimited lattice point variant. It can be associated with newly created types of Hardy–Landau lattice point identities of analytic number theory which, in the language of sampling, give a deepened insight, e.g., into over- and undersampling and the representation of the aliasing error.

The specific objective for the book was to write a contribution that would be a companion to the work of W. Freeden, *Metaharmonic Lattice Point Theory* (CRC Press, Taylor & Francis, Boca Raton, 2011). The final outcome of our work, presented under the title *Lattice Point Identities and Shannon-Type Sampling*, is not a textbook; it is a research monograph. Notably, the research intention of the authors from the beginning was to fill up a scientific niche, where it was envisioned that the book should provide a comprehensive coverage of application of lattice point theory to sampling and reconstruction/representation of signals, particularly on geoscientifically relevant regions in Euclidean spaces [such as a ball, an ellipsoid, the geoid (cf. the cover illustration), the real Earth, etc.] In fact, our approach shows a tremendous potential for geoscientific applications. This undoubtly is the reason why the book, besides the number-theoretically reflected and sampling-related fields of research, can be used in an advanced graduate course for geoscience-oriented content. It is applicable in inverse problems, e.g., downward continuation problems of satellite technology and downward calculation problems in exploration geophysics, such as gravimetry and magnetometry using lattice point grids on regions as data systems.

The first obligation of this research monograph is to work out some preparatory Shannon-type relevant aspects of lattice point summation. We start with a multi-dimensional generalization of the Euler summation formula by suitably interpreting its constituents, i.e., the classical "Bernoulli polynomials" as Euler–Green functions relative to arbitrary lattices Λ and specifically chosen operators L, namely the second-order differential operators $L = \Delta + 4\pi i a \cdot \nabla$, $a \in \mathbb{R}^q$, with Δ the Laplace operator and ∇ the gradient operator in $\mathbb{R}^q, q \geq 1$. The generalization of the classical Euler summation formula is obtained by the conversion of a "wave number" λ-dependent Helmholtz-type $(L + \lambda)$-differential equation into an associated integral equation. In doing so, we are able to compare weighted sums of functional values for a prescribed system of lattice points in a regular, e.g., "potato-like" region $\mathcal{G} \subset \mathbb{R}^q$ with $\mathcal{G}$-bandlimited Fourier transform integrals, plus a remainder term reflecting the wave number–dependent $\mathcal{G}$-frequency band in $\mathbb{R}^q$. The remainder term is particularly useful for two applicabilities of multivariate lattice point theory, viz. to investigate the convergence of multi-dimensional alternating series, such as number-theoretic Hardy–Landau-type series by specifying the wave number λ appropriately to the oscillation properties of the summand and to characterize multivariate alternating Shannon-type cardinal series for the explicit description of under- and oversampling obligations.

The second obligation of this research book can be regarded as the operator $(L+\lambda)$-induced verification of weighted Hardy–Landau-type lattice point identities. This part of the book shows a close relationship to the analytic and geometric theory of numbers in Euclidean spaces (e.g., Gaussian circle problem, Minkowski's theorem), essentially connected with methods and procedures of mathematical physics. A particular result of the Hardy–Landau-type lattice point identities is that the number of weighted lattice points both inside a regular region $\mathcal{G}$ and (under certain conditions on the weight functions at infinity) inside the whole Euclidean space $\mathbb{R}^q$ may be expressed as a sum extended over Fourier transform integrals resulting in bandlimited and non-bandlimited approaches, respectively. More concretely, variants of the Euler and Poisson summation formulas are applied successively to balls with increasing radii in order to develop certain limit criteria imposed on the functions generating the Fourier transform integrals for the purpose of sampling. All in all, it turns out that, for certain choices of wave numbers $\lambda \in \mathbb{R}$, $(L + \lambda)$-resulting Poisson-type summation formulas exist for both finite (regular) regions $\mathcal{G} \subset \mathbb{R}^q$ and the (infinite) Euclidean space $\mathbb{R}^q$, so that bandlimited and non-bandlimited aspects of Shannon-type sampling are natural consequences of a multivariate integration over the associated Poisson-type summation formulas. An important task for the Shannon-type framework is to derive explicit expressions of over- and undersampling, thereby specifying not only the occurrence, but also the type of aliasing in a thorough mathematical realization.

The third obligation of this work is to explain significant aspects of constructive approximation in the reproducing kernel Paley–Wiener Hilbert space context resulting from the multivariate bandlimited Shannon-type sampling theorem corresponding to regular regions $\mathcal{G} \subset \mathbb{R}^q$. Minimum norm (spline) interpolation and/or smoothing are applied, e.g., to the multivariate antenna problem.

The roots of this book are fivefold: (i) the basic results due to L.J. Mordell [1928a,b, 1929] on one-dimensional Euler and Poisson summation formulas, (ii) the work by C. Müller [1954a,b,c] on two-dimensional periodic Euler–Green functions with respect to the Laplacian Δ and their representation in the framework of complex analysis, (iii) the work of W. Freeden [1975], W. Freeden [1978a,b] on multi-dimensional generalizations of the Euler summation formula to $(\Delta + \lambda)$-elliptic partial differential operators and $(\Delta + \lambda)$-adapted multi-dimensional Poisson summation formulas over regular, i.e., "potato-like" regions and their application in number theory (W. Freeden [2011]) and cubature (W. Freeden, M. Gutting [2017]), (iv) the work on multi-dimensional bandlimited lattice point Shannon-type sampling (W. Freeden, M.Z. Nashed [2015, 2017a]), and (v) the work of M.Z. Nashed [1981] on the one-dimensional antenna problem and its relationship to today's inverse problems (see also W. Freeden, M.Z. Nashed [2018a,b]).

Willi Freeden and M. Zuhair Nashed
August 2019, Kaiserslautern and Orlando

Authors

Willi Freeden received his Philosophicum in 1970, diploma in mathematics in 1971, Staatsexamen in mathematics and geography in 1972, PhD in mathematics in 1975, and Habilitation in mathematics in 1979, all from RWTH Aachen University, Germany. He has served for many years as a professor at the RWTH Aachen University and the University of Kaiserslautern. He has held various visiting professor positions, including the Ohio State University, Columbus (Department of Geodetic Science and Surveying). He is recipient of the RWTH Borchers Award, the Eurasian Association on Inverse Problems (EAIP) Award, and the Fellowship of the International Association of Geodesy (IAG). Since 1996, he has been a member of the German Geodetic Commission of the Bavarian Academy of Sciences, Munich. He held the position of Vice President for Research and Technology at the University of Kaiserslautern from 2002–2006. He is author, editor, and coeditor of 20 books, and has published more than 225 papers, several expository papers, and book chapters. He is founding editor-in-chief of Springer's *GEM International Journal on Geomathematics*, editor-in-chief of Springer's *Handbook of Geomathematics*, editor-in-chief of the (German) Springer-Spektrum *Handbuch Tiefe Geothermie* and *Handbuch Oberflächennahe Geothermie*, editor-in-chief of the Birkhäuser book series, *Geosystems Mathematics*, editor-in-chief of the Birkhäuser lecture notes, *Geosystems Mathematics and Computing*, 2015 editor-in-chief of the second edition of Springer's *Handbook of Geomathematics*, 2018 editor-in-chief of Birkhäuser's *Handbook of Mathematical Geodesy*, editor-in-chief of the Springer-Spektrum *Handbuch der Geodäsie*. He is member of the editorial board of a large number of international journals. He is the organizer of several Oberwolfach conferences, mini-symposia, and special sessions at meetings of the American Mathematical Society.

M. Zuhair Nashed received his SB and SM degrees in electrical engineering from MIT and his PhD in mathematics from the University of Michigan. He has served for many years as a professor at Georgia Tech and the University of Delaware and has held visiting professor positions at the University of Michigan, University of Wisconsin, AUB, and KFUPM. He has held distinguished visiting scholar positions at many universities around the world. He joined the University of Central Florida as Chair of the Department of Mathematics (2002 - 2006). Dr. Nashed is a fellow of the American Mathematical Society (Inaugural Class of 2013) and the recipient of the Lester Ford Award of the Mathematical Association of America, the Sigma Xi Faculty Research Award, and Sustained Research Award in Science from Georgia Tech, Dr. Zakir Husain Award of the Indian Society of Industrial and Applied Mathematics, as well as several other international awards. He has published over 140 papers in mathematics and 30 papers in applied sciences, physics, and engineering, has written 30 expository papers and book chapters, and authored and edited 15 books. He is editor-in-chief of *Numerical Functional Analysis and Optimization*, executive editor of *Sampling Theory in Signal and Image Processing*, founding and past coeditor of the *Journal of Integral Equations*, and a member of the editorial board of 36 journals. He is also editor-in-chief of Springer's *Handbook of Geomathematics*, editor-in-chief of the Birkhäuser book series, *Geosystems Mathematics* and of Birkhäuser's *Lecture Notes on Geosystems Mathematics and Computing*, and editor-in-chief of *Handbook of Mathematical Geodesy*. He has served for 40 years as executive editor of the book series, *Pure and Applied Mathematics: A Program of Monographs, Textbooks, and Lecture Notes*, published by Marcel Dekker, Inc., and later by Chapman & Hall/CRC. He is editor-in-chief of the book series, *Contemporary Mathematics and Its Applications*, launched recently by World Scientific. He gave an invited hour address at a meeting of the American Mathematical Society, three plenary lectures at meetings of the Mathematical Association of America and plenary lectures at meetings of the French, Tunisian, and Lebanese Mathematical Societies; Indian Society of Industrial and Applied Mathematics; and Japan Society of Mechanical Engineers. He has given over 400 plenary and invited talks at conferences and colloquia. He has organized over 40 conferences, mini-symposia, and special sessions.

Acknowledgments

The authors would like to thank CRC for giving them the opportunity to publish this research monograph. Particular thanks go to Sunil Nair, Sarfraz Khan, and Callum Fraser for their obligingness and cooperation.

The authors also express their gratitude to Teena Lawrence, Nova Techset, Chennai, India, who had been assigned to oversee the production of our book. Furthermore, the authors thank Varun Gopal, Editorial Assistant, CRC Press, for his assistance to solve some LaTex problems.

The cover illustration includes a model of the geoid (i.e., the equipotential surface of the Earth, at sea level). The "geoidal potato" constitutes a typical geophysically-relevant regular region as discussed in this work. We are obliged to Dr. Roger Haagmans, Head of Solid Earth Unit Mission Science Division, Earth Observations Department of Science, Applications and Future Technologies European Space Agency, Noordwijk, The Netherlands, for providing us with the image (ESA ID number SEMLXEOA90).

Part I

Central Theme

Chapter 1

From Lattice Point to Shannon-Type Sampling Identities

In the year 1765, J.L. Lagrange proposed a sampling theorem for bandlimited periodic functions by stating that the knowledge of functional values at $2n+1$ equidistant points within a period is sufficient to represent uniquely a periodic function by sine and cosine terms. A.L. Cauchy [1841] found an interpolation formula for a bandlimited function, i.e., a function representable by a complex Fourier expansion showing no more than a certain number of coefficients. These essential results may be seen as the prehistory of sampling.

1.1 Classical Framework of Shannon Sampling

Following P.L. Butzer and R.L. Stens [1992], the history of today's understanding of the sampling theory can be traced back to the interpolation theory using equidistant nodes published by the Belgian mathematician Charles–Jean Baron de la Vallée Poussin (1866-1962). He probably was the first person to consider sampling for not necessarily bandlimited functions already in 1908. In fact, Ch.–J. de la Vallée Poussin [1908] treated the particular case of non-bandlimited (more concretely, spacelimited) functions, a class of functions that actually represent signals occurring often in engineering practice (note that these signals cannot be bandlimited simultaneously due to the uncertainty principle). The German mathematician M. Theis [1919] continued de la Vallée Poussin's work in a considerable way.

Early in one-dimensional theory, variants of bandlimited lattice point sampling were given by E.T. Whittaker [1915], E.T. Whittaker [1923], E.T. Whittaker [1929], and K. Ogura [1920]. In later years, it became known that the classical sampling theorem had been presented before C.E. Shannon [1949a,b] to the Russian communication community by V.A. Kotel'nikov [1933]. In a more implicit verbal form, it had also been described in the German literature by H. Raabe [1939] (for more details concerning the history see H.D. Lüke [1999]). As a consequence of the discovery of the several independent introductions of the sampling theorem, people started to refer to the theorem by including the names of the aforementioned authors, resulting in such phrases as the Whittaker-Kotel'nikov-Shannon (WKS) sampling theorem.

The most common form of sampling today is lattice point sampling of a band-limited signal $F^{\wedge}_{\mathcal{F}_\Lambda}$ given by

$$F^{\wedge}_{\mathcal{F}_\Lambda}(t) = \int_{\mathcal{F}_\Lambda} F(x)\, e^{-2\pi i x\cdot t}\, dx,\ t \in \mathbb{R}, \tag{1.1}$$

over the fundamental cell $\mathcal{F}_\Lambda = [-\frac{\sigma}{2}, \frac{\sigma}{2})$, $\sigma > 0$, of the lattice $\Lambda = \sigma\mathbb{Z}$ for a square-integrable function F on $\mathcal{F}_\Lambda$. Already in the first half of the last century, H. Nyquist [1924, 1928] was led to the following 1D-statement: If a 1D-function (signal) $F^{\wedge}_{\mathcal{F}_\Lambda}$ (dependent on the time variable t) contains no frequencies higher than $\frac{\sigma}{2}$, it is completely determined by its functional values at a series of points spaced $\frac{1}{\sigma}$ apart. Hence, a sufficient rate amounts to σ samples, or anything larger. In fact, a more detailed study in the sampling area due to C.E. Shannon [1949a,b] tells us that, for a given *sample rate* $\tau = \frac{1}{T}$, perfect reconstruction is guaranteed for any *bandwidth* $\sigma \leq \tau$. If the bandwidth σ is too high, the reconstruction exhibits imperfections known as *aliasing*. The two thresholds, T and τ, respectively, are called the *Nyquist rate* and the *Nyquist frequency*.

There are different ways to the *1D-Shannon sampling theorem* (based, e.g., on tools already presented by G.H. Hardy [1941]). Continuing the nomenclature used in this work, we are able to explain the sampling theorem as follows: Let $F^{\wedge}_{\mathcal{F}_\Lambda}$ be the Fourier transform $F^{\wedge}_{\mathcal{F}_\Lambda}$ given by (1.1) for some $F \in \mathrm{L}^2([-\frac{\sigma}{2}, \frac{\sigma}{2}))$ extended Λ-periodically to the real line $\mathbb{R}$. We expand the extension in a series

$$F(x) = \sum_{h\in\Lambda^{-1}} c_h\ \Phi_h(x), \tag{1.2}$$

where the system $\{\Phi_h\}_{h\in\Lambda^{-1}}$ of 1D periodic polynomials is given by

$$\Phi_h(x) = \frac{1}{\sqrt{\|\mathcal{F}_\Lambda\|}}\ e(h\cdot x) = \frac{1}{\sqrt{\|\mathcal{F}_\Lambda\|}}\ e^{2\pi i h\cdot x}, \tag{1.3}$$

and $\Lambda^{-1} = \sigma^{-1}\mathbb{Z}$ is the inverse lattice of $\Lambda = \sigma\mathbb{Z}$, $\|\mathcal{F}_\Lambda\| = \sigma$ is the length of the fundamental cell $\mathcal{F}_\Lambda$, and $\Sigma_{h\in\Lambda^{-1}} \ldots$ means that the sum is extended over all h of the inverse lattice Λ^{-1}.

It is not hard to see that

$$\begin{aligned} \int_{\mathcal{F}_\Lambda} \Phi_h(x)\overline{\Phi_{h'}(x)}\, dx &= \int_{\mathcal{F}_\Lambda} \frac{1}{\sqrt{\|\mathcal{F}_\Lambda\|}}\, e^{2\pi i h\cdot x} \frac{1}{\sqrt{\|\mathcal{F}_\Lambda\|}}\, e^{-2\pi i h'\cdot x}\, dx \\ &= \delta_{h,h'} \\ &= \begin{cases} 1 & , \quad h = h' \\ 0 & , \quad h \neq h'. \end{cases} \end{aligned} \tag{1.4}$$

Hence, the system $\{\Phi_h\}_{h\in\Lambda^{-1}}$ is orthonormal with respect to the inner product:

$$\int_{\mathcal{F}_\Lambda} \Phi_h(x)\ \overline{\Phi_{h'}(x)}\ dx = \delta_{h,h'}. \tag{1.5}$$

The Fourier coefficients c_h are given in the form

$$c_h = \int_{\mathcal{F}_\Lambda} F(x)\ \Phi_h(x)\ dx = \frac{1}{\sqrt{\|\mathcal{F}_\Lambda\|}} \int_{\mathcal{F}_\Lambda} F(x)\ e^{2\pi i h\cdot x}\ dx. \tag{1.6}$$

Comparing (1.1) and (1.6) we are led to the coefficients

$$c_h = \frac{1}{\sqrt{\|\mathcal{F}_\Lambda\|}}\ F^{\wedge}_{\mathcal{F}_\Lambda}(h). \tag{1.7}$$

Substituting (1.2) into (1.1) we obtain the identity

$$F^{\wedge}_{\mathcal{F}_\Lambda}(t) = \int_{\mathcal{F}_\Lambda} \sum_{h\in\Lambda^{-1}} c_h \frac{1}{\sqrt{\|\mathcal{F}_\Lambda\|}} e^{2\pi i h\cdot x}\ e^{-2\pi i x\cdot t}\ dx, \quad t\in\mathbb{R}, \tag{1.8}$$

such that by interchanging the order of integration and summation

$$F^{\wedge}_{\mathcal{F}_\Lambda}(t) = \sum_{h\in\Lambda^{-1}} F^{\wedge}_{\mathcal{F}_\Lambda}(h) \frac{1}{\|\mathcal{F}_\Lambda\|} \int_{\mathcal{F}_\Lambda} e^{2\pi i x\cdot(h-t)}\ dx. \tag{1.9}$$

This identity yields the celebrated classical expansion of a bandlimited signal

$$F^{\wedge}_{\mathcal{F}_{\sigma\mathbb{Z}}}(t) = \sum_{h\in\Lambda^{-1}} F^{\wedge}_{\mathcal{F}_{\sigma\mathbb{Z}}}(h)\ \frac{\sin(2\pi(h-t)\frac{\sigma}{2})}{2\pi(h-t)\frac{\sigma}{2}}. \tag{1.10}$$

Equivalently, we find in the standard nomenclature of sampling that

$$F^{\wedge}_{\mathcal{F}_{\sigma\mathbb{Z}}}(t) = \sum_{n\in\mathbb{Z}} F^{\wedge}_{\mathcal{F}_{\sigma\mathbb{Z}}}\left(\frac{n}{\sigma}\right)\ \frac{\sin(\pi\sigma(\frac{n}{\sigma}-t))}{\pi\sigma(\frac{n}{\sigma}-t)}. \tag{1.11}$$

So, in the lattice point terminology used in this work, Shannon's sampling result (1.10) can be rephrased as follows: Any function $F^{\wedge}_{\mathcal{F}_\Lambda}$ bandlimited to the fundamental cell $\mathcal{F}_\Lambda = \left[-\frac{\sigma}{2}, \frac{\sigma}{2}\right)$ of the lattice $\Lambda = \sigma\mathbb{Z}$, i.e., the Fourier transform integral (1.1) may be reconstructed from the sequence of samples

$$\{F^{\wedge}_{\mathcal{F}_\Lambda}(h)\}_{h\in\Lambda^{-1}} \tag{1.12}$$

via the *cardinal series*

$$F^{\wedge}_{\mathcal{F}_\Lambda}(t) = \sum_{h\in\Lambda^{-1}} F^{\wedge}_{\mathcal{F}_\Lambda}(h) \frac{1}{\|\mathcal{F}_\Lambda\|} \int_{\mathcal{F}_\Lambda} e^{2\pi i x\cdot(h-t)}\ dx, \quad t\in\mathbb{R}, \tag{1.13}$$

where the series is absolutely and uniformly convergent on any compact set of the real line $\mathbb{R}$. Obviously, the Fourier transform $F^{\wedge}_{\mathcal{F}_\Lambda}$ (considered as a

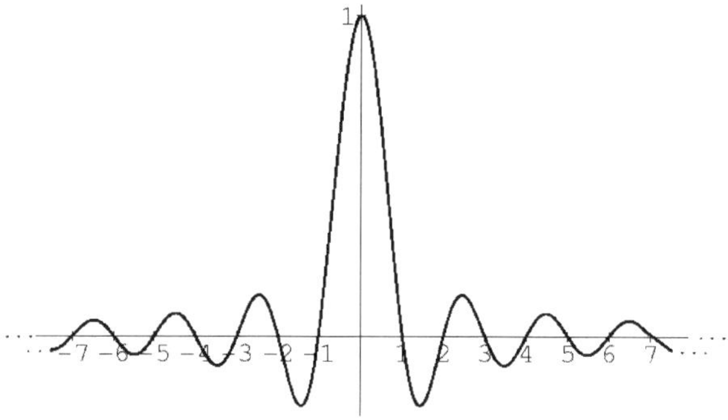

FIGURE 1.1: Illustration of the function $t \mapsto \operatorname{sinc}(\pi t) = \frac{\sin(\pi t)}{\pi t}$, $t \in \mathbb{R}$.

1D signal) is an infinitely repeated replication of the samples at points h of the inverse lattice Λ^{-1}. The portion of the Fourier transform integral between $-\frac{\sigma}{2}$ and $\frac{\sigma}{2}$, i.e., the fundamental cell $\mathcal{F}_{\sigma\mathbb{Z}}$ of the lattice $\sigma\mathbb{Z}$ is called the base band.

The interpolatory background of bandlimited sampling can be recapitulated briefly as follows: A mathematical way to interpolate samples $F^{\wedge}_{\mathcal{F}_\Lambda}(h'), h' \in \Lambda^{-1}$, $\Lambda^{-1} = \frac{1}{\sigma}\mathbb{Z}$, may be realized by use of sinc-functions (cf. Figure 1.1)

$$t \mapsto \operatorname{sinc}(\pi\sigma\,(h-t)) = \frac{\sin(\pi\sigma(h-t))}{\pi\sigma(h-t)}, \quad t \in \mathbb{R}. \tag{1.14}$$

The sequence

$$\{\operatorname{sinc}(\pi\sigma\,(h-\cdot))\}_{h\in\Lambda^{-1}} \tag{1.15}$$

shows the discrete orthogonality property:

$$\operatorname{sinc}(\pi\sigma(h-h')) = \delta_{h,h'} = \begin{cases} 1, & h = h', \\ 0, & h \neq h'. \end{cases} \tag{1.16}$$

As a consequence, $F^{\wedge}_{\mathcal{F}_\Lambda}$ given by the so-called *cardinal series*

$$F^{\wedge}_{\mathcal{F}_\Lambda}(t) = \sum_{h\in\Lambda^{-1}} F^{\wedge}_{\mathcal{F}_\Lambda}(h)\,\operatorname{sinc}(\pi\sigma\,(h-t)), \quad t \in \mathbb{R}, \tag{1.17}$$

satisfies the interpolatory property

$$F^{\wedge}_{\mathcal{F}_\Lambda}(t)|_{t=h'} = \sum_{h\in\Lambda^{-1}} F^{\wedge}_{\mathcal{F}_\Lambda}(h)\,\operatorname{sinc}(\pi\sigma\,(h-t))|_{t=h'} = F^{\wedge}_{\mathcal{F}_\Lambda}(h'), \quad h' \in \Lambda^{-1}. \tag{1.18}$$

1.2 Transition From Shannon to Shannon-Type Sampling

It should be noted that the classical proof of the Shannon sampling theorem as recapitulated in Section 1.1 is indeed rigorous. The interchange of integration and summation from (1.7) to (1.9) can be easily justified. However, the proof is very revealing (as already pointed out by M.Z. Nashed [2010]): We perform this interchange and a theorem pops up, but what are its essential core and its perspective consequences?

As a matter of fact, many *questions* arise in future-oriented sampling obligations:

- Is there a close interrelationship between the Shannon sampling and the Dirac distributional concept?
- Are there straightforward counterparts and generalizations of the univariate sampling theorem to general lattices not only by 1D iteration but also in a specifically multivariate framework?
- Is a specifically multivariate Shannon sampling variant for general lattices also valid for a class of bandlimited sample functions related not only to the base band determined by the fundamental cell of the lattice, but also to general geometric bands, such as certain "potato-like" (i.e., regular) regions $\mathcal{G}$ in Euclidean space $\mathbb{R}^q$?
- How can we describe resulting procedures of over- and undersampling in bandlimited Shannon sampling related to $\mathcal{G}$? How can we specify the explicit aliasing error between the cardinal series and the (Fourier transformed) functional value in bandlimited sampling?
- Are there certain aspects in multivariate bandlimited Shannon sampling for general lattices that are applicable to a class of not necessarily bandlimited functions?
- How can we characterize the asymptotic properties of the functions at infinity reflecting non-bandlimited sample manifestations and explicit representations of the aliasing error?
- Is there a particular way to organize a variant of Shannon sampling for locally supported functions? What are the resulting imperfections of the locally supported sampling expressions?

These questions will be discussed and answered in this research book by choosing a non-standard and novel approach to sampling involving specific tools of lattice point theory called *Shannon-type sampling.*

1.3 Novel Framework of Shannon-Type Sampling

Conceptually written in univariate shorthand formulas over $\mathbb{Z}$ (note that the proofs of the univariate theory can be found in Chapter 4), the transition from the 1D-Shannon sampling identity to the 1D-Shannon-type sampling identities for a (suitably given) function (signal) F on an interval $[c,d] \subset \mathbb{R}, c < d$ leads to the following specification:

The *classical univariate approach* (CUA) is well-known and characterized by the following identities:

(CUA 1) *Euler Summation Formula:*

$$\begin{aligned} \sum_{\substack{g\in[c,d]\\ g\in\mathbb{Z}}}{}' F(g) &= \int_c^d F(x)\, dx \\ &\quad + \int_c^d G(\Delta;x)\, \Delta F(x)\, dx \\ &\quad + \{F(x)\, \nabla G(\Delta;x)\}\,\big|_c^d \\ &\quad - \{G(\Delta;x)\ (\nabla F(x))\}\,\big|_c^d, \end{aligned} \tag{1.19}$$

where $G(\Delta;\cdot)$ is apart from a factor the Bernoulli function B_2 of degree 2

$$G(\Delta;\cdot) = -\frac{1}{2}B_2(x) = -\frac{1}{2}\,(x-\lfloor x\rfloor)^2 - (x-\lfloor x\rfloor) + \frac{1}{6},\ x\in\mathbb{R} \tag{1.20}$$

(note that $\lfloor x\rfloor = \text{floor}(x)$) and

$$\sum_{\substack{g\in[c,d]\\ g\in\mathbb{Z}}}{}' F(g) = \sum_{\substack{g\in(c,d)\\ g\in\mathbb{Z}}} F(g) + \frac{1}{2}\sum_{\substack{g=c,d\\ g\in\mathbb{Z}}} F(g). \tag{1.21}$$

(CUA 2) *Hardy–Landau Lattice Point Identity:*

$$\sum_{\substack{g\in[c,d]\\ g\in\mathbb{Z}}}{}' 1 = (d-c) + \sum_{\substack{h\neq 0\\ h\in\mathbb{Z}}} \frac{1}{2\pi i h}\left(e^{2\pi i h\cdot d} - e^{2\pi i h\cdot c}\right). \tag{1.22}$$

(CUA 2a) *Special Hardy–Landau Lattice Point Identity:* c>0

$$\sum_{\substack{g\in[-\frac{c}{2},\frac{c}{2})\\ g\in\mathbb{Z}}}{}' 1 = c + \sum_{\substack{h\neq 0\\ h\in\mathbb{Z}}} \frac{\sin(\pi h c)}{\pi h}. \tag{1.23}$$

(CUA 3) *Poisson Summation Formula:*

$$\sum_{\substack{g\in[c,d]\\ g\in\mathbb{Z}}}{}' F(g) = \lim_{N\to\infty}\sum_{\substack{|h|\le N\\ h\in\mathbb{Z}}}\int_c^d F(x)\, e^{2\pi ih\cdot x}\, dx. \tag{1.24}$$

(CUA 4) *Shannon Sampling Formula:* $y\in\mathbb{R}$

$$F^{\wedge}_{\mathcal{F}_{\mathbb{Z}}}(y) = \sum_{h\in\mathbb{Z}} F^{\wedge}_{\mathcal{F}_{\mathbb{Z}}}(h)\,\frac{\sin(\pi(h-y))}{\pi(h-y)}. \tag{1.25}$$

The *novel univariate approach* (NUA) is characterized by the following identities:

(NUA 1) *Euler-Type Summation Formula for L:*

$$\begin{aligned}\sum_{\substack{g\in[c,d]\\ g\in\mathbb{Z}}}{}' e^{2\pi ia\cdot g}F(g) &= \sum_{\substack{L^\wedge(h)=0\\ h\in\mathbb{Z}}}\int_{[c,d]} e^{2\pi ia\cdot x}F(x)\, e^{2\pi ih\cdot x}\, dx\\ &+ \int_{[c,d]} G(L;x)\,(\overline{L}+\lambda)\, e^{2\pi ia\cdot x}F(x)\, dx\\ &+ \{e^{2\pi ia\cdot x}F(x)\,(\nabla_x+4\pi ia)\,G(L;x)\}\big|_c^d\\ &- \{G(L;x)\nabla_x e^{2\pi ia\cdot x}F(x)\}\big|_c^d,\end{aligned} \tag{1.26}$$

where

$$L = \Delta + 4\pi ia\cdot\nabla,\ a\in\mathbb{R}, \tag{1.27}$$

$$L^\wedge(h) = 4\pi^2((h+a)^2-a^2),\ h\in\mathbb{Z}, \tag{1.28}$$

and $G(L;\cdot)$ is the Euler–Green function (or $\mathbb{Z}$-lattice function) for the operator L (cf. Definition 4.2).

(NUA 2) *Hardy–Landau-Type Identity:* $y\in\mathbb{R}$

$$\sum_{\substack{a+g\in[c,d]\\ g\in\mathbb{Z}}}{}' e^{2\pi iy\cdot(a+g)} = \lim_{N\to\infty}\sum_{\substack{|h-y|\le N\\ h\in\mathbb{Z}}} e^{2\pi ia\cdot h}\int_c^d e^{2\pi i(h-y)\cdot x}\, dx. \tag{1.29}$$

(NUA 3) *Poisson-Type Summation Formula:* $y\in\mathbb{R}$

$$\begin{aligned}&\sum_{\substack{a+g\in[c,d]\\ g\in\mathbb{Z}}}{}' e^{2\pi iy\cdot(a+g)}F(a+g)\\ &= \lim_{N\to\infty}\sum_{\substack{|h-y|\le N\\ h\in\mathbb{Z}}} e^{2\pi ia\cdot h}\int_{\substack{x\in[c,d]\\ x\in\mathbb{R}}} F(x)\, e^{-2\pi ix\cdot(h-y)}\, dx.\end{aligned} \tag{1.30}$$

(NUA 4) *Shannon-Type Sampling Formula:* $y \in \mathbb{R}$, $[u,v] \subset \mathbb{R}$ arbitrary

$$\sum_{\substack{(\mathcal{F}_{\mathbb{Z}}+\{g'\})\cap[u,v]\neq\emptyset \\ g'\in\mathbb{Z}}} F^{\wedge}_{[c,d]\cap \bigcup_{g\in\mathbb{Z}}((([u,v]\cap(\mathcal{F}_{\mathbb{Z}}+\{g'\}))-\{g'\})+\{g\})}(y) \tag{1.31}$$
$$= \sum_{\substack{(\mathcal{F}_{\mathbb{Z}}+\{g'\})\cap[u,v]\neq\emptyset \\ g'\in\mathbb{Z}}} \int_{[c,d]\cap \bigcup_{g\in\Lambda}((([u,v]\cap(\mathcal{F}_{\mathbb{Z}}+\{g'\}))-\{g'\})+\{g\})} F(x)\, e^{-2\pi i y\cdot x}\, dx$$
$$= \sum_{h\in\mathbb{Z}} \underbrace{\int_{[c,d]} F(x)\, e^{-2\pi i h\cdot x}\, dx}_{=F^{\wedge}_{[c,d]}(h)} \underbrace{\int_{[u,v]} e^{2\pi i x\cdot(h-y)}\, dx}_{=K_{[u,v]}(h-y)}.$$

(NUA 4a) *Special Shannon-Type Sampling Formula:* $y \in \mathbb{R}$, $[c,d] \subset [u,v] \subset \overline{\mathcal{F}_{\mathbb{Z}}}$

$$F^{\wedge}_{[c,d]}(y) = \sum_{h\in\mathbb{Z}} F^{\wedge}_{[c,d]}(h)\;\; K_{[u,v]}(h-y), \tag{1.32}$$

(NUA 4b) *Special Shannon-Type Sampling Formula:* $y \in \mathbb{R}$, $[u,v] \subset [c,d] \subset \overline{\mathcal{F}_{\mathbb{Z}}}$
,

$$F^{\wedge}_{[u,v]}(y) = \sum_{h\in\mathbb{Z}} F^{\wedge}_{[c,d]}(h)\; K_{[u,v]}(h-y). \tag{1.33}$$

(NUA 4c) *Shannon Sampling Formula:* $y \in \mathbb{R}$, $[u,v] = [c,d] = \overline{\mathcal{F}_{\mathbb{Z}}}$,

$$F^{\wedge}_{\mathcal{F}_{\mathbb{Z}}}(y) = \sum_{h\in\mathbb{Z}} F^{\wedge}_{\mathcal{F}_{\mathbb{Z}}}(h) \underbrace{\int_{[-\frac{1}{2},\frac{1}{2}]} e^{2\pi i x\cdot(h-y)}\, dx}_{=K_{\mathcal{F}_{\mathbb{Z}}}(h-y)} \tag{1.34}$$

with

$$K_{\mathcal{F}_{\mathbb{Z}}}(h-y) = \int_{[-\frac{1}{2},\frac{1}{2}]} e^{2\pi i x\cdot(h-y)}\, dx \;=\; \frac{\sin(\pi(h-y))}{\pi(h-y)}, \quad h\in\mathbb{Z}. \tag{1.35}$$

In other words, Shannon-type sampling restricted to the (closure of the) fundamental cell $\mathcal{F}_{\mathbb{Z}} = [-\frac{1}{2},\frac{1}{2})$ of the lattice $\mathbb{Z}$ leads to ordinary Shannon sampling.

Chapter 2

Obligations, Ingredients, Achievements, and Innovations

2.1 Obligations and Ingredients

The essential obligations of this work are as follows: We first explain the lattice point approach to Shannon sampling for the univariate case. Even in this 1D context, our number theoretically-motivated framework will be able to produce new sampling aspects, for example, in characterizing explicit over- and undersampling Shannon-type identities. Unfortunately, we are confronted with the serious difficulty that the presented 1D context does not allow an immediate generalization to higher dimensions. The reason is that the multivariate counterparts of the 1D-Bernoulli polynomials, also called Euler–Green or lattice functions, constituting the essential ingredients of the Euler summation formula show a much stronger singularity behavior in higher dimensions than in the 1D-theory. Actually, classical 1D-Bernoulli polynomials are lattice periodic Green functions to ordinary 1D-derivatives and the boundary condition of periodicity. They allow explicit elementary representations and are continuous functions with a "finite jump" singularity in lattice points for its derivative. Moreover, their Fourier series is uniformly convergent. Sampling-type purposes as intended here, however, will be based on a different operator, namely the aforementioned L-operator $L = \Delta + 4\pi i a \cdot \nabla$. Nonetheless, in the 1D framework, its handling is straightforward and quite analogous to the Euler–Green function with respect to the Laplace operator Δ, i.e., the second-order derivative. In higher dimensions the situation, however, is completely different. For example, the Euler–Green function with respect to the differential operator L does not imply a uniformly convergent Fourier series as for the univariate theory. Roughly speaking, we have to overcome the following obstacles: The discussion of the dimension $q = 2$, i.e., the bivariate case, is "not far away" from the univariate context, whereas the handling of the higher dimensions $q \geq 3$ becomes more and more critical. As a consequence, lattice point identities in the plane turn out to be attackable (almost) similarly to the 1D case, while the technical procedure becomes more and more problematic for increasing dimensions (as already pointed out for number-theoretic obligations by W. Freeden [2011]). Another difficulty in higher dimensions is the geometric variety. General regions $\mathcal{G}$ admitting the validity of the Gaussian theorem of vector analysis in $\mathbb{R}^q$, i.e., so-called regular regions (cf. the cover illustration), come into play. Furthermore, multivariate lattices are of higher complexity. The most demanding problem, however, is that the $(L+\lambda)$-Euler–Green function of the sampling reflected $(L+\lambda)$-Euler-type summation formula is not available as elementary function in closed form. So, our approach has to use auxiliary tools of mathematical physics to guarantee, for example, the existence of a multivariate $(L+\lambda)$-Euler–Green function. More concretely, we have to generalize the fundamentals of the classical one-dimensional Hilbert theory of Green functions (D. Hilbert [1912], R. Courant, D. Hilbert [1924]) to the multivariate lattice periodic case and the Shannon sampling reflecting iterated Helmholtz-type differential operators $(L+\lambda)^m, m \in \mathbb{N}, \lambda \in \mathbb{R}$. By specifying the defining properties of multivariate periodic Euler–Green function, i.e., boundary condition of periodicity, $(L+\lambda)$-differential equation, $(L+\lambda)$-characteristic singularity, normalization condition, we are able to make essential scientific progress in both overcoming convergence problems for bilinear expansions of Euler–Green functions to iterated operators and formulating Euler-type summation formulas of theoretical significance for sampling. Seen from a structural point of view, every generalization of the Euler summation formula such as the required Euler-type variant involving the differential operator $L+\lambda$ rests on the basic idea to relate a sum of values of a function at finitely many lattice points to finite sums of Fourier transform integrals of this function with respect to certain frequencies dependent on the specific operator $(L+\lambda)$. On the one hand, this makes things complicated because of the occurrence of non-standard $(L+\lambda)$-derivatives in the remainder of the Euler summation formula. On the other hand, it opens a perspective to handle the remainder term explicitly in integral form, so that the non-availability of the Euler–Green function in elementary form does not matter for purposes of sampling.

Actually, in addition to the bandlimited theory of Shannon sampling, our approach based on Euler-type summation additionally attempts to come close to Shannon-type sampling of non-bandlimited functions (note that the non-bandlimited theory is *not* treated in the recently published contribution W. Freeden, M.Z. Nashed [2015], which exclusively deals with aspects of the bandlimited theory).

The essential ingredients of our Shannon-type sampling results are based on a number of auxiliary means and tools, which in principle are well-known, but to some extent unusual and innovative in the context of Shannon-type sampling, such as

- fundamental solutions for the iterated $(L+\lambda)$-operators,
- asymptotic laws for spherical integrals involving the $(L+\lambda)$-Euler–Green function,

- an adequate nomenclature of the theories of spherical harmonics and cylinder functions (Bessel, Kelvin functions, etc.), providing the canonical multivariate counterparts to, e.g., the sinc-function,
- integral transforms, e.g., Gauss–Weierstrass transform, pointwise Fourier inversion formula, Hankel transform involving discontinuous integrals over regular regions,
- functional equation for the Theta function as the basic tool for Gauss–Weierstrass summability,
- asymptotic criteria of the Fourier theory for the formulation of weighted "lattice point and lattice ball discrepancies" as preparation for the explicit study of alternating cardinal series and the phenomena of aliasing in sampling problems,
- reproducing kernel Paley–Wiener Hilbert space framework and spline approximation techniques, such as interpolation, smoothing, best approximation.

2.2 Achievements and Innovative Results

This monograph provides a bundle of far-reaching achievements and a cornucopia of sampling results:

- The transition to lattice point Shannon-type sampling, as presented in this work, provides the fruits of cross-fertilization of various subjects, namely Green's concept of elliptic partial differential theory, Fourier inversion theory, special function theory, constructive approximation involved with Euler and Poisson summation formulas, and aspects of analytic and geometric number theory. All these mathematical disciplines put together apparently disparate features as closely related building blocks of a common core, i.e., lattice point identities, ready to obtain bandlimited as well as non-bandlimited lattice point Shannon-type sampling identities and cardinal series expansions.
- Euler-type and Poisson-type summation formulas associated to iterated Helmholtz-type $(L + \lambda)$-operators in the particular manifestation of Hardy–Landau-type identities do not only provide an alternative to prove multivariate generalizations of lattice point identities (as in the research monograph by W. Freeden [2011] for the operators $\Delta + \lambda$), but also lead to new aspects in Shannon-type sampling and explicit over- as well as undersampling characterizations. An important result worth mentioning in this context is the explicit availability of expressions for aliasing errors.

- The choice of the "wave number" as an adaptive parameter to the oscillation characteristics of the weight function generating the cardinal series is a decisive instrument of our sampling approach involving the (infinite) structure of Euclidean spaces.

- Non-bandlimited Fourier transfoms of certain locally-supported functions result in the astonishing sampling statement that, for arbitrary lattices, each lattice point value inside the local support (in our case always the closure of a given regular region) can be calculated from the other lattice point values of the function inside the local support, under explicit knowledge of the aliasing error.

Evidently, a great number of questions still remain unanswered by our approach. A striking confinement is the restriction to the special Helmholtz-type elliptic differential operator $L + \lambda$. It helps to apply the classical Fredholm theory for (singular) integral equations. Moreover, it avoids distributional characterizations of the Euler–Green function, but it also prevents us, for instance, from incorporating the specific treatment of hyperbolic and/or parabolic differential equations in lattice point theory. In the opinion of the authors, the transition to these types of differential operators demands its own nature which by no means can be studied in parallel wih the elliptic-based theory of this work.

2.3 Methods and Tools

The essential tool of this book is an Euler-type summation formula corresponding to differential operators $L = \Delta + 4\pi i a \cdot \nabla$, $a \in \mathbb{R}^q$, with Δ the Laplace operator and ∇ the gradient operator in $\mathbb{R}^q, q \geq 1$, over y-shifted lattices, $y \in \mathbb{R}^q$. The list of significant topics and innovative results is broad. It enables us to

- interpret "Bernoulli polynomials" as Green's functions to differential operators corresponding to the boundary condition of periodicity,

- develop multivariate Euler-type summation formulas with respect to operators $L + \lambda$ to specify convergence criteria for alternating multi-dimensional series such as the Hardy–Landau-type series and Shannon-type cardinal series,

- deduce Poisson-type summation formulas for regular, i.e., "potato-like" regions, in the sense of Gauss–Weierstrass summability and in an ordinary, i.e., non-averaging sense,

- outline Hardy–Landau-type identities for regular regions in the sense of Gauss–Weierstrass summability and, under certain circumstances, in an ordinary non-averaging sense,
- integrate Poisson-type summation formulas over regular regions to obtain bandlimited Shannon-type sampling theorems,
- investigate important aspects of constructive approximation in Paley–Wiener spaces including Paley–Wiener spline approximations and their application to the multi-dimensional antenna problem,
- devise $(L + \lambda)$-conditions for the validity of multi-dimensional Poisson-type summation formulas in Euclidean spaces by adaption of the wave number λ to the specific properties of the alternating functions, e.g., sinc- and jinc-expressions,
- derive specific $(L+\lambda)$-conditions (at infinity) to guarantee wave number-dependent Poisson-type summation formulas over Euclidean spaces for application to multivariate alternating functions,
- integrate Poisson-type summation formulas over Euclidean spaces to formulate non-bandlimited Shannon-type sampling procedures,
- characterize over- and undersampling in non-bandlimited Shannon-type sampling and to give explicit representations of the aliasing error.

In conclusion, this book demonstrates that significant roots of today's facets of Shannon's sampling theorem rest on number-theoretical results starting from famous contributions of C.F. Gauss [1801, 1826]. In fact, our excursion

- going out from the Gaussian circle problem of the early-nineteenth century
- via the classical Hardy–Landau lattice point identity of the first half of the twentieth century
- and the Shannon sampling theorem of the second half of the twentieth century

has resulted in new multivariate extensions of lattice point identities and Shannon-type sampling procedures of high practical applicability, thereby providing new types of cardinal series, explicit under- and oversampling error expressions, and general reproducing kernel Hilbert space structure of an associated Paley–Wiener theory.

Chapter 3

Layout

The *objectives of this book* are achieved in accordance with the following oganisational plan, which provides a subdivision into ten parts and twenty chapters.

3.1 Structural Organisation

Part I is concerned with the central theme of this book.

- Chapter 1 gives an overview of the univariate Shannon sampling theory and its history. The specific features for choosing a non-standard and novel approach to sampling based methods and tools involving lattice point theory are outlined, and the resulting innovative approach is summarized under the label Shannon–type sampling.
- Chapter 2 gives a survey about the achievement and innovation in lattice point and Shannon-type sampling theory. Furthermore, the layout of the book is presented.
- Chapter 3 presents the organisational structure and the layout of the book.

Part II starts with the univariate sampling theory.

- Chapter 4 relates the classical Shannon sampling theorem to an Euler-type summation formula. An essential tool is the Green function with respect to the special differential operator $L = \Delta + 4\pi i\ a \cdot \nabla + \lambda$, $a \in \mathbb{R}$, $\lambda \in \mathbb{R}$, and the "boundary condition" of Λ-lattice periodicity. Shannon-type sampling including all manifestions of over- and undersampling for arbitrary lattices $\Lambda \subset \mathbb{R}$ is provided for arbitrary finite reference intervals and the whole Euclidean space $\mathbb{R}$ (under certain conditions at infinity for the functions under consideration). The applicability of the results is studied by specifying certain cases of generalized sinc-functions and associated cardinal series expansions.

Part III presents preparatory material that is needed for the multivariate context. Almost all results of Part III are well-known. Therefore we decided to omit most of the proofs but to supply the reader with references instead, in order to keep the book to a manageable size.

- Chapter 5 gives an introduction of the basic nomenclature in Cartesian as well as spherical notation. Basic integral theorems of vector analysis are studied for regular regions (such as the cover illustration of the geoid).
- Chapter 6 provides necessary tools of the theory of special functions in a notation especially suitable for the development and formulation of lattice point and Shannon-type sampling identities in Euclidean spaces $\mathbb{R}^q$.
- Chapter 7 presents helpful material of analytic number theory and the geometry of numbers.
- Chapter 8 is devoted to the Fourier theory for the Λ-lattice periodic case and the Fourier transform in Euclidean spaces $\mathbb{R}^q$. Particular topics include the method of stationary phase, the pointwise Fourier inversion formula (in ordinary sense and in Gauss–Weierstrass summability form), and the Hankel transform of discontinuous integrals (in the sense of the Bessel function theory).

Part IV studies the decisive instrument for the book in order to handle lattice point and resulting Shannon-type identities, viz. the Euler-type summation formula for regular regions $\mathcal{G} \subset \mathbb{R}^q$ and the differential operators $L+\lambda$, $\lambda \in \mathbb{R}$, and general y-shifted lattices $\Lambda + \{y\} \subset \mathbb{R}^q$, $y \in \mathbb{R}^q$.

- Chapter 9 presents an introduction of the Green function with respect to the operator L and the "boundary condition" of Λ-lattice periodicity, which forms the constituting ingredient of the Euler-type summation formula. Subsequently, the Green functions and associated Euler-type summation formulas are investigated for iterated operators L^m, $m \in \mathbb{N}$. The Euler-type summation formulas open the perspective to derive lattice point and Shannon-type identities based on their fundamental structural framework.

Part V shows essential aspects of both Hardy–Landau lattice point and band-limited Shannon-type identities by use of bivariate variants of the Euler-type summation formula as essential tool.

- Chapter 10 provides Λ-lattice point Hardy–Landau identities for the spherical context and constant lattice point weights. Basic Λ-discrepancy asymptotics are recapitulated (leading to the famous Hardy conjuncture).

- Chapter 11 deals with the extension of the Hardy–Landau identities to regular regions $\mathcal{G}$ and general Λ-lattice point weights. It is shown via the bivariate Euler-type summation formula that the validity of weighted Hardy–Landau identites may be verified for regular regions $\mathcal{G}$ with boundaries $\partial\mathcal{G}$, for which the technique of the stationary phase may be applied on their parameter representations.

- Chapter 12 is concerned with the transition from bivariate Hardy-Landau-type identities for general lattice point weights to bandlimited bivariate Shannon–type sampling theorems. The transition is obtained by integrating the Hardy-Landau-type identity over a regular region of the variable a with y fixed, respectively, the variable y with a fixed. In both cases, the aliasing error between the functional value and the cardinal series is explicitly documented for over – and undersampling techniques.

- Chapter 13 generalizes the bivariate Λ-lattice point concept to a Λ-lattice ball concept. As a consequence, the Λ-lattice ball cardinal series shows an improved convergence behavior.

Part VI discusses the equivalence of Euler-type and the Poisson-type summation formula for regular regions $\mathcal{G} \in \mathbb{R}^q$. The main difficulty in this context is the pointwise convergence of the occurring alternating series. A remedy against divergence is Gauss–Weierstrass summability. It is pointed out that the convergence of lattice point identities over regular regions $\mathcal{G}$ in the ordinary pointwise sense is realizable only under additional requirements, mainly under the assumption of vanishing weights on the boundary $\partial\mathcal{G}$, whereas Shannon-type sampling identities over $\mathcal{G}$ obtained by integration show ordinary convergent cardinal series.

- Chapter 14 describes Gauss–Weierstrass averages of the multivariate Euler-type summation formula and first representations of Shannon-type theorems for regular regions $\mathcal{G}$ without any geometric restriction, however, under smoothness assumptions originated by the ordinary Euler-type summation formula in $\mathbb{R}^q$.

- Chapter 15 uses as auxiliary tool the functional equation of the multivariate Theta function in order to develop the Poisson-type summation formula for regular regions and continuous weights via the Fourier inversion formula understood in Gauss–Weierstrass nomenclature.

Part VII demonstrates that two different variants of Shannon-type sampling formulas over regular regions $\mathcal{G}$ can be canonically obtained from the Poisson-type summation formula in $\mathbb{R}^q$ by reflecting functional values–based and Fourier transformed values–based weight functions, respectively, for purposes of sampling.

- Chapter 16 provides a different entry point to Shannon-type sampling over regular regions $\mathcal{G}$. Firstly the cardinal series is constructed in such a way that it sums up functional values of the Fourier transform of the weight function over $\mathcal{G}$ in y-shifted lattice points, secondly the cardinal series sums up functional values of the weight function over $\mathcal{G}$ in a-shifted lattice points.

- Chapter 17 presents multivariate Paley–Wiener space theory. Its reproducing Hilbert space kernel structure is used to develop spline interpolatory sampling. The antenna problem as inverse problem to regular regions $\mathcal{G}$ as reference pattern is discussed in an appropriate spline framework.

Part VIII provides adaptive sufficient conditions to guarantee the convergence of multivariate alternating series and to ensure the validity of the Poisson-type summation formulas over the whole Euclidean spaces $\mathbb{R}^q$.

- Chapter 18 starts from the Euler-type summation formula over increasing balls in $\mathbb{R}^q$. Spherically-reflected convergence criteria are formulated in adaptation to the alternating summands. Hardy–Landau-type lattice point identities are derived from Poisson-type summation formulas in $\mathbb{R}^q$.

Part IX is concerned with multivariate Shannon-type formulas over Euclidean spaces $\mathbb{R}^q$ and a Paley–Wiener space manifestation over $\mathbb{R}^q$.

- Chapter 19 offers the entry points for Shannon-type sampling over Euclidean spaces. Essential tools are the Poisson-type summation formulas in $\mathbb{R}^q$. Cardinal series are investigated over functional values in y-shifted lattice points as well as functional values of the Fourier transform in a-shifted lattice points. In all cases, over- and undersampling errors are explicitly available from the Poisson-type summation formulas. Finally, Shannon-type sampling for locally supported functions is investigated in more detail, and a study of the aliasing error is given.

Part X gives some concluding remarks on sampling and perspective comments about its role in the theory of recovery problems.

- Chapter 20 presents a collection of publications showing that Shannon sampling is the point of departure for many develpments and far-reaching progress in sampling theory. Moreover, it is pointed out that sampling methods play pivotal roles in the recovery of an object from discrete information, where the bridging role of sampling in the theory of recovery problems (i.e., signal analysis, image processing, inverse problems) is an auspicious perspective for the future.

3.2 Relationship to Other Monographs

Finally some remarks should be made: The book is written under the auspices that the essence of Shannon-type sampling theory is its *applicability to other branches in science and engineering*. It is useful not only in number theory, but also in Fourier analysis, constructive approximation, numerical integration, data handling, and so on. Moreover, the ideas, concepts, and structures in lattice point theory strongly influence a variety of areas, such as inverse problems, signal analysis, image analysis, e.g., in geomathematics and mathematical physics. As a consequence, the unusual selection of the material and its presentation in the framework of the $(L + \lambda)$-Euler summation formula makes this book *an addendum to the research monographs on lattice point and sampling theory.*

Part II

Univariate Poisson-Type Summation Formulas and Shannon-Type Sampling

Chapter 4

Euler/Poisson-Type Summation Formulas and Shannon-Type Sampling

Our aim in this preparative chapter is to demonstrate that the 1D Shannon sampling theory may be obtained by variants of (1D) Euler/Poisson summation formulas, namely lattice point identities resulting in so-called Hardy–Landau identities.

For the reason of readability, we first explain the classical Euler summation formula based on the the Euler–Green function $G(\Delta; \cdot)$ relative to the (1D) Laplace operator Δ, i.e., the second-order derivative, in its equivalence to the standard Poisson summation formula.

4.1 Classical Euler Summation Formula

Under the assumption of second-order continuous derivatives of a function F on the interval $[0, n]$, L. Euler (1707-1783) succeeded in finding the *summation formula*

$$\sum_{\substack{0 \leq g \leq n \\ g \in \mathbb{Z}}} F(g) - \frac{1}{2}\left(F(0) + F(n)\right) \tag{4.1}$$

$$= \int_0^n F(x)\, dx + \frac{1}{12}\left(F'(n) - F'(0)\right) + \int_0^n \underbrace{\left(-\frac{1}{2}B_2(x)\right)}_{=G(\Delta;x)} F''(x)\, dx,$$

where B_2 given by

$$B_2(x) = (x - \lfloor x \rfloor)^2 - (x - \lfloor x \rfloor) + \frac{1}{6} \tag{4.2}$$

is known as the *Bernoulli function of degree 2* in today's nomenclature.

Seen from the mathematical point of view, the Euler summation formula (4.1) is a fine illustration of how a generalization can lead to the solution of seemingly independent problems. The particular structure of the summation formula also captures the delicate details of the connection between integration, i.e., continuous summation, and its various discretizations. Obviously, it subsumes and resolves the appropriate bridge between continuous and discrete summation within a single exposition. So, it provides the canonical bridge to modern sampling. At this stage, however, it should be pointed out that L. Euler himself used this interrelation between continuous and discrete sums only for estimating sums and series by virtue of integrals. It was actually C. Maclaurin (1698-1746), who discovered the summation formula (4.1) independently in 1742, to use it for the evaluation of integrals in terms of sums. Hence, C. Maclaurin's understanding is much closer to Shannon sampling in the sense of expressing a function in integral form by a cardinal series taken over discrete functional values at lattice points.

The essential idea of the Euler summation formula as used in our context is different from standard procedures, which can be found in the literature (see, e.g., T.M. Apostol [1999] and the references therein). Our approach is based on the physically motivated interpretation of the Bernoulli function B_2 occurring in the classical (1D) Euler summation formula by means of mathematical physics as the Green function $G(\Delta;\cdot)$ with respect to the (1D) Laplace operator $\Delta = \nabla\cdot\nabla$ corresponding to boundary conditions of $\mathbb{Z}$–periodicity. In doing so, the Bernoulli function, i.e., in the jargon of mathematical physics, the *Euler–Green function*, acts as the connecting tool to convert a differential equation involving the 1D Laplace operator corresponding to periodic boundary conditions into an associated integral equation, i.e., the summation formula. More concretely, the $\mathbb{Z}$-periodical Euler–Green function $G(\Delta;\cdot)$ for the Laplace operator Δ is constructed so as to have the bilinear expansion

$$G(\Delta; x-y) \;=\; \sum_{\substack{\Delta^\wedge(h)\neq 0\\ h\in\mathbb{Z}}} \frac{e^{2\pi i h\cdot x}e^{-2\pi i h\cdot y}}{-\Delta^\wedge(h)}, \qquad x,y\in\mathbb{R}, \tag{4.3}$$

where the sequence $\{\Delta^\wedge(h)\}_{h\in\mathbb{Z}}$ forms the spectrum $\mathrm{Spect}_\Delta(\mathbb{Z})$ of the Laplace operator Δ, i.e., $(\Delta+\Delta^\wedge(h))\; e^{2\pi i h\cdot x} = 0, x\in\mathbb{R}$, so that $\Delta^\wedge(h) = 4\pi^2h^2,\ h\in\mathbb{Z}$. Observing the special values

$$G(\Delta;0) \;=\; G(\Delta;n) \;=\; \sum_{\substack{\Delta^\wedge(h)\neq 0\\ h\in\mathbb{Z}}} \frac{1}{-\Delta^\wedge(h)} \;=\; -\frac{1}{12} \tag{4.4}$$

and the explicit representation (4.2) of the Fourier series expansion (4.3) we are able to reformulate the Euler summation formula (4.1). Integration by parts yields

$$\begin{aligned}
&\sum_{\substack{0\le g\le n\\ g\in\mathbb{Z}}} F(g) - \frac{1}{2}\left(F(0)+F(n)\right) \qquad (4.5)\\
&= \int_0^n F(x)\,dx + \sum_{\substack{h\neq 0\\ h\in\mathbb{Z}}} \left(\frac{\nabla F(n) - \nabla F(0)}{4\pi^2 h^2} - \int_0^n \frac{e^{2\pi i h\cdot x}}{4\pi^2 h^2}\,\Delta F(x)\,dx \right)\\
&= \int_0^n F(x)\,dx - \lim_{N\to\infty} \sum_{\substack{|h|\le N\\ h\neq 0}} \frac{1}{2\pi i h} \int_0^n \nabla F(x)\, e^{2\pi i h\cdot x}\,dx,
\end{aligned}$$

such that the classical *Poisson summation formula* comes into play in

$$\sum_{\substack{0\le g\le n\\ g\in\mathbb{Z}}} F(g) - \frac{1}{2}\left(F(0)+F(n)\right) = \lim_{N\to\infty} \sum_{\substack{|h|\le N\\ h\in\mathbb{Z}}} \int_0^n F(x)\, e^{2\pi i h\cdot x}\,dx. \qquad (4.6)$$

In spite of their apparent dissimilarity, the Euler summation formula (4.1) and the Poisson summation formula (4.6) are equivalent for twice continuously differentiable functions on the interval $[0, n]$.

The "building blocks" of the bridge between the two equivalent formulas (4.1) and (4.6) are the *defining constituents of the Euler–Green function* $G(\Delta;\cdot)$, which can be characterized as follows:

(*Periodicity*) $G(\Delta;\cdot)$ is continuous in $\mathbb{R}$ and $\mathbb{Z}$–periodical

$$G(\Delta;x) \;=\; G(\Delta;x+g),\ x\in\mathbb{R},\ g\in\mathbb{Z}. \qquad (4.7)$$

(*Differential equation*) $\Delta G(\Delta;\cdot)$ "coincides" apart from an additive constant with the Dirac function(al)

$$\Delta_x G(\Delta;x) \;=\; -1,\ \ x\in\mathbb{R}\backslash\mathbb{Z}. \qquad (4.8)$$

(*Characteristic singularity*) $G(\Delta;\cdot)$ possesses the singularity of the 1D-fundamental solution of the Laplace operator

$$G(\Delta;x) \;-\; \frac{1}{2}\,|x| \;=\; O(1),\ \ x\to 0. \qquad (4.9)$$

(*Normalization*) $G(\Delta;\cdot)$ integrated over a whole period interval of length 1 is assumed to be zero, i.e.,

$$\int_{\mathcal{F}_{\mathbb{Z}}} G(\Delta;x)\,dx \;=\; \int_{-\frac{1}{2}}^{\frac{1}{2}} G(\Delta;x)\,dx \;=\; 0. \qquad (4.10)$$

The Euler–Green function $G(\Delta;\cdot)$ is uniquely determined by the properties (4.7)–(4.10).

Even more generally, for *arbitrary* intervals $[c,d] \subset \mathbb{R}$, $c < d$, and *arbitrary* twice continuously differentiable (weight) functions F on $[c,d]$, the constituents (4.7)–(4.10) of the Euler–Green function $G(\Delta;\cdot)$ enable us to guarantee the equivalence of the *Euler summation formula*

$$\sum_{\substack{g\in[c,d]\\ g\in\mathbb{Z}}}{}' F(g) = \int_c^d F(x)\,dx + \int_c^d G(\Delta;x)\,\Delta F(x)\,dx + \{F(x)\,(\nabla G(\Delta;x)) - G(\Delta;x)\,(\nabla F(x))\}\,\Big|_c^d \tag{4.11}$$

and the *Poisson summation formula*

$$\sum_{\substack{g\in[c,d]\\ g\in\mathbb{Z}}}{}' F(g) = \lim_{N\to\infty} \sum_{\substack{|h|\leq N\\ h\in\mathbb{Z}}} \int_c^d F(x)\,e^{2\pi ih\cdot x}\,dx, \tag{4.12}$$

where we have used the convention

$$\sum_{\substack{g\in[c,d]\\ g\in\mathbb{Z}}}{}' F(g) = \sum_{\substack{c<g<d\\ g\in\mathbb{Z}}} F(g) + \frac{1}{2}\sum_{\substack{g=c,d\\ g\in\mathbb{Z}}} F(g), \tag{4.13}$$

and the second sum on the right side of (4.13) only occurs if c and/or d is an integer.

Univariate Hardy–Landau Identity. Of particular interest in lattice point theory is the special case of a constant weight function (i.e., $F = 1$), i.e., the *1D Hardy–Landau identity*

$$\sum_{\substack{g\in[c,d]\\ g\in\mathbb{Z}}}{}' 1 = (d-c) + \sum_{\substack{h\neq 0\\ h\in\mathbb{Z}}} \int_c^d e^{2\pi ih\cdot x}\,dx \tag{4.14}$$

$$= (d-c) + \sum_{\substack{h\neq 0\\ h\in\mathbb{Z}}} \frac{1}{2\pi ih}\, e^{2\pi ih\cdot x}\Big|_c^d . \tag{4.15}$$

The formula (4.14) compares the number of $\mathbb{Z}$–lattice points inside an interval $[c,d]$ with the length $d-c$ of the interval $[c,d]$ under the explicit knowledge of the remainder term (usually called, the *discrepancy*) as a 1D alternating series.

In the literature, the 1D Hardy–Landau identity is usually formulated on an interval $[-R,R]$, $R > 0$, that is symmetric to the origin. Under this particular choice we have

$$\int_{-R}^{R} e^{2\pi ih\cdot x}\,dx = 2R\,\mathrm{sinc}(2\pi hR), \tag{4.16}$$

and the sinc-term given by

$$\operatorname{sinc}(2\pi Rh) = \frac{\sin(2\pi Rx)}{2\pi Rh}, \quad x \in \mathbb{Z},\ R > 0, \tag{4.17}$$

enables us to write

$$\sum_{\substack{|g|\leq R\\ g\in\mathbb{Z}}}{}' \ 1 = 2R + 2R \sum_{\substack{h\neq 0\\ h\in\mathbb{Z}}} \operatorname{sinc}(2\pi hR). \tag{4.18}$$

Note that the 1D Poisson summation formula yields the identity

$$\lim_{N\to\infty} \sum_{\substack{|g|\leq N\\ g\in\mathbb{Z}}} \frac{\sin(2\pi Rg)}{2\pi Rg} = \sum_{h\in\mathbb{Z}} \int_{\mathbb{R}} \frac{\sin(2\pi Rx)}{2\pi Rx} e^{-2\pi ih\cdot x}\, dx. \tag{4.19}$$

The (discontinuous) integral on the right side of (4.19) is explicitly known. By virtue of the Fourier inversion formula (see, e.g., G.N. Watson [1944]), we have

$$\lim_{N\to\infty} \int_{-N}^{N} \frac{\sin(2\pi Rx)}{\pi x} e^{-2\pi ih\cdot x}\, dx = \begin{cases} 1 & , \quad |h| < R, \\ \frac{1}{2} & , \quad |h| = R, \\ 0 & , \quad |h| > R. \end{cases} \tag{4.20}$$

so that the 1D Hardy–Landau identity allows an immediate interpretation as a Poisson summation-type formula, too. In addition, the close relation between the Hardy–Landau summation and the metaharmonicity of the summands (cf. W. Freeden [2011]) becomes obvious in the identity (4.14), since the function $e(h\cdot) = e^{2\pi ih\cdot}, h \in \mathbb{Z}$, satisfies *the 1D Helmholtz equation* $(\Delta + \Delta^\wedge(h))\, e(h\cdot x) = 0, x \in \mathbb{R}$, corresponding to the wave number $\lambda = \Delta^\wedge(h) = 4\pi^2h^2$, $h \in \mathbb{Z}$; i.e., $e(h\cdot)$, $h \in \mathbb{Z}$, is *metaharmonic* in $\mathbb{R}$.

As a consequence, the Euler and Poisson summation formulas offer two significant perspectives, which turn out to be of decisive importance in lattice point sampling:

- to specify convergence criteria for alternating infinite series and to verify limits and asymptotic relations of infinite lattice point sums (see, e.g., W. Freeden [2011] for more details).

- to evaluate integrals constituting bandlimited as well as non-bandlimited functions and to estimate, to optimize the error, and to provide settings of constructive approximation (see, e.g., W. Freeden, M. Gutting [2017] for more details on integration and numerical cubature).

4.2 Variants of the Euler Summation Formula

Our aim next is to introduce the Euler–Green function $G(L+\lambda;\cdot)$ with respect to the operator

$$L+\lambda = \Delta + 4\pi i\ a\cdot\nabla + \lambda \tag{4.21}$$

and the "boundary condition" of Λ-periodicity, where $\lambda\in\mathbb{R}, a\in\mathbb{R}, \Lambda = \sigma\mathbb{Z}, \sigma>0, \Lambda^{-1}=\sigma^{-1}\mathbb{Z}$. To this end, we notice that

$$(L+L^\wedge(h))\ e^{2\pi ih\cdot x} = 0,\ x\in\mathbb{R}, \tag{4.22}$$

where

$$L^\wedge(h) = 4\pi^2h^2 - 4\pi ia\cdot 2\pi ih = 4\pi^2\left((h+a)^2-a^2\right),\ h\in\Lambda^{-1}. \tag{4.23}$$

Remark 4.1. *Keeping these facts in mind we are formally able to interpret the expression $L\ G(L;\cdot)$ as a modified Dirac functional δ_L given by*

$$\delta_L(x) = \frac{1}{\|\mathcal{F}_\Lambda\|}\sum_{\substack{L^\wedge(h)\neq 0\\ h\in\Lambda^{-1}}} e^{2\pi ih\cdot x},\quad x\in\mathbb{R}, \tag{4.24}$$

such that

$$\delta(x) = \frac{1}{\|\mathcal{F}_\Lambda\|}\sum_{\substack{L^\wedge(h)=0\\ h\in\Lambda^{-1}}} e^{2\pi ih\cdot x}\ +\ \delta_L(x),\quad x\in\mathbb{R}. \tag{4.25}$$

Moreover, we have

$$\delta(x) = \frac{1}{\|\mathcal{F}_\Lambda\|}\sum_{\substack{L^\wedge(h)=\lambda\\ h\in\Lambda^{-1}}} e^{2\pi ih\cdot x}\ +\ \delta_{L+\lambda}(x),\quad x\in\mathbb{R}, \tag{4.26}$$

where, as always, the sum $\sum_{L^\wedge(h)=\lambda, h\in\Lambda^{-1}}\dots$ is assumed to be zero if there is no lattice point $h\in\Lambda$ satisfying $L^\wedge(h)=\lambda$. In this case, $\delta(x)=\delta_{L+\lambda}(x)$. Thus we are confronted with a striking difference between the case of eigenvalues and non-eigenvalues of the operator L, which reflects the situation, which we have to expect for the structural context of Euler summation formulas.

The formal background provided by the Dirac equations (4.24), (4.25), and (4.26) motivate the following constituting properties of the 1D-Euler–Green function function:

Definition 4.2. *$G(L+\lambda;\cdot):\mathbb{R}\setminus\Lambda\to\mathbb{C}$, $\lambda\in\mathbb{R}$ fixed, is called Euler–Green function (or Λ-lattice function) for the operator $L+\lambda, L=\Delta+4\pi ia\cdot\nabla$, $a\in\mathbb{R}$, if it satisfies the following properties:*

(i) For all $x\in\mathbb{R}\setminus\Lambda$ and $g\in\Lambda$

$$G(L+\lambda;x+g) = G(L+\lambda;x). \tag{4.27}$$

(ii) For all $x \in \mathbb{R}\backslash\Lambda$

$$\overline{G(L+\lambda;x)} = G(L+\lambda;-x). \tag{4.28}$$

(iii) $G(L+\lambda;\cdot)$ *is twice continuously differentiable for all* $x \in \mathbb{R}\setminus\Lambda$. *For* $\lambda \notin \mathrm{Spect}_L(\Lambda)$, *i.e.,* $\lambda \neq 4\pi^2\left((h+a)^2 - a^2\right)$, $h \in \Lambda^{-1}$,

$$(L+\lambda)G(L+\lambda;x) = 0. \tag{4.29}$$

For $\lambda \in \mathrm{Spect}_L(\Lambda)$, *i.e.,* $\lambda = 4\pi^2\left((h+a)^2 - a^2\right)$

$$(L+\lambda)G(L+\lambda;x) = -\frac{1}{\|\mathcal{F}_\Lambda\|}\sum_{\substack{(L+\lambda)^\wedge(h)=0\\ h\in\Lambda^{-1}}} e^{2\pi i h\cdot x}. \tag{4.30}$$

(iv) In a neighborhood of the origin

$$G(L;x) \; - \; \frac{1}{2}\,|x| \; = \; O(1), \;\; x \to 0, \tag{4.31}$$

(so that $\nabla G(L;x)$ *shows a finite jump as singularity in the origin, hence, in all lattice points* h *of* Λ *because of condition (ii)).*

(v) For all $h \in \Lambda^{-1}$ *with* $(L+\lambda)^\wedge(h) = 0$

$$\int_{\mathcal{F}_\Lambda} G(L+\lambda;x)\; e^{-2\pi i h\cdot x}\; dx = 0. \tag{4.32}$$

For all $h \in \Lambda^{-1}$ and $\lambda \in \mathbb{R}$ we readily see by a double integration by parts, that

$$\int_{\mathcal{F}_\Lambda} G(L+\lambda;x)\; e^{-2\pi i h\cdot x}\; dx = \begin{cases} 0, & (L+\lambda)^\wedge(h) = 0, \\ -\frac{1}{\|\mathcal{F}_\Lambda\|}\frac{1}{(L+\lambda)^\wedge(h)}, & (L+\lambda)^\wedge(h) \neq 0. \end{cases} \tag{4.33}$$

As a consequence, for each number $\lambda \in \mathbb{R}$, there exists one and only one Euler–Green function $G(L+\lambda;\cdot)$ satisfying the constituting conditions *(i)* – *(v)* listed under Definition 4.2 (the proof follows from standard arguments of the Fourier theory). Moreover, it is easy to see that $G(L+\lambda;\cdot)$ has the absolutely and uniformly convergent Fourier series expansion

$$G(L+\lambda;x) \; = \; \frac{1}{\|\mathcal{F}_\Lambda\|}\sum_{\substack{L^\wedge(h)\neq\lambda\\ h\in\Lambda^{-1}}} \frac{e^{2\pi i h\cdot x}}{-(L+\lambda)^\wedge(h)}, \quad x \in \mathbb{R} \tag{4.34}$$

(note that $-(L+\lambda)^\wedge(h) = \lambda - L^\wedge(h) = \lambda - 4\pi^2\left((h+a)^2 - a^2\right)$).

The Euler–Green function (or Λ-lattice function) $G(L+\lambda;\cdot)$ for the operator $L+\lambda$ acts as the connecting tool to convert the differential equation involving the Helmholtz-type operator $L + \lambda$ corresponding to Λ–periodic boundary conditions into an associated integral equation, which turns out to be the counterpart of the Euler summation formula (4.1) in the particular manifestation for the differential operator $L + \lambda$.

Euler-Type Summation Formula. Observing the defining properties of the Euler–Green function (or Λ-lattice function) $G(L + \lambda; \cdot)$ for the operator $L + \lambda$ we are immediately able to formulate an Euler summation formula analogously to the case of the Helmholtz operator $\Delta + \lambda$ (cf. W. Freeden [2011]) by an ordinary twofold integration by parts.

Theorem 4.3. *(Euler-Type Summation Formula for the Operator $L + \lambda$). Let Λ be an arbitrary lattice in $\mathbb{R}$. Let $(c, d) \subset \mathbb{R}, c < d$, be a finite interval. Let H be a twice continuously differentiable function on $[c, d]$. Then, for $\lambda \notin \mathrm{Spect}_L(\Lambda)$, we have*

$$\begin{aligned}
\sum_{\substack{g\in[c,d]\\ g\in\Lambda}}{}' H(g) & \\
= & \int_{[c,d]} G(L+\lambda; x)\, (\overline{L}+\lambda)H(x)\, dx \\
& + \{H(x)\, (\nabla_x + 4\pi i a)\, G(L+\lambda; x)\}\big|_c^d \\
& - \{G(L+\lambda; x)\nabla_x H(x)\}\big|_c^d,
\end{aligned} \tag{4.35}$$

while, for $\lambda \in \mathrm{Spect_L}(\Lambda)$, we have

$$\begin{aligned}
\sum_{\substack{g\in[c,d]\\ g\in\Lambda}}{}' H(g) &= \frac{1}{\|\mathcal{F}_\Lambda\|} \sum_{\substack{(L+\lambda)^\wedge(h)=0\\ h\in\Lambda^{-1}}} \int_{[c,d]} H(x)\, e^{2\pi i h\cdot x}\, dx \\
&+ \int_{[c,d]} G(L+\lambda; x)\, (\overline{L}+\lambda)H(x)\, dx \\
&\quad + \{H(x)\, (\nabla_x + 4\pi i a)\, G(L+\lambda; x)\}\big|_c^d \\
&\quad - \{G(L+\lambda; x)\nabla_x H(x)\}\big|_c^d,
\end{aligned} \tag{4.36}$$

where a is an arbitrary real number and $\overline{L}$ is the conjugate complex of L, i.e.,

$$\overline{L} = \Delta - 4\pi i a \cdot \nabla, \quad a \in \mathbb{R}. \tag{4.37}$$

and

$$\sum_{\substack{g\in[c,d]\\ g\in\Lambda}}{}' H(g) = \sum_{\substack{g\in(c,d)\\ g\in\Lambda}} H(g) + \frac{1}{2} \sum_{\substack{g=c,d\\ g\in\Lambda}} H(g). \tag{4.38}$$

Once more, $\sum_{L^\wedge(h)=\lambda, h\in\Lambda^{-1}} \cdots$ means that the summation has to be extended over all $h \in \Lambda^{-1}$ satisfying $L^\wedge(h) = \lambda$. In the case, that there is no lattice point $h \in \Lambda^{-1}$ satisfying $L^\wedge(h) = \lambda$, the sum is assumed to be zero.

The summation formulas (4.35) and (4.36) express a sum of functional values in terms of Fourier integrals over $[c, d]$ and associated boundary terms, thereby involving the derivatives of the function H up to the second order. The difference between the two cases (4.35) and (4.36) shows that "resonance phenomena" occur in the case of eigenvalues.

From Theorem 4.3 we immediately obtain

Corollary 4.4. *Let Λ be an arbitrary lattice in $\mathbb{R}$. Let $(c,d) \subset \mathbb{R}, c < d$, be a finite interval. Let H be a twice continuously differentiable function on $[c, d]$. Then we have*

$$\begin{aligned}
\sum_{\substack{g\in[c,d]\\ g\in\Lambda}}{}' e^{2\pi i a\cdot g} H(g) &= \frac{1}{\|\mathcal{F}_\Lambda\|} \sum_{\substack{(L+\lambda)^\wedge(h)=0\\ h\in\Lambda^{-1}}} \int_{[c,d]} e^{2\pi i a\cdot x} H(x)\; e^{2\pi i h\cdot x}\; dx \\
&+ \int_{[c,d]} G(L+\lambda; x)\; (\overline{L}+\lambda)\; e^{2\pi i a\cdot x} H(x)\; dx \\
&\quad + \{e^{2\pi i a\cdot x} H(x)\, (\nabla_x + 4\pi i a)\, G(L+\lambda; x)\}\big|_c^d \\
&\quad - \{G(L+\lambda; x) \nabla_x e^{2\pi i a\cdot x} H(x)\}\big|_c^d . \qquad (4.39)
\end{aligned}$$

Remark 4.5. *The Euler-type summation formula (4.39) coincides with (4.11), i.e., the classical Euler summation formula, by letting $\Lambda = \mathbb{Z}$, $\lambda = 0$, and $a = y = 0$.*

4.3 Poisson-Type Summation Formula over Finite Intervals

Next we choose the 1D-ball $\mathbb{B}^1_N(y) = (-N+y, N+y)$ as the reference interval in the identity (4.39). Furthermore, we replace the term $H(x)$ especially by $F^\wedge_{\mathbb{B}^1_R}(x-y), y \in \mathbb{R}$, with F being of class $\mathrm{C}^{(2)}(\overline{\mathbb{B}^1_R})$, $R > 0$, where the Fourier transform $F^\wedge_{\mathbb{B}^1_R}$ over $\mathbb{B}^1_R$ is given by the integral

$$F^\wedge_{\mathbb{B}^1_R}(x) = \int_{\mathbb{B}^1_R} F(z)\; e^{-2\pi i x\cdot z}\; dz = \int_{\substack{|z|\le R\\ z\in\mathbb{R}}} F(z)\; e^{-2\pi i x\cdot z}\; dz, \quad x \in \mathbb{R}. \qquad (4.40)$$

Then we are led to the following summation formula.

Corollary 4.6. *For given $y \in \mathbb{R}$, let F be a twice continuously differentiable function on $\overline{\mathbb{B}^1_N(y)}$. Then we have*

$$\sum_{\substack{|g-y|\le N\\ g\in\Lambda}}{}' e^{2\pi i a\cdot g} F^{\wedge}_{\mathbb{B}^1_R}(g-y) \tag{4.41}$$

$$= \frac{1}{\|\mathcal{F}_\Lambda\|} \sum_{\substack{4\pi^2((h+a)^2-a^2)=\lambda\\ h\in\Lambda^{-1}}} e^{2\pi i y\cdot(h+a)} \int_{\substack{|x|\le N\\ x\in\mathbb{R}}} F^{\wedge}_{\mathbb{B}^1_R}(x)\; e^{2\pi i x\cdot(h+a)}\; dx$$

$$+ \int_{\substack{|x-y|\le N\\ x\in\mathbb{R}}} G(L+\lambda;x)\;(\overline{L}_x+\lambda)\; e^{2\pi i a\cdot x} F^{\wedge}_{\mathbb{B}^1_R}(x-y)\; dx$$

$$- \left\{ G(L+\lambda;x)\; \nabla_x e^{2\pi i a\cdot x} F^{\wedge}_{\mathbb{B}^1_R}(x-y) \right\} \Big|_{-N+y}^{N+y}$$

$$+ \left\{ e^{2\pi i a\cdot x} F^{\wedge}_{\mathbb{B}^1_R}(x-y)\,(\nabla_x + 4\pi i a)\, G(L+\lambda;x) \right\} \Big|_{-N+y}^{N+y}.$$

Obviously, the identity (4.41) compares a finite sum of functional values in lattice points of Λ with a finite sum of Fourier transforms corresponding to lattice points $h \in \Lambda^{-1}$ with $4\pi^2((h+a)^2 - a^2) = \lambda$ under explicit knowledge of the remainder term.

Observing the elementary identities

$$\Delta_x e^{2\pi i a\cdot x} F^{\wedge}_{\mathbb{B}^1_R}(x-y) = e^{2\pi i a\cdot x}\left(\Delta_x + 4\pi i a\cdot\nabla_x - 4\pi^2 a^2\right) F^{\wedge}_{\mathbb{B}^1_R}(x-y), \tag{4.42}$$

$$a\cdot\nabla_x e^{2\pi i a\cdot x} F^{\wedge}_{\mathbb{B}^1_R}(x-y) = e^{2\pi i a\cdot x}\left(a\cdot\nabla_x + 2\pi i a^2\right) F^{\wedge}_{\mathbb{B}^1_R}(x-y), \tag{4.43}$$

and

$$(\Delta_x - 4\pi i a\cdot\nabla_x)\, e^{2\pi i a\cdot x} F^{\wedge}_{\mathbb{B}^1_R}(x-y) \;=\; e^{2\pi i a\cdot x}\;\left(\Delta_x + 4\pi^2 a^2\right) F^{\wedge}_{\mathbb{B}^1_R}(x-y), \tag{4.44}$$

we obtain

$$\overline{L}_x\; e^{2\pi i a x} F^{\wedge}_{\mathbb{B}^1_R}(x-y) = e^{2\pi i a\cdot x}(\Delta_x + 4\pi^2 a^2) F^{\wedge}_{\mathbb{B}^1_R}(x-y). \tag{4.45}$$

Thus we arrive at the following Euler-type summation formula:

Corollary 4.7. *For $a, y \in \mathbb{R}$ and $N > 0$ we have*

$$\sum_{\substack{|g-y|\le N\\ g-y\in\Lambda}}{}' \; e^{2\pi i a\cdot g} F^{\wedge}_{\mathbb{B}^1_R}(g-y) \tag{4.46}$$

$$= \frac{1}{\|\mathcal{F}_\Lambda\|} \sum_{\substack{4\pi^2((h+a)^2-a^2)=\lambda\\ h\in\Lambda^{-1}}} e^{2\pi i y\cdot(h+a)} \int_{\substack{|x|\le N\\ x\in\mathbb{R}}} F^{\wedge}_{\mathbb{B}^1_R}(x)\; e^{2\pi i x\cdot(h+a)}\; dx$$

$$+ \int_{\substack{|x-y|\leq N \\ x\in\mathbb{R}}} G(L+\lambda;x)\; e^{2\pi i a\cdot x} \left(\Delta_x + \lambda + 4\pi^2 a^2\right) F^{\wedge}_{\mathbb{B}^1_R}(x-y)\; dx$$

$$+ \left\{ e^{2\pi i a\cdot x} G(L+\lambda;x) \nabla_x F^{\wedge}_{\mathbb{B}^1_R}(x-y) \right\}\Big|_{-N+y}^{N+y}$$

$$- \left\{ F^{\wedge}_{\mathbb{B}^1_R}(x-y) e^{2\pi i a\cdot x} (\nabla_x + 4\pi i a) G(L+\lambda;x) \right\}\Big|_{-N+y}^{N-+y}.$$

Now, for $x \neq y$, we get by a twofold integration by parts, i.e., the 1D Green theorem,

$$\begin{aligned} F^{\wedge}_{\mathbb{B}^1_R}(x-y) &= -\frac{1}{4\pi^2(x-y)^2} \int_{\substack{|w|\leq R \\ w\in\mathbb{R}}} \Delta_w F(w)\; e^{-2\pi i w\cdot(x-y)}\; dw \qquad (4.47) \\ &- \frac{1}{4\pi^2(x-y)^2} \left\{ (-2\pi i(x-y))\; e^{-2\pi i w\cdot(x-y)} F(w) \right\}\Big|_{-R}^{R} \\ &+ \frac{1}{4\pi^2(x-y)^2} \left\{ (\nabla_w F(w))\; e^{-2\pi i w\cdot(x-y)} \right\}\Big|_{-R}^{R}. \end{aligned}$$

So, we are confronted with the fact that the second term of the right side of (4.47) is of order $O(|x-y|^{-1})$, while the other terms of the right side of (4.47) are of order $O(|x-y|^{-2})$ for $|x-y| \to \infty$, because of the continuity properties of F and its derivatives (note that this observation is not critical in the univariate context, but it becomes problematic in higher dimensions, where methods of the stationary phase and/or asymptotic relations of the metaharmonic theory have to be applied).

In connection with the continuity of $G(L+\lambda;\cdot)$ and the piecewise continuity and boundedness of its first derivative we obtain for all wave numbers λ from Corollary 4.7

$$\begin{aligned} &\sum_{\substack{|g-y|\leq N \\ g-y\in\Lambda}} e^{2\pi a\cdot g} F^{\wedge}_{\mathbb{B}^1_R}(g-y) \qquad (4.48) \\ &= \frac{1}{\|\mathcal{F}_\Lambda\|} \sum_{\substack{4\pi^2((a+h)^2-a^2)=\lambda \\ h\in\Lambda^{-1}}} e^{2\pi i y\cdot(a+h)} \int_{\substack{|x|\leq N \\ x\in\mathbb{R}}} F^{\wedge}_{\mathbb{B}^1_R}(x)\; e^{2\pi i x\cdot(a+h)}\; dx \\ &+ \int_{\substack{|x-y|\leq N \\ x\in\mathbb{R}}} G(L+\lambda;x)\; e^{2\pi i a\cdot x} \left(\Delta_x + \lambda + 4\pi^2 a^2\right) F^{\wedge}_{\mathbb{B}^1_R}(x-y)\; dx \\ &+ o(1) \end{aligned}$$

for $N \to \infty$.

Now, the key step of our considerations is as follows: Choosing the wave number in a special way, i.e.,

$$\lambda = 4\pi^2(R^2 - a^2) \tag{4.49}$$

we see that

$$\left(\Delta_x + \lambda + 4\pi^2 a^2\right) F^{\wedge}_{\mathbb{B}^1_R}(x-y) = \int_{\substack{|w|\leq R\\ w\in\mathbb{R}}} \underbrace{4\pi^2(R^2 - w^2)F(z)}_{=\tilde{F}(\omega)} \; e^{-2\pi i w\cdot(x-y)}\, dw. \tag{4.50}$$

Therefore, because of the factor $4\pi^2(R^2 - w^2)$, twofold integration by parts yields the asymptotic relation

$$(\Delta_x + \lambda + 4\pi^2 a^2) F^{\wedge}_{\mathbb{B}^1_R}(x-y) = O\left(|x-y|^{-2}\right), \quad |x-y| \to \infty. \tag{4.51}$$

Consequently, the indefinite integral

$$\lim_{N\to\infty} \int_{\substack{|x-y|\leq N\\ x\in\mathbb{R}}} G(L+\lambda; x)\, e^{2\pi i a\cdot x} \left(\Delta_x + \lambda + 4\pi^2 a^2\right) F^{\wedge}_{\mathbb{B}^1_R}(x-y)\, dx \tag{4.52}$$

is absolutely convergent for the special choice $\lambda = 4\pi^2(R^2 - a^2)$.

Considering the Fourier series expansion (4.34) with $\lambda = 4\pi^2(R^2 - a^2)$ we obtain

$$\int_{\substack{|x-y|\leq N\\ x\in\mathbb{R}}} G(L+\lambda; x)\, e^{2\pi i a\cdot x} \; F^{\wedge}_{\mathbb{B}^1_R}(x-y)\, dx \tag{4.53}$$

$$= \frac{1}{\|\mathcal{F}_\Lambda\|} \sum_{\substack{|h+a|\neq R\\ h\in\Lambda^{-1}}} \frac{e^{2\pi i y\cdot(a+h)}}{4\pi^2(R^2 - (a+h)^2)} \int_{|z|\leq N} e^{2\pi i z\cdot(a+h)} \int_{\substack{|w|\leq R\\ w\in\mathbb{R}}} \tilde{F}(w) e^{-2\pi i w\cdot z}\, dw\, dz,$$

where we have introduced the abbreviation

$$\tilde{F}(w) = 4\pi^2(R^2 - w^2)F(w), \quad w \in \mathbb{R}. \tag{4.54}$$

The application of the inversion formula of the 1D Fourier theory (see, e.g., M.A. Pinsky [2002]) yields

$$\lim_{N\to\infty} \int_{|z|\leq N} e^{2\pi i z\cdot(a+h)} \int_{\substack{|w|\leq R\\ w\in\mathbb{R}}} \tilde{F}(w) e^{-2\pi i w\cdot z}\, dw\, dz \tag{4.55}$$

$$= \begin{cases} \tilde{F}(h+a), & |h+a| < R, \\ \frac{1}{2}\tilde{F}(h+a), & |h+a| = R, \\ 0, & |h+a| > R. \end{cases}$$

After some obvious manipulations we therefore obtain

$$\frac{1}{\|\mathcal{F}_\Lambda\|} \sum_{\substack{|a+h|\leq R\\ h\in\Lambda^{-1}}}{}' e^{2\pi i y\cdot(a+h)} F(a+h) = \lim_{N\to\infty} \sum_{\substack{|g-y|\leq N\\ g\in\Lambda}} e^{2\pi i a\cdot g} F^{\wedge}_{\mathbb{B}^1_R}(g-y). \tag{4.56}$$

Writing out the identity as indicated by (4.40) and replacing Λ by its inverse lattice Λ^{-1} we obtain as a variant of the 1D Poisson summation formula the following theorem.

Theorem 4.8. *(Poisson-Type Summation Formula.) The lattice point identity*

$$\sum_{\substack{|a+g|\le R\\ g\in\Lambda}}{}' F(a+g)\; e^{2\pi i y\cdot(a+g)} \tag{4.57}$$

$$= \lim_{N\to\infty} \frac{1}{\|\mathcal{F}_\Lambda\|} \sum_{\substack{|h-y|\le N\\ h\in\Lambda^{-1}}} e^{2\pi i a\cdot h} \int_{\substack{|x|\le R\\ x\in\mathbb{R}}} F(x)\; e^{-2\pi i x\cdot(h-y)}\, dx$$

holds true for arbitrary $R>0$ and all $a,y\in\mathbb{R}$, provided that F is a twice continuously differentiable function on the interval $\overline{\mathbb{B}^1_R}=[-R,R]$.

Fourier Theoretic Interpretation of Theorem 4.8: On the one hand, it is not difficult to see that

$$\sum_{\substack{|a+g+g'|\le R\\ g\in\Lambda}}{}' e^{2\pi i y\cdot(a+g+g')} F(a+g+g') = \sum_{\substack{|a+g|\le R\\ g\in\Lambda}}{}' e^{2\pi i y\cdot(a+g)} F(a+g) \tag{4.58}$$

holds true for all $g'\in\Lambda$. Hence, for every $R>0$ and fixed $y\in\mathbb{R}$, the function

$$a \mapsto \sum_{\substack{|a+g|\le R\\ g\in\Lambda}}{}' e^{2\pi i y\cdot(a+g)} F(a+g), \quad a\in\mathbb{R} \tag{4.59}$$

is Λ-periodic. Moreover, the function

$$a \mapsto \lim_{N\to\infty} \frac{1}{\|\mathcal{F}_\Lambda\|} \sum_{\substack{|h-y|\le N\\ h\in\Lambda^{-1}}} e^{2\pi i a\cdot h} \int_{\substack{|x|\le R\\ x\in\mathbb{R}}} F(x)\; e^{-2\pi i x\cdot(h-y)}\, dx, \quad a\in\mathbb{R} \tag{4.60}$$

is Λ-periodic. Theorem 4.8 therefore means that the piecewise continuous function (4.59) is representable by its Fourier series (4.60) in the pointwise sense, provided that F is twice continuously differentiable on $\overline{\mathbb{B}^1_R}$. In other words, the Fourier coefficients of the function (4.59) are given by

$$\begin{aligned}
&\int_{\mathcal{F}_\Lambda} \sum_{\substack{|a+g|\le R\\ g\in\Lambda}}{}' e^{2\pi i y\cdot(a+g)} F(a+g)\; e^{-2\pi i a\cdot h}\, da \qquad (4.61)\\
&\quad= \int_{\mathcal{F}_\Lambda} \sum_{\substack{|a+g|\le R\\ g\in\Lambda}}{}' e^{2\pi i y\cdot(a+g)} F(a+g) e^{-2\pi i (a+g)\cdot h}\, da\\
&\quad= \int_{\mathcal{F}_\Lambda} \sum_{\substack{|a+g|\le R\\ g\in\Lambda}}{}' F(a+g) e^{-2\pi i (a+g)\cdot(h-y)}\, da\\
&\quad= \int_{\substack{|x|\le R\\ x\in\mathbb{R}}} F(x)\; e^{-2\pi i x\cdot(h-y)}\, dx
\end{aligned}$$

for all $h\in\Lambda^{-1}$.

On the other hand, Theorem 4.8 can be rewritten in the equivalent form as follows:

Corollary 4.9. *The lattice point identity*

$$\sum_{\substack{|a+h|\leq R \\ h\in\Lambda^{-1}}}{}' F(a+h)\ \frac{1}{\|\mathcal{F}_\Lambda\|}\ e^{2\pi i y\cdot h} \tag{4.62}$$

$$= \lim_{N\to\infty} \sum_{\substack{|g-y|\leq N \\ g\in\Lambda}} e^{2\pi i a\cdot(g-y)} \int_{\substack{|x|\leq R \\ x\in\mathbb{R}}} F(x)\ e^{-2\pi i x\cdot(g-y)}\ dx$$

holds true for arbitrary $R>0$ and all $a, y \in \mathbb{R}$, provided that F is a twice continuously differentiable function on the interval $\overline{\mathbb{B}_R^1}$.

For $a \in \mathbb{R}$, the functions

$$y \mapsto \sum_{\substack{|a+h|\leq R \\ h\in\Lambda^{-1}}}{}' \frac{1}{\|\mathcal{F}_\Lambda\|}\ e^{2\pi i y\cdot h}\ F(a+h),\ \ y\in\mathbb{R} \tag{4.63}$$

and

$$y \mapsto \lim_{N\to\infty} \sum_{\substack{|g-y|\leq N \\ g\in\Lambda}} e^{2\pi i a\cdot(g-y)} \int_{\substack{|x|\leq R \\ x\in\mathbb{R}}} F(x)\ e^{-2\pi i x\cdot(g-y)}\ dx,\ \ y\in\mathbb{R}, \tag{4.64}$$

are periodic with respect to the lattice Λ. The Fourier coefficients of the function (4.64) are given by

$$\int_{\mathcal{F}_\Lambda} \lim_{N\to\infty} \sum_{\substack{|g-y|\leq N \\ g\in\Lambda}} e^{2\pi i a\cdot(g-y)} \int_{\substack{|x|\leq R \\ x\in\mathbb{R}}} F(x)\ e^{-2\pi i x\cdot(g-y)}\ dx\ e^{-2\pi i y\cdot h}\ dy \tag{4.65}$$

$$= \int_{\mathcal{F}_\Lambda} \lim_{N\to\infty} \sum_{\substack{|g-y|\leq N \\ g\in\Lambda}} e^{2\pi i (a+h)\cdot(g-y)} \int_{\substack{|x|\leq R \\ x\in\mathbb{R}}} F(x)\ e^{-2\pi i x\cdot(g-y)}\ dx\ dy$$

$$= \int_{\mathbb{R}} e^{2\pi i (a+h)\cdot w} \int_{\substack{|x|\leq R \\ x\in\mathbb{R}}} F(x)\ e^{-2\pi i x\cdot w}\ dx\ dw$$

for all $h \in \Lambda^{-1}$, where the Fourier inversion formula (see, e.g., M.A. Pinsky [2002]) tells us that

$$\int_{\mathbb{R}} e^{2\pi i (a+h)\cdot w} \int_{\substack{|x|\leq R \\ x\in\mathbb{R}}} F(x)\ e^{-2\pi i x\cdot w}\ dx\ dw = \begin{cases} F(a+h), & |a+h|<R, \\ \frac{1}{2}F(a+h), & |a+h|=R, \\ 0, & |a+h|>R. \end{cases} \tag{4.66}$$

Again, Theorem 4.8 in the manifestation of Corollary 4.9 means that the function (4.64) is representable by its Fourier series (4.63) in the pointwise sense, provided that F is twice continuously differentiable on $\overline{\mathbb{B}_R^1}$.

Preparatory Number Theoretic Interpretation of Theorem 4.8: Theorem 4.8 can be regarded as an extended 1D realization of the famous Hardy–Landau identity. Indeed, for $a = y = 0$, and constant weight function $F = 1$, the identity (4.57) expresses the number of Λ–lattice points in $\overline{\mathbb{B}^1_R}$ in terms of an alternating sinc-series

$$\sum_{\substack{|g|\leq R\\ g\in\Lambda}}{}' 1 = 2R\frac{1}{\|\mathcal{F}_\Lambda\|} \lim_{N\to\infty} \sum_{\substack{|h|\leq N\\ h\in\Lambda^{-1}}} \operatorname{sinc}(2\pi hR). \tag{4.67}$$

Thus, the Λ *lattice discrepancy*

$$\sum_{\substack{|g|\leq R\\ g\in\Lambda}}{}' 1 - 2R\frac{1}{\|\mathcal{F}_\Lambda\|} \tag{4.68}$$

allows the following representation in terms of an alternating series expansion

$$2R \lim_{N\to\infty} \frac{1}{\|\mathcal{F}_\Lambda\|} \sum_{\substack{|h|\leq N\\ h\in\Lambda^{-1}\setminus\{0\}}} \frac{\sin(2\pi hR)}{2\pi hR}, \tag{4.69}$$

i.e.,

$$\sum_{\substack{|g|\leq R\\ g\in\Lambda}}{}' 1 - 2R\frac{1}{\|\mathcal{F}_\Lambda\|} = 2R \lim_{N\to\infty} \frac{1}{\|\mathcal{F}_\Lambda\|} \sum_{\substack{|h|\leq N\\ h\in\Lambda^{-1}\setminus\{0\}}} \frac{\sin(2\pi hR)}{2\pi hR}. \tag{4.70}$$

Moreover, it follows from (4.57) that

$$\sum_{\substack{|a+g|\leq R\\ g\in\Lambda}}{}' e^{2\pi i y\cdot(a+g)} = \lim_{N\to\infty} \frac{1}{\|\mathcal{F}_\Lambda\|} \sum_{\substack{|h-y|\leq N\\ h\in\Lambda^{-1}}} e^{2\pi i a\cdot h} \frac{\sin(2\pi(h-y)R)}{\pi(h-y)}, \quad y\in\mathbb{R}, \tag{4.71}$$

holds true for all $a, y \in \mathbb{R}$.

In particular, if $g = 0$ is the only lattice point in $\mathbb{B}^\wedge_R(a)$, we obtain the identity

$$e^{2\pi i y\cdot a} = \lim_{N\to\infty} \sum_{\substack{|h-y|\leq N\\ h\in\Lambda^{-1}}} e^{2\pi i a\cdot h} \frac{1}{\|\mathcal{F}_\Lambda\|} \frac{\sin(2\pi(h-y)R)}{\pi(h-y)}, \quad y\in\mathbb{R}, \tag{4.72}$$

which can be regarded as one of the keystones for Papoulis's approach (cf. A. Papoulis [1971]) to 1D sampling. It demonstrates that, for every $y \in \mathbb{R}$, the function $a \mapsto e^{2\pi i y\cdot a}, a \in \mathbb{B}^\wedge_R(a)$, can be obtained from its samples $e^{2\pi i a\cdot h}, h \in \Lambda^{-1}$, by use of the kernel $\operatorname{sinc}(2\pi(h-y)R)$ as connecting tool. This observation also is a strong link to a close relationship of lattice point summation and sampling.

4.4 Shannon Sampling Based on the Poisson Summation-Type Formula

Next our purpose is to recognize the classical Shannon sampling theory in our Euler/Poisson-type summation reflected approach.

Our point of departure for realizing Shannon lattice point sampling actually starts from the Poisson summation-type formula

$$\begin{aligned} &\sum_{\substack{|a+g|\leq R \\ g\in\Lambda}}{}' F(a+g)\, e^{2\pi i y\cdot(a+g)} \\ &= \lim_{N\to\infty} \frac{1}{\|\mathcal{F}_\Lambda\|} \sum_{\substack{|h-y|\leq N \\ h\in\Lambda^{-1}}} e^{2\pi i a\cdot h} \int_{\substack{|x|\leq R \\ x\in\mathbb{R}}} F(x)\, e^{-2\pi i x\cdot(h-y)}\, dx \end{aligned} \tag{4.73}$$

From (4.73) we are immediately led to

$$F^{\wedge}_{\mathbb{B}^1_R}(y) = \int_{\mathbb{B}^1_R} F(a)\, e^{-2\pi i y\cdot a}\, da = \int_{\mathcal{F}_\Lambda} \sum_{\substack{|a+g|\leq R \\ g\in\Lambda}}{}' F(a+g) e^{-2\pi i y\cdot(a+g)}\, da. \tag{4.74}$$

provided that $\overline{\mathbb{B}^1_R} \subset \overline{\mathcal{F}_\Lambda}$. In other words, it follows from (4.85) that

$$F^{\wedge}_{\mathbb{B}^1_R}(-y) = \lim_{N\to\infty} \sum_{\substack{|h-y|\leq N \\ h\in\Lambda^{-1}}} F^{\wedge}_{\mathbb{B}^1_R}(h-y) \frac{1}{\|\mathcal{F}_\Lambda\|} \frac{\sin(\pi h)}{\pi h}, \tag{4.75}$$

for all $\overline{\mathbb{B}^1_R} \subset \overline{\mathcal{F}_\Lambda}$. Consequently, going over to the usual Shannon context $\overline{\mathbb{B}^1_R} = \overline{\mathcal{F}_\Lambda}$, we obtain for the problem of reconstructing a bandlimited signal $F^{\wedge}_{\mathcal{F}_\Lambda}$ of the form

$$F^{\wedge}_{\mathcal{F}_\Lambda}(y) = \int_{\mathcal{F}_\Lambda} F(a)\, e^{-2\pi i a\cdot y}\, da, \quad F \in \mathrm{C}^{(2)}\left(\overline{\mathcal{F}_\Lambda}\right), \quad y \in \mathbb{R}, \tag{4.76}$$

from (4.75) the sampling identity

$$F^{\wedge}_{\mathcal{F}_\Lambda}(-y) = \sum_{h\in\Lambda^{-1}} F^{\wedge}_{\mathcal{F}_\Lambda}(h) \frac{1}{\|\mathcal{F}_\Lambda\|} \frac{\sin(\pi(h+y))}{\pi(h+y)}, \quad y \in \mathbb{R}. \tag{4.77}$$

Hence,

$$F^{\wedge}_{\mathcal{F}_\Lambda}(y) = \sum_{h\in\Lambda^{-1}} F^{\wedge}_{\mathcal{F}_\Lambda}(h) \frac{1}{\|\mathcal{F}_\Lambda\|} \frac{\sin(\pi(h-y))}{\pi(h-y)} \tag{4.78}$$

for all $y \in \mathbb{R}$. The formula (4.78), indeed, is the classical *Shannon sampling formula.* It tells us that $F^{\wedge}_{\mathcal{F}_\Lambda}$, as defined by (4.76), can be recovered by its points on the lattice $\mathbb{Z}$. In fact, it should be noted that the proof of our Shannon approach can be provided by a 1D integration of a weighted Hardy–Landau-type lattice point identity.

Summarizing in the lattice point nomenclature as used in this work we are able to rephrase Shannon's sampling result (4.77) as follows:

Let $\Lambda = \sigma\mathbb{Z}$, $\sigma > 0$, designate the lattice in $\mathbb{R}$ consisting of all points σg, $g \in \mathbb{Z}$. Moreover, denote by $\Lambda^{-1} = \frac{1}{\sigma}\mathbb{Z}$ the inverse lattice to Λ. Then, any function $F^{\wedge}_{\mathcal{F}_\Lambda}$ bandlimited to the fundamental cell $\mathcal{F}_\Lambda = \left[-\frac{\sigma}{2}, \frac{\sigma}{2}\right)$ of the lattice $\Lambda = \sigma\mathbb{Z}$, i.e., the Fourier transform over the finite interval $\mathcal{F}_\Lambda$

$$F^{\wedge}_{\mathcal{F}_\Lambda}(y) = \int_{\mathcal{F}_\Lambda} F(x) e^{-2\pi i x\cdot y}\, dx, \quad y \in \mathbb{R} \tag{4.79}$$

may be reconstructed from the sequence of samples

$$\{F^{\wedge}_{\mathcal{F}_\Lambda}(h)\}_{h\in\Lambda^{-1}} \tag{4.80}$$

via its *cardinal series*

$$F^{\wedge}_{\mathcal{F}_\Lambda}(y) = \sum_{h\in\Lambda^{-1}} F^{\wedge}_{\mathcal{F}_\Lambda}(h) \frac{1}{\|\mathcal{F}_\Lambda\|} \underbrace{\int_{\mathcal{F}_\Lambda} e^{2\pi i x\cdot(h-y)}\, dx}_{=\,K_{\mathcal{F}_\Lambda}(h-y)}, \quad y \in \mathbb{R}, \tag{4.81}$$

where the symbol $\sum_{h\in\Lambda^{-1}} \ldots$, as always, means that the sum is extended over all lattice points h of the inverse lattice Λ^{-1}. Clearly, for $\Lambda = \sigma\mathbb{Z}$, $\sigma > 0$, we have

$$\|\mathcal{F}_{\Lambda^{-1}}\| = \frac{1}{\|\mathcal{F}_\Lambda\|} = \frac{1}{\sigma} \tag{4.82}$$

and $K_{\mathcal{F}_\Lambda}$ represents the sinc-kernel

$$K_{\mathcal{F}_\Lambda}(h-y) = \int_{\mathcal{F}_\Lambda} e^{2\pi i x\cdot(h-y)}\, dx = \frac{\sin(\pi\sigma(h-y))}{\pi(h-y)}, \quad y \in \mathbb{R},\ h \in \Lambda^{-1}. \tag{4.83}$$

So, the Fourier transform $F^{\wedge}_{\mathcal{F}_\Lambda}$ (considered as a 1D signal) is an infinitely repeated replication of the samples at points h of the inverse lattice Λ^{-1}. Combining (4.81) and (4.83) we equivalently obtain the well-known sampling formula

$$F^{\wedge}_{\mathcal{F}_{\sigma\mathbb{Z}}}(y) = \sum_{n=-\infty}^{\infty} F^{\wedge}_{\mathcal{F}_{\sigma\mathbb{Z}}}\left(\frac{n}{\sigma}\right) \frac{\sin\left(\pi\sigma(\frac{n}{\sigma} - y)\right)}{\pi\sigma(\frac{n}{\sigma} - y)}, \quad y \in \mathbb{R}, \tag{4.84}$$

such that, as announced earlier, Shannon sampling provides the bridge between continuous and discrete versions of a $\mathcal{F}_{\sigma\mathbb{Z}}$-bandlimited signal $F^{\wedge}_{\mathcal{F}_{\sigma\mathbb{Z}}}$.

4.5 Shannon-Type Sampling Based on Poisson Summation-Type Formulas

The univariate variant (Theorem 4.8) of the *Poisson summation formula*, i.e., in our nomenclature, the *Poisson summation-type formula*, allows the following two equivalent writings

$$\sum_{\substack{|a+g|\leq R\\ g\in\Lambda}}{}' F(a+g)\ e^{2\pi i y\cdot(a+g)} \tag{4.85}$$

$$= \lim_{N\to\infty}\frac{1}{\|\mathcal{F}_\Lambda\|}\sum_{\substack{|h-y|\leq N\\ h\in\Lambda^{-1}}} e^{2\pi i a\cdot h}\int_{\substack{|x|\leq R\\ x\in\mathbb{R}}} F(x)\ e^{-2\pi i x\cdot(h-y)}\ dx$$

and

$$\sum_{\substack{|a+h|\leq R\\ h\in\Lambda^{-1}}}{}' F(a+h)\frac{1}{\|\mathcal{F}_\Lambda\|}e^{2\pi i y\cdot h} \tag{4.86}$$

$$= \lim_{N\to\infty}\sum_{\substack{|g-y|\leq N\\ g\in\Lambda}} e^{2\pi i a\cdot(g-y)}\int_{\substack{|x|\leq R\\ x\in\mathbb{R}}} F(x)\ e^{-2\pi i x\cdot(g-y)}\ dx.$$

In spite of their equivalence, the Poisson-type summation formulas (4.85) and (4.86) offer two different entry points for Shannon-type sampling. In fact, it can be easily seen that two different cardinal series can be specified from the same Poisson-type summation formula (Theorem 4.8), namely

- a cardinal series resulting from the right side of (4.85), summing up functional values of the Fourier transform $F^{\wedge}_{\mathbb{B}^1_R}$ of $F\in \mathrm{C}^{(2)}(\mathbb{B}^1_R)$, $R>0$, in y-shifted lattice points,
- a cardinal series resulting from the left side of (4.86), summing up functional values of $F\in \mathrm{C}^{(2)}(\mathbb{B}^1_R)$, $R>0$, in a-shifted lattice points.

Clearly, the role of functional values and Fourier transformed values of F within Theorem 4.8 is exchangeable, since the Fourier transformed values of the Fourier transform canonically lead back to the functional values by virtue of the Fourier inversion formula. Nevertheless, it turns out that the two different entry points (4.85) and (4.86) provided by the same Poisson-type summation formula (Theorem 4.8) subsequently imply different Shannon-type sampling variants.

4.6 Fourier Transformed Values–Based Shannon-Type Sampling (Finite Intervals)

Firstly, our goal is to go over from the Poisson-type summation formula (4.85) to a *Shannon-type sampling theorem*:

Choosing $\Lambda = \sigma\mathbb{Z}$, $\sigma > 0$, we obtain by integration of both sides of (4.85) over $\mathbb{B}^1_\tau$, $\tau > 0$, the following Shannon-type sampling result (for the proof the reader is referred the multivariate context presented in this work):

Theorem 4.10. *(Fourier Transformed Values–Based Shannon-Type Sampling Theorem). Assume that* Λ *is the lattice* $\sigma\mathbb{Z}$, $\sigma > 0$. *Suppose that* F *is a member of the class* $\mathrm{C}^{(2)}(\overline{\mathbb{B}^1_\rho})$. *Then, for* $\mathbb{B}^1_\rho$, $\mathbb{B}^1_\tau$ *with* $\rho, \tau > 0$ *arbitrary, we have*

$$\sum_{\substack{(\mathcal{F}_{\sigma\mathbb{Z}}+\{g'\})\cap\mathbb{B}^1_\tau\neq\emptyset \\ g'\in\sigma\mathbb{Z}}} F^\wedge_{\mathbb{B}^1_\rho\cap\bigcup_{g\in\sigma\mathbb{Z}}(((\mathbb{B}^1_\tau\cap(\mathcal{F}_{\sigma\mathbb{Z}}+\{g'\}))-\{g'\})+\{g\})}(y) \tag{4.87}$$

$$= \sum_{\substack{(\mathcal{F}_{\sigma\mathbb{Z}}+\{g'\})\cap\mathbb{B}^1_\tau\neq\emptyset \\ g'\in\sigma\mathbb{Z}}} \int_{\mathbb{B}^1_\rho\cap\bigcup_{g\in\Lambda}(((\mathbb{B}^1_\tau\cap(\mathcal{F}_{\sigma\mathbb{Z}}+\{g'\}))-\{g'\})+\{g\})} F(x)\, e^{-2\pi i y\cdot x}\, dx$$

$$= \sum_{h\in\frac{1}{\sigma}\mathbb{Z}} \underbrace{\int_{\mathbb{B}^1_\rho} F(x)\, e^{-2\pi i h\cdot x}\, dx}_{=F^\wedge_{\mathbb{B}^1_\rho}(h)} \; \underbrace{\frac{1}{\|\mathcal{F}_{\sigma\mathbb{Z}}\|}\int_{\mathbb{B}^1_\tau} e^{2\pi i x\cdot(h-y)}\, dx}_{=K_{\mathbb{B}^1_\tau}(h-y)}.$$

Actually, the identity (4.87) forms a quantification of the earlier mentioned Nyquist-context, i.e., they explicitly include all manifestations of over- and undersampling involving the intervals $\mathbb{B}^1_\rho, \mathbb{B}^1_\tau$ for the lattice $\Lambda = \sigma\mathbb{Z}, \sigma > 0$: *Oversampling* means that the signal can be sampled, where "superfluous information" can be removed (see, e.g., M.Z. Nashed, Q. Sun [2010]) or used in an appropriate way to accelerate the convergence of the series (see, e.g., R.J. Marks II [1991]). Otherwise, we speak of *undersampling.* In the case of undersampling, we are confronted with the well-known phenomenon of *aliasing,* that the signal reconstructed from the samples is different from the original continuous signal.

Some particular geometric configurations, i.e., special choices of intervals $\mathbb{B}^1_\rho, \mathbb{B}^1_\tau$ should be studied in more detail: We begin with $\overline{\mathbb{B}^1_\tau} \subset [-\frac{\sigma}{2}, \frac{\sigma}{2}]$ and

$\mathbb{B}^1_\rho$ arbitrary. In this case, we have

$$\begin{aligned}
&\sum_{\substack{(\mathcal{F}_{\sigma\mathbb{Z}}+\{g'\})\cap\mathbb{B}^1_\tau\neq\emptyset \\ g'\in\sigma\mathbb{Z}}} F^\wedge_{\mathbb{B}^1_\rho\cap\bigcup_{g\in\sigma\mathbb{Z}}(((\mathbb{B}^1_\tau\cap(\mathcal{F}_{\sigma\mathbb{Z}}+\{g'\}))-\{g'\})+\{g\})}(y) \\
&\qquad = F^\wedge_{\mathbb{B}^1_\rho\cap\bigcup_{g\in\sigma\mathbb{Z}}(\mathbb{B}^1_\tau+\{g\})}(y) \\
&\qquad = \sum_{h\in\frac{1}{\sigma}\mathbb{Z}} F^\wedge_{\mathbb{B}^1_\rho}(h)\ \underbrace{\frac{1}{\|\mathcal{F}_{\sigma\mathbb{Z}}\|}}_{=\frac{1}{\sigma}}\ K_{\mathbb{B}^1_\tau}(h-y).
\end{aligned} \tag{4.88}$$

As an immediate consequence, for $\overline{\mathbb{B}^1_\tau}=[-\frac{\sigma}{2},\frac{\sigma}{2}]=\overline{\mathcal{F}_{\sigma\mathbb{Z}}}$ and $\mathbb{B}^1_\rho$ arbitrary, we obtain

$$\begin{aligned}
&\sum_{\substack{(\mathcal{F}_{\sigma\mathbb{Z}}+\{g'\})\cap\mathbb{B}^1_\tau\neq\emptyset \\ g'\in\sigma\mathbb{Z}}} F^\wedge_{\mathbb{B}^1_\rho\cap\bigcup_{g\in\sigma\mathbb{Z}}(((\mathbb{B}^1_\tau\cap(\mathcal{F}_{\sigma\mathbb{Z}}+\{g'\}))-\{g'\})+\{g\})}(y) \\
&\qquad = F^\wedge_{\mathbb{B}^1_\rho}(y) \\
&\qquad = \sum_{h\in\frac{1}{\sigma}\mathbb{Z}} F^\wedge_{\mathbb{B}^1_\rho}(h)\ \frac{1}{\sigma}K_{\mathbb{B}^1_\tau}(h-y).
\end{aligned} \tag{4.89}$$

We continue with $\overline{\mathbb{B}^1_\rho}=[-\frac{\sigma}{2},\frac{\sigma}{2}]$ and $\mathbb{B}^1_\tau$ arbitrary. This yields the identity

$$\begin{aligned}
&\sum_{\substack{(\mathcal{F}_{\sigma\mathbb{Z}}+\{g'\})\cap\mathbb{B}^1_\tau\neq\emptyset \\ g'\in\sigma\mathbb{Z}}} F^\wedge_{\mathbb{B}^1_\rho\cap\bigcup_{g\in\sigma\mathbb{Z}}(((\mathbb{B}^1_\tau\cap(\mathcal{F}_{\sigma\mathbb{Z}}+\{g'\}))-\{g'\})+\{g\})}(y) \\
&\qquad = \sum_{\substack{\mathcal{F}_{\sigma\mathbb{Z}}+\{g'\}\cap\mathbb{B}^1_\tau\neq\emptyset \\ g'\in\sigma\mathbb{Z}}} F^\wedge_{\mathbb{B}^1_\rho\cap((\mathbb{B}^1_\tau\cap(\mathcal{F}_\Lambda+\{g'\}))-\{g'\})}(y) \\
&\qquad = \sum_{h\in\frac{1}{\sigma}\mathbb{Z}} F^\wedge_{\mathbb{B}^1_\rho}(h)\ \frac{1}{\sigma}K_{\mathbb{B}^1_\tau}(h-y).
\end{aligned} \tag{4.90}$$

For $\overline{\mathbb{B}^1_\rho},\overline{\mathbb{B}^1_\tau}\subset[-\frac{\sigma}{2},\frac{\sigma}{2}]$ we have

$$\begin{aligned}
&\sum_{\substack{(\mathcal{F}_{\sigma\mathbb{Z}}+\{g'\})\cap\mathbb{B}^1_\tau\neq\emptyset \\ g'\in\sigma\mathbb{Z}}} F^\wedge_{\mathbb{B}^1_\rho\cap\bigcup_{g\in\sigma\mathbb{Z}}(((\mathbb{B}^1_\rho\cap(\mathcal{F}_{\sigma\mathbb{Z}}+\{g'\}))-\{g'\})+\{g\})}(y) \\
&\qquad = F^\wedge_{\mathbb{B}^1_\rho\cap\mathbb{B}^1_\tau}(y) \\
&\qquad = \sum_{h\in\frac{1}{\sigma}\mathbb{Z}} F^\wedge_{\mathbb{B}^1_\rho}(h)\ \frac{1}{\sigma}K_{\mathbb{B}^1_\tau}(h-y).
\end{aligned} \tag{4.91}$$

For $\overline{\mathbb{B}^1_\rho}\subset\overline{\mathbb{B}^1_\tau}\subset[-\frac{\sigma}{2},\frac{\sigma}{2}]$ we have

$$F^\wedge_{\mathbb{B}^1_\rho}(y)=\sum_{h\in\frac{1}{\sigma}\mathbb{Z}} F^\wedge_{\mathbb{B}^1_\rho}(h)\ \frac{1}{\sigma}K_{\mathbb{B}^1_\tau}(h-y), \tag{4.92}$$

whereas, for $\overline{\mathbb{B}^1_\tau} \subset \overline{\mathbb{B}^1_\rho} \subset [-\frac{\sigma}{2}, \frac{\sigma}{2}]$,

$$F^{\wedge}_{\mathbb{B}^1_\tau}(y) = \sum_{h \in \frac{1}{\sigma}\mathbb{Z}} F^{\wedge}_{\mathbb{B}^1_\rho}(h) \, \frac{1}{\sigma} \, K_{\mathbb{B}^1_\tau}(h - y). \tag{4.93}$$

In particular, for $\overline{\mathbb{B}^1_\rho} = \overline{\mathbb{B}^1_\tau} \subset [-\frac{\sigma}{2}, \frac{\sigma}{2}]$, we are able to formulate the identity

$$F^{\wedge}_{\mathbb{B}^1_\rho}(y) = \sum_{h \in \frac{1}{\sigma}\mathbb{Z}} F^{\wedge}_{\mathbb{B}^1_\rho}(h) \, \frac{1}{\sigma} K_{\mathbb{B}^1_\rho}(h - y). \tag{4.94}$$

Finally it should be noted that the standard form of Shannon sampling (cf. (4.84)) is provided by taking $\overline{\mathbb{B}^1_\rho} = \overline{\mathbb{B}^1_\tau} = [-\frac{\sigma}{2}, \frac{\sigma}{2}] = \overline{\mathcal{F}_{\sigma\mathbb{Z}}}$, so that

$$F^{\wedge}_{\mathcal{F}_{\sigma\mathbb{Z}}}(y) = \sum_{h \in \frac{1}{\sigma}\mathbb{Z}} F^{\wedge}_{\mathcal{F}_{\sigma\mathbb{Z}}}(h) \, \frac{1}{\sigma} K_{\mathcal{F}_{\sigma\mathbb{Z}}}(h - y). \tag{4.95}$$

Obviously, by choosing a small sampling density in (4.94) such that the interval $[-\frac{\sigma}{2}, \frac{\sigma}{2}]$ covers the compact support $\overline{\mathbb{B}^1_\rho}$ of the original signal $F^{\wedge}_{\mathbb{B}^1_\rho}$, the number of samples $F^{\wedge}_{\mathbb{B}^1_\rho}(h), h \in \frac{1}{\sigma}\mathbb{Z}$, for reconstruction is high, and vice versa. In practice, we are therefore required to find a compromise between the sampling density and the total number of samples. This compromise can be achieved by a choice of $\sigma\mathbb{Z}$ such that $[-\frac{\sigma}{2}, \frac{\sigma}{2}]$ covers $\overline{\mathbb{B}^1_\rho}$ tightly.

4.7 Functional Values–Based Shannon-Type Sampling (Finite Intervals)

Secondly, our goal is to go over from the Poisson-type summation formula (4.86) to a *Shannon-type sampling theorem:*

Choosing $\Lambda = \sigma\mathbb{Z}$ we obtain by integration of both sides of (4.86) over $\mathbb{B}^1_\tau$, $\tau > 0$, the following Shannon-type sampling result (for the proof the reader is referred the multivariate context presented later in this work).

Theorem 4.11. *(Functional Values–Based Shannon-Type Sampling Theorem). Suppose that F is a member of the class $\mathrm{C}^{(2}(\overline{\mathbb{B}^1_\rho})$. Then, for $a \in \mathbb{R}$, we have*

$$\sum_{\substack{(\mathcal{F}_{\sigma\mathbb{Z}}+\{g'\})\cap\mathbb{B}^1_\tau \neq \emptyset \\ g' \in \sigma\mathbb{Z}}} \int_{\bigcup\limits_{g \in \sigma\mathbb{Z}} (((\mathbb{B}^1_\rho \cap (\mathcal{F}_{\sigma\mathbb{Z}}+\{g'\}))-\{g'\})+\{g\})} e^{2\pi i a \cdot x} \; F^{\wedge}_{\mathbb{B}^1_\rho}(x) \;\; dx$$
$$= \sum_{\substack{a+h \in \overline{\mathbb{B}^1_\rho} \\ h \in \frac{1}{\sigma}\mathbb{Z}}}{}' \; F(a+h) \, \frac{1}{\|\mathcal{F}_{\sigma\mathbb{Z}}\|} K_{\mathbb{B}^1_\tau}(-h). \tag{4.96}$$

We only discuss the following three cases $\overline{\mathbb{B}^1_\tau} \subsetneq \overline{\mathcal{F}_{\sigma\mathbb{Z}}}$, $\overline{\mathbb{B}^1_\tau} = \overline{\mathcal{F}_\Lambda}$, and $\overline{\mathbb{B}^1_\tau} \supsetneq \overline{\mathcal{F}_{\sigma\mathbb{Z}}}$:

For $\overline{\mathbb{B}^1_\tau} \subsetneq \overline{\mathcal{F}_\Lambda}$ we find that

$$\int_{\bigcup\limits_{g\in\sigma\mathbb{Z}} (\mathbb{B}^1_\rho+\{g\})} e^{2\pi i a\cdot x}\ F^\wedge_{\mathbb{B}^1_\rho}(x)\ \ dx = \sum_{\substack{a+h\in\overline{\mathbb{B}^1_\rho}\\ h\in\frac{1}{\sigma}\mathbb{Z}}}{}' \ F(a+h)\ \frac{1}{\sigma}\ K_{\mathbb{B}^1_\tau}(-h). \tag{4.97}$$

As a consequence we are able to see by application of the Fourier inversion formula that

$$\begin{aligned} &\sum_{\substack{a+h\in\overline{\mathbb{B}^1_\rho}\\ h\in\frac{1}{\sigma}\mathbb{Z}}}{}' \ F(a+h)\ \frac{1}{\sigma}\ K_{\mathbb{B}^1_\tau}(-h) \\ &\qquad = \begin{cases} F(a), & |a|<\rho \\ \frac{1}{2}F(a), & |a|=\rho \\ 0, & |a|>\rho \end{cases} \\ &\qquad - \int_{\bigcup\limits_{g\in\sigma\mathbb{Z}} ((\overline{\mathcal{F}_{\sigma\mathbb{Z}}}\setminus\mathbb{B}^1_\rho)+\{g\})} e^{2\pi i a\cdot x}\ F^\wedge_{\mathbb{B}^1_\rho}(x)\ \ dx, \end{aligned} \tag{4.98}$$

so that the aliasing error between the functional values on the local support $\overline{\mathbb{B}^1_\rho}$ and the cardinal series are given by the integral on the left side.

For $\overline{\mathbb{B}^1_\tau} = \overline{\mathcal{F}_{\sigma\mathbb{Z}}}$ we obtain by application of the Fourier inversion formula

$$\begin{aligned} &\int_{\mathbb{R}} e^{2\pi i a\cdot x}\ F^\wedge_{\mathbb{B}^1_\rho}(x)\ \ dx \\ &\qquad = \begin{cases} F(a), & |a|<\rho \\ \frac{1}{2}F(a), & |a|=\rho \\ 0, & |a|>\rho \end{cases} \\ &\qquad = \sum_{\substack{a+h\in\overline{\mathbb{B}^1_\rho}\\ h\in\frac{1}{\sigma}\mathbb{Z}}}{}' \ F(a+h)\ \frac{1}{\sigma}\ K_{\mathcal{F}_{\sigma\mathbb{Z}}}(-h). \end{aligned} \tag{4.99}$$

For $\overline{\mathbb{B}^1_\tau} \supsetneq \overline{\mathcal{F}_{\sigma\mathbb{Z}}}$ we find that

$$\begin{aligned} &\sum_{\substack{(\mathcal{F}_{\sigma\mathbb{Z}}+\{g'\})\cap\mathbb{B}^1_\tau\neq\emptyset\\ g'\in\sigma\mathbb{Z}}} \int_{\bigcup\limits_{g\in\sigma\mathbb{Z}} (((\mathbb{B}^1_\rho\cap(\mathcal{F}_{\sigma\mathbb{Z}}+\{g'\}))-\{g'\})+\{g\})} e^{2\pi i a\cdot x}\ F^\wedge_{\mathbb{B}^1_\rho}(x)\ \ dx \\ &\qquad = \sum_{\substack{a+h\in\overline{\mathbb{B}^1_\rho}\\ h\in\frac{1}{\sigma}\mathbb{Z}}}{}' \ F(a+h)\ \frac{1}{\|\mathcal{F}_{\sigma\mathbb{Z}}\|} K_{\mathbb{B}^1_\tau}(-h). \end{aligned} \tag{4.100}$$

Thus we are confronted with the following situation: The closer $\overline{\mathbb{B}^1_\tau}$ is to a finite (closed) union of lattice cells of $\Lambda = \sigma\mathbb{Z}$, the smaller is the aliasing error by this variant of Shannon-type sampling.

In conclusion, Theorem 4.8 opens two different manifestations of Shannon-type sampling. The Fourier integral–based context provides diverse sampling capabilities. The functional based Shannon-type sampling, however, does not provide a palette of results comparable to the Fourier integral–based case. Nevertheless, also in this framework, a way is found to detect an explicit aliasing error for non-standard geometric configurations.

4.8 Paley–Wiener Reproducing Kernel Hilbert Spaces

Looking at the context of our Fourier integral–based Shannon-type sampling we critically notice that the classical Shannon sampling formula (4.77) for $\Lambda = \mathbb{Z}$ is formulated on the reference set

$$\mathrm{B}^{(2)}(\mathcal{F}_\mathbb{Z}) = \left\{ y \mapsto F^\wedge_{\mathcal{F}_\mathbb{Z}}(y) = \int_{\mathcal{F}_\mathbb{Z}} e^{-2\pi i a\cdot y} F(a)\, da,\ y \in \mathbb{R} : F \in \mathrm{C}^{(2)}(\overline{\mathcal{F}_\mathbb{Z}}) \right\}, \tag{4.101}$$

that is a strict subset of the *Paley–Wiener space*

$$\mathrm{B}(\mathcal{F}_\mathbb{Z}) = \{ y \mapsto F^\wedge_{\mathcal{F}_\mathbb{Z}} = \int_{\mathcal{F}_\mathbb{Z}} e^{-2\pi i a\cdot y} F(a)\, da,\ y \in \mathbb{R} : F \in \mathrm{L}^2(\mathcal{F}_\mathbb{Z}) \}. \tag{4.102}$$

This observation, however, does not bother us very much, since every $F \in \mathrm{L}^2(\mathcal{F}_\mathbb{Z})$ can be approximated (in $\mathrm{L}^2(\mathcal{F}_\mathbb{Z})$-sense) by a function $F_\varepsilon \in \mathrm{C}^{(2)}(\overline{\mathcal{F}_\mathbb{Z}})$ in ε-accuracy such that

$$\begin{aligned} \sup_{y\in\mathbb{R}} \left| \int_{\mathcal{F}_\mathbb{Z}} e^{-2\pi i a\cdot y} (F(a) - F_\varepsilon(a))\, da \right| & \qquad\qquad (4.103)\\ \leq \left(\int_{\mathcal{F}_\mathbb{Z}} |F(a) - F_\varepsilon(a)|^2\, da \right)^{1/2} \left(\int_{\mathcal{F}_\mathbb{Z}} |e^{-2\pi i a\cdot y}|^2\, da \right)^{1/2} & \\ \leq \varepsilon. & \end{aligned}$$

The *Paley–Wiener space* $\mathrm{B}(\mathcal{F}_\mathbb{Z})$ has many interesting properties. All of these properties have been collected and generalized, e.g., in M.Z. Nashed [2010], M.Z. Nashed, Q. Sun [2013]:

(i) $\mathrm{B}(\mathcal{F}_\mathbb{Z})$ consists of all functions $F^\wedge_{\mathcal{F}_\mathbb{Z}} = \int_{\mathcal{F}_\mathbb{Z}} F(a)\, e^{-2\pi i a\cdot}\, da$ with $F \in \mathrm{L}^2(\mathcal{F}_\mathbb{Z})$:

$$\mathrm{B}(\mathcal{F}_\mathbb{Z}) = \left\{ F^\wedge_{\mathcal{F}_\mathbb{Z}} = \int_{\mathcal{F}_\mathbb{Z}} F(a)\, e^{-2\pi i a\cdot}\, da : F \in \mathrm{L}^2(\mathcal{F}_\mathbb{Z}) \right\}. \tag{4.104}$$

(ii) $\mathrm{B}(\mathcal{F}_{\mathbb{Z}})$ is a reproducing kernel Hilbert space with the uniquely-determined reproducing kernel

$$\text{sinc}\ (\pi(g-y)) = \frac{\sin(\pi(g-y))}{\pi(g-y)}. \tag{4.105}$$

(iii) The sequence

$$\{\text{sinc}\ (\pi(g-y))\}_{g\in\mathbb{Z}} \tag{4.106}$$

is an orthonormal basis for $\mathrm{B}(\mathcal{F}_{\mathbb{Z}})$.

(iv) The sequence (4.106) shows the discrete orthogonality property:

$$\text{sinc}\ (\pi(g'-g)) = \delta_{g',g} = \begin{cases} 1, & g' = g, \\ 0, & g' \neq g. \end{cases} \tag{4.107}$$

(v) $F^{\wedge}_{\mathcal{F}_{\mathbb{Z}}}(\cdot - c) \in \mathrm{B}(\mathcal{F}_{\mathbb{Z}})$ and $\|F^{\wedge}_{\mathcal{F}_{\mathbb{Z}}}(\cdot - c)\|_{\mathrm{L}^2(\mathbb{R})} = \|F^{\wedge}_{\mathcal{F}_{\mathbb{Z}}}\|_{\mathrm{L}^2(\mathbb{R})}$ for all $S \in \mathrm{B}(\mathcal{F}_{\mathbb{Z}})$ and $c \in \mathbb{R}$, such that $\mathrm{B}(\mathcal{F}_{\mathbb{Z}})$ is a unitarily translation-invariant subspace of $\mathrm{L}^2(\mathbb{R})$.

(vi) $\mathrm{B}(\mathcal{F}_{\mathbb{Z}})$ is a shift-invariant subspace of $\mathrm{L}^2(\mathbb{R})$ generated by the sinc-function:

$$\mathrm{B}(\mathcal{F}_{\mathbb{Z}}) = \left\{ \sum_{g\in\mathbb{Z}} c(g)\ \text{sinc}\ \pi(\cdot - g) : \sum_{g\in\mathbb{Z}} |c(g)|^2 < \infty \right\}. \tag{4.108}$$

4.9 Poisson-Type Summation Formula over the Euclidean Space

Next we apply the Euler-type summation formula with respect to the operator $L+\lambda$, $L = \Delta + 4\pi i a \cdot \nabla$, $a \in \mathbb{R}$, especially to sums of spherical type, i.e., to intervals $\overline{\mathbb{B}^1_N(y)}$ with arbitrarily chosen $y \in \mathbb{R}$ and $N > 0$. From Corollary 4.7 we obtain

$$\sum_{\substack{|g-y|\leq N \\ g\in\Lambda}}{}' e^{2\pi i a\cdot g} F^{\wedge}_{\mathbb{B}^1_R}(g-y) \tag{4.109}$$

$$\begin{aligned}
&= \sum_{\substack{4\pi^2((h+a)^2-a^2)=\lambda \\ h\in\Lambda^{-1}}} e^{2\pi i y\cdot(h+a)} \int_{\substack{|x|\leq N \\ x\in\mathbb{R}}} F^{\wedge}_{\mathbb{B}^1_R}(x)\ e^{2\pi i x\cdot(h+a)}\ dx \\
&\quad + \int_{\substack{|x-y|\leq N \\ x\in\mathbb{R}}} G(L+\lambda;x)\ e^{2\pi i a\cdot x} \left(\Delta_x + \lambda + 4\pi^2 a^2\right) F^{\wedge}_{\mathbb{B}^1_R}(x-y)\ dx \\
&\quad + \left\{ e^{2\pi i a\cdot x} G(L+\lambda;x) \nabla_x F^{\wedge}_{\mathbb{B}^1_R}(x-y) \right\}\Big|_{-N+y}^{N+y} \\
&\quad - \left\{ F^{\wedge}_{\mathbb{B}^1_R}(x-y)\ e^{2\pi i a\cdot x} (\nabla_x + 4\pi i a) G(L+\lambda;x) \right\}\Big|_{-N+y}^{N-+y}.
\end{aligned}$$

We are interested in investigating the asymptotic behavior with increasing radii N to derive sufficient criteria for the validity of the Poisson-type summation formula over the whole Euclidean space $\mathbb{R}$.

Questions of the convergence of (4.109) as $N \to \infty$ require asymptotic estimates of the constituting elements of the remainder term $R(N)$ given by

$$\begin{aligned} R(N) & \qquad\qquad (4.110) \\ = & \ \frac{1}{2} \sum_{\substack{|g-y|=N \\ g\in\Lambda}} e^{2\pi i a\cdot g} F(g-y) \\ & + \left\{ e^{2\pi i a\cdot x} G(L+\lambda; x) \nabla_x F^{\wedge}_{\mathbb{B}^1_R}(x \quad y) \right\} \Big|_{-N+y}^{N+y} \\ & - \left\{ F^{\wedge}_{\mathbb{B}^1_R}(x-y) e^{2\pi i a\cdot x} (\nabla_x + 4\pi i a) G(L+\lambda; x) \right\} \Big|_{-N+y}^{N-+y}. \end{aligned}$$

The following estimates, which essentially follow from the boundedness of $G(L+\lambda;\cdot)$, play an important role in the development of convergence criteria:

$$|G(L+\lambda;x)| \ |_{-N+y}^{N+y} = O(1), \quad N \to \infty \qquad (4.111)$$

$$|\nabla_x G(L+\lambda;x)| \ |_{-N+y}^{N+y} = O(1), \quad N \to \infty. \qquad (4.112)$$

In the sequel, the properties (4.111) and (4.112) are used to formulate 1D-convergence criteria for spherically reflected series.

Proposition 4.12. *A function F is called of class $\mathrm{X}^{(2)}(\mathbb{R})$, if F is of class $\mathrm{C}^{(2)}(\mathbb{R})$ satisfying*

$$|F(x)| \ = \ o(1), \quad |x| \to \infty \qquad (4.113)$$

and

$$|\nabla_x F(x)| \ = \ o(1), \quad |x| \to \infty. \qquad (4.114)$$

By virtue of Proposition 4.12 we are immediately led to

Lemma 4.13. *For $N \to \infty$*

$$R(N) = o(1), \qquad (4.115)$$

so that

$$\begin{aligned} & \sum_{\substack{|g-y|\le N \\ g-y\in\Lambda}} e^{2\pi i a\cdot g} F(g-y) \qquad\qquad (4.116) \\ = & \ \frac{1}{\|\mathcal{F}_\Lambda\|} \ e^{2\pi i a\cdot y} \sum_{\substack{4\pi^2((h+a)^2-a^2)=\lambda \\ h\in\Lambda^{-1}}} e^{2\pi i h\cdot y} \int_{|x|\le N} F(x) e^{2\pi i x\cdot(a+h)} dx \\ & + \int_{\substack{|x-y|\le N \\ x\in\mathbb{R}^q}} G(L+\lambda;x) \ e^{2\pi i a\cdot x} \left(\Delta_x + \lambda + 4\pi^2 a^2\right) F(x-y) \ dx \\ & + \ o(1). \end{aligned}$$

Next we discuss the second term of the right side of (4.116).

Proposition 4.14. *For an appropriately specified* $\lambda \in \mathbb{R}$, F *is called of class* $\mathrm{Y}_\lambda^{(2)}(\mathbb{R})$, *if* F *is of class* $\mathrm{C}^{(2)}(\mathbb{R})$ *satisfying*

$$(\Delta_x + \lambda + 4\pi^2 a^2) F(x) = O\left(|x|^{-(1+\varepsilon)}\right), \quad \varepsilon > 0, \ |x| \to \infty. \tag{4.117}$$

In connection with Proposition 4.14 we obtain the following lemma.

Lemma 4.15. *The integral*

$$\int_{\mathbb{R}} G(L+\lambda; x)\, e^{2\pi i a\cdot x} \left(\Delta_x + \lambda + 4\pi^2 a^2\right) F(x-y)\, dx \tag{4.118}$$

is absolutely convergent.

Proof. By virtue of (4.117) we are immediately able to guarantee that, for suitable positive constants M, N,

$$\left| \int_{\substack{M \le |x-y| \le N \\ x \in \mathbb{R}^q}} G(L+\lambda; x)\, e^{2\pi i a\cdot x} \left(\Delta_x + \lambda + 4\pi^2 a^2\right) F(x-y)\, dx \right| = O\left(\int_M^N \frac{1}{(1+x)^{1+\varepsilon}}\, dx \right). \tag{4.119}$$

Consequently, the absolute convergence of the integral (4.119) is guaranteed. □

Combining Lemma 4.13 and Lemma 4.15 we are able to formulate the following limit relation.

Theorem 4.16. *Suppose that, for a preassigned* $\lambda \in \mathbb{R}$, *the function* F *is of class* $\mathrm{X}^{(2)}(\mathbb{R}) \cap \mathrm{Y}_\lambda^{(2)}(\mathbb{R})$. *Then, the limit*

$$\lim_{N\to\infty} \left(\sum_{\substack{|g-y|\le N \\ g\in\Lambda}} e^{2\pi i a\cdot g} F(g-y) - \frac{1}{\|\mathcal{F}_\Lambda\|}\, e^{2\pi i a\cdot y} \sum_{\substack{4\pi^2((h+a)^2-a^2)=\lambda \\ h\in\Lambda^{-1}}} e^{2\pi i h\cdot y} \int_{\substack{|x|\le N \\ x\in\mathbb{R}}} F(x)\, e^{2\pi i x\cdot(a+h)} dx \right) \tag{4.120}$$

exists, and we have

$$\lim_{N\to\infty}\left(\sum_{\substack{|g-y|\le N\\ g\in\Lambda}} e^{2\pi i a\cdot g}F(g-y)\right.$$
$$\left. -\frac{1}{\|\mathcal{F}_\Lambda\|}\; e^{2\pi i a\cdot y}\sum_{\substack{4\pi^2((h+a)^2-a^2)=\lambda\\ h\in\Lambda^{-1}}} e^{2\pi i h\cdot y}\int_{\substack{|x|\le N\\ x\in\mathbb{R}}} F(x)\, e^{2\pi i x\cdot(a+h)}dx\right)$$
$$= \int_{\mathbb{R}} G(L+\lambda;x)\; e^{2\pi i a\cdot x}\left(\Delta_x+\lambda+4\pi^2a^2\right)F(x-y)\;dx, \tag{4.121}$$

where, as usual, the sum

$$\sum_{\substack{4\pi^2((h+a)^2-a^2)=\lambda\\ h\in\Lambda^{-1}}}\cdots \tag{4.122}$$

has to be extended over all $h\in\Lambda^{-1}$ satisfying $4\pi^2\left((h+a)^2-a^2\right)=\lambda$.

In the case that the parameter λ is given in such a way that there is no lattice point $h\in\Lambda^{-1}$ satisfying $4\pi^2\left((h+a)^2-a^2\right)=\lambda$, the finite lattice point sum (4.122) is understood to be zero, and we simply arrive at the following corollary.

Corollary 4.17. *Suppose that there is no lattice point $h\in\Lambda^{-1}$ satisfying the equation $4\pi^2\left((h+a)^2-a^2\right)=\lambda$, i.e., $\lambda\notin \mathrm{Spect}_L(\Lambda)$. Then, under the assumption of Theorem 4.16,*

$$\lim_{N\to\infty}\sum_{\substack{|g-y|\le N\\ g\in\Lambda}} e^{2\pi i a\cdot g}F(g-y)$$
$$= \int_{\mathbb{R}} G(L+\lambda;x)\; e^{2\pi i a\cdot x}\left(\Delta_x+\lambda+4\pi^2a^2\right)F(x-y)\;dx. \tag{4.123}$$

Theorem 4.16 demonstrates that the convergence of 1D-sums (understood in spherical summation) is closely related to the spectrum. In fact, for the class of non-eigenvalues, i.e., for values $\lambda\notin\mathrm{Spect}_L(\Lambda)$, Theorem 4.16 immediately guarantees the convergence of the infinite series.
In order to ensure the convergence of the sum (4.122) in the case of eigenvalues, however, we have to require an additional convergence condition for the remaining volume integrals:

Proposition 4.18. *For a given $\lambda\in\mathrm{Spect}_L(\Lambda)$, $F\in \mathrm{C}^{(2)}(\mathbb{R})$ is said to be of class $\mathrm{Z}_\lambda^{(2)}(\mathbb{R})$, if the integrals*

$$\int_{\mathbb{R}} F(x)\; e^{2\pi i x\cdot(a+h)}dx \tag{4.124}$$

exist for all $h \in \Lambda^{-1}$ *with* $4\pi^2\left((h+a)^2 - a^2\right) = \lambda$ *in the (spherical) sense*

$$\int_{\mathbb{R}} F(x)\ e^{2\pi i x\cdot(a+h)}dx = \lim_{N\to\infty} \int_{\substack{|x|\le N \\ x\in\mathbb{R}}} F(x)\ e^{2\pi i x\cdot(a+h)}dx. \tag{4.125}$$

Observing Proposition 4.18 we are able to formulate the following convergence theorem.

Theorem 4.19. *For a preassigned* $\lambda \in \mathrm{Spect}_L(\Lambda)$, *let* F *be a member of class* $\mathrm{X}^{(2)}(\mathbb{R}) \cap \mathrm{Y}^{(2)}_\lambda(\mathbb{R}) \cap \mathrm{Z}^{(2)}_\lambda(\mathbb{R})$. *Then the series*

$$\lim_{N\to\infty} \sum_{\substack{|g-y|\le N \\ g\in\Lambda}} e^{2\pi i a\cdot g}F(g-y) \tag{4.126}$$

is convergent (in the indicated spherical sense).
In more detail, for given $\lambda \notin \mathrm{Spect}_L(\Lambda)$ *and* $F \in \mathrm{X}^{(2)}(\mathbb{R}) \cap \mathrm{Y}^{(2)}_\lambda(\mathbb{R})$, *we have*

$$\begin{aligned}\lim_{N\to\infty} \sum_{\substack{|g-y|\le N \\ g\in\Lambda}} & e^{2\pi i a\cdot g}F(g-y) \\ &= \int_{\mathbb{R}} G(L+\lambda;x)\ e^{2\pi i a\cdot x}\left(\Delta_x + \lambda + 4\pi^2 a^2\right) F(x-y)\ dx,\end{aligned} \tag{4.127}$$

whereas, for given $\lambda \in \mathrm{Spect}_L(\Lambda)$ *and* $F \in \mathrm{X}^{(2)}(\mathbb{R}) \cap \mathrm{Y}^{(2)}_\lambda(\mathbb{R}) \cap \mathrm{Z}^{(2)}_\lambda(\mathbb{R})$, *we have*

$$\begin{aligned}\lim_{N\to\infty} & \sum_{\substack{|g-y|\le N \\ g\in\Lambda}} e^{2\pi i a\cdot g}F(g-y) \\ = & \lim_{N\to\infty} \frac{1}{\|\mathcal{F}_\Lambda\|}\ e^{2\pi i a\cdot y} \sum_{\substack{4\pi^2\left((h+a)^2-a^2\right)=\lambda \\ h\in\Lambda^{-1}}} e^{2\pi i h\cdot y} \int_{\substack{|x|\le N \\ x\in\mathbb{R}}} F(x)\ e^{2\pi i x\cdot(a+h)}dx \\ & + \int_{\mathbb{R}} G(L+\lambda;x)\ e^{2\pi i a\cdot x}\left(\Delta_x + \lambda + 4\pi^2 a^2\right) F(x-y)\ dx.\end{aligned} \tag{4.128}$$

As a consequence, our approach canonically implies to the following useful convention.

Proposition 4.20. *For a given* $\lambda \in \mathbb{R}$, *a function* $F \in \mathrm{C}^{(2)}(\mathbb{R})$ *is said to belong to the class* $\mathrm{C}^{(2)}_\lambda(\mathbb{R})$, *if one of the alternative conditions is satisfied:*

(i) F *is of class* $\mathrm{X}^{(2)}(\mathbb{R}) \cap \mathrm{Y}^{(2)}_\lambda(\mathbb{R})$, *provided that* $\lambda \notin \mathrm{Spect}_L(\Lambda)$,

(ii) F *is of class* $\mathrm{X}^{(2)}(\mathbb{R}) \cap \mathrm{Y}^{(2)}_\lambda(\mathbb{R}) \cap \mathrm{Z}^{(2)}_\lambda(\mathbb{R})$, *provided that* $\lambda \in \mathrm{Spect}_L(\Lambda)$.

The convergence criteria, which led to the formulation of Theorem 4.19, enable us to derive an extension of the classical Poisson summation formula. As the Λ–Euler–Green function $G((L+\lambda);\cdot)$ permits an absolutely and uniformly convergent Fourier series in $\mathbb{R}$, Lebesgue's theorem of dominated convergence allows us to interchange summation and integration such that

$$\begin{aligned}&\int_{\mathbb{R}} G((L+\lambda);x)\, e^{2\pi i a\cdot x}\left(\Delta_x+\lambda+4\pi^2a^2\right)F(x-y)\,dx \\ &\quad= \frac{1}{\|\mathcal{F}_\Lambda\|}\sum_{\substack{4\pi^2((h+a)^2-a^2)\neq\lambda\\ h\in\Lambda^{-1}}}\frac{1}{\lambda-4\pi^2\left((h+a)^2-a^2\right)} \\ &\qquad\times\int_{\mathbb{R}} e^{2\pi i(a+h)\cdot x}\left(\Delta_x+\lambda+4\pi^2a^2\right)F(x-y)\,dx\end{aligned} \tag{4.129}$$

holds true for a function $F\in \mathrm{C}^{(2)}_\lambda(\mathbb{R})$. Moreover, integration by parts, i.e., Green's theorem yields the identity

$$\begin{aligned}&\int_{\mathbb{R}} e^{2\pi i(a+h)\cdot x}\left(\Delta_x+\lambda+4\pi^2a^2\right)F(x-y)\,dx \\ &\quad=\int_{\mathbb{R}}\left(\Delta_x+\lambda+4\pi^2a^2\right)e^{2\pi i(a+h)\cdot x}\,F(x-y)\,dx \\ &\quad=\left(\lambda-4\pi^2\left((h+a)^2-a^2\right)\right)\int_{\mathbb{R}} e^{2\pi i(a+h)\cdot x}\,F(x-y)\,dx.\end{aligned} \tag{4.130}$$

Inserting (4.130) into (4.129) we therefore find that

$$\begin{aligned}&\int_{\mathbb{R}} G((L+\lambda);x)\, e^{2\pi i a\cdot x}\left(\Delta_x+\lambda+4\pi^2a^2\right)F(x-y)\,dx \\ &\quad=\frac{1}{\|\mathcal{F}_\Lambda\|}\,e^{2\pi i a\cdot y}\sum_{\substack{4\pi^2((h+a)^2-a^2)\neq\lambda\\ h\in\Lambda^{-1}}} e^{2\pi i h\cdot y}\int_{\mathbb{R}}F(x)\,e^{2\pi i x\cdot(a+h)}dx.\end{aligned} \tag{4.131}$$

In combination with Theorem 4.19 this finally yields the *1D-Poisson-type summation formula.*

Theorem 4.21. *For a preassigned $\lambda\in\mathbb{R}$, suppose that the function F is of class $\mathrm{C}^{(2)}_\lambda(\mathbb{R})$. Then the limit*

$$\lim_{N\to\infty}\sum_{\substack{|g-y|\le N\\ g\in\Lambda}} e^{2\pi i a\cdot g}F(g-y) \tag{4.132}$$

is convergent, and we have

$$\begin{aligned}&\lim_{N\to\infty}\sum_{\substack{|g-y|\le N\\ g\in\Lambda}} e^{2\pi i a\cdot g}F(g-y) \\ &\quad=\sum_{h\in\Lambda^{-1}}\int_{\mathbb{R}}F(x)\,e^{2\pi i x\cdot(a+h)}dx\,\frac{1}{\|\mathcal{F}_\Lambda\|}e^{2\pi i(a+h)\cdot y}.\end{aligned} \tag{4.133}$$

It is not hard to see that Theorem 4.21 allows two equivalent representations, namely

$$\lim_{N\to\infty} \frac{1}{\|\mathcal{F}_\Lambda\|} \sum_{\substack{|h-y|\le N\\ h\in\Lambda^{-1}}} e^{2\pi i a\cdot h}\, F(h-y) \tag{4.134}$$
$$= \sum_{g\in\Lambda} e^{2\pi i y\cdot(a+g)} \int_{\mathbb{R}^q} F(x)\, e^{2\pi i x\cdot(a+g)} dx$$

and

$$\lim_{N\to\infty} \sum_{\substack{|g+y|\le N\\ g\in\Lambda}} e^{-2\pi i a\cdot(g+y)} F(g+y) \tag{4.135}$$
$$= \frac{1}{\|\mathcal{F}_\Lambda\|} \sum_{h\in\Lambda^{-1}} e^{2\pi i h\cdot y} \int_{\mathbb{R}} F(x)\, e^{-2\pi i x\cdot(h+a)} dx, \quad a,y\in\mathbb{R}.$$

Analogously to the context implied by the Poisson-type summation formula for finite intervals (Theorem 4.8), the identities (4.134) and (4.135), respectively, offer two different entry points for Shannon-type sampling, i.e., functional values–based Shannon-type sampling and Fourier transformed values–based Shannon-type sampling characterized, respectively, by

- the summation of the resulting cardinal series on the left side of (4.134) is extended over functional values of $F \in \mathrm{C}^{(2)}_\lambda(\mathbb{R})$ in y-shifted lattice points of Λ,
- the summation of the resulting cardinal series on the right side of (4.135) is extended over functional values of the Fourier transform of $F \in \mathrm{C}^{(2)}_\lambda(\mathbb{R})$ in a-shifted lattice points of Λ^{-1}.

4.10 Functional Values–Based Shannon-Type Sampling (Euclidean Space)

From the identity (4.134) we obtain by integration over $\mathbb{B}^1_\tau$ the following Shannon-type sampling theorem (the proof is analogous to that presented in the multivariate part of this book).

Theorem 4.22. *Assume that Λ is an arbitrary lattice in $\mathbb{R}$. Suppose that F is a member of the class $\mathrm{C}^{(2)}_\lambda(\mathbb{R}^q)$. Then, for all $y\in\mathbb{R}$, we have*

$$\sum_{\substack{(\mathcal{F}_\Lambda+\{g'\})\cap\mathbb{B}^1_\tau\neq\emptyset\\ g'\in\Lambda}} \int_{\bigcup_{g\in\Lambda}(((\mathbb{B}^1_\tau\cap(\mathcal{F}_\Lambda+\{g'\}))-\{g'\})+\{g\})} F^\wedge_{\mathbb{R}}(w)\, e^{2\pi i y\cdot w}\, dw \tag{4.136}$$
$$= \lim_{N\to\infty} \sum_{\substack{|h|\le N\\ h\in\Lambda^{-1}}} F(h)\, \frac{1}{\|\mathcal{F}_\Lambda\|}\, K_{\mathbb{B}^1_\tau}(h+y).$$

We choose $\Lambda = \sigma\mathbb{Z}$ and distinguish the three cases $\overline{\mathbb{B}^1_\tau} \subsetneq \overline{\mathcal{F}_{\sigma\mathbb{Z}}}$, $\overline{\mathbb{B}^1_\tau} = \overline{\mathcal{F}_{\sigma\mathbb{Z}}}$, and $\overline{\mathbb{B}^1_\tau} \supsetneq \overline{\mathcal{F}_{\sigma\mathbb{Z}}}$:

For $\overline{\mathbb{B}^1_\tau} \subsetneq \overline{\mathcal{F}_{\sigma\mathbb{Z}}}$ we find that

$$\int_{\bigcup\limits_{g\in\sigma\mathbb{Z}} (\mathbb{B}^1_\tau+\{g\})} F^\wedge_{\mathbb{R}}(x)\, e^{2\pi i y\cdot x}\, dx = \lim_{N\to\infty} \sum_{\substack{|h|\le N \\ h\in\frac{1}{\sigma}\mathbb{Z}}} F(h)\, \frac{1}{\sigma}\, K_{\mathbb{B}^1_\tau}(h+y). \tag{4.137}$$

Thus we are confronted with the following observation: The closer $\overline{\mathbb{B}^1_\tau}$ is to the fundamental cell of $\sigma\mathbb{Z}$, the smaller is the aliasing error by this variant of Shannon-type sampling.

For $\overline{\mathbb{B}^1_\tau} = \overline{\mathcal{F}_{\sigma\mathbb{Z}}}$, we obtain by application of the Fourier inversion formula

$$\begin{aligned} \int_{\bigcup\limits_{g\in\sigma\mathbb{Z}} (\mathcal{F}_{\sigma\mathbb{Z}}+\{g\})} e^{2\pi i y\cdot x}\, F^\wedge_{\mathbb{R}}(x)\; dx &= \int_{\mathbb{R}} e^{2\pi i y\cdot x}\, F^\wedge_{\mathbb{R}}(x)\, dx \\ &= F(y) \\ &= \lim_{N\to\infty} \sum_{\substack{|h|\le N \\ h\in\frac{1}{\sigma}\mathbb{Z}}} F(h)\, \frac{1}{\sigma}\, K_{\mathcal{F}_{\sigma\mathbb{Z}}}(h+y). \end{aligned} \tag{4.138}$$

In other words, the functions F of the class $\mathrm{C}^{(2}_\lambda(\mathbb{R})$ can be sampled in the standard way by choosing the fundamental cell $\mathcal{F}_{\sigma\mathbb{Z}}$ to act as the kernel generator $K_{\mathcal{F}_{\sigma\mathbb{Z}}}$ of the cardinal series in Shannon sampling

$$F(y) = \lim_{N\to\infty} \sum_{\substack{|h|\le N \\ h\in\frac{1}{\sigma}\mathbb{Z}}} F(h)\, \frac{1}{\sigma}\, K_{\mathcal{F}_{\sigma\mathbb{Z}}}(h+y), \quad y\in\mathbb{R}. \tag{4.139}$$

It should be noted that F is not necessarily bandlimited.

For $\overline{\mathbb{B}^1_\tau} \supsetneq \overline{\mathcal{F}_{\sigma\mathbb{Z}}}$

$$\begin{aligned} &\sum_{\substack{(\mathcal{F}_{\sigma\mathbb{Z}}+\{g'\})\cap\mathbb{B}^1_\tau\neq\emptyset \\ g'\in\Lambda}} \int_{\bigcup\limits_{g\in\Lambda} (((\mathbb{B}^1_\tau\cap(\mathcal{F}_\Lambda+\{g'\}))-\{g'\})+\{g\})} F^\wedge_{\mathbb{R}}(x)\, e^{2\pi i y\cdot x}\, dx \\ &= \lim_{N\to\infty} \sum_{\substack{|h|\le N \\ h\in\frac{1}{\sigma}\mathbb{Z}}} F(h)\, \frac{1}{\sigma}\, K_{\mathbb{B}^1_\tau}(h-y). \end{aligned} \tag{4.140}$$

Altogether, in the case of a lattice cell (or finite unions of them) aliasing can be avoided. In all other cases, a way is found in not-necessarily bandlimited sampling to detect the explicit aliasing error.

4.11 Fourier Transformed Values–Based Shannon-Type Sampling (Euclidean Space)

Next we start from the *Poisson-type summaton formula* in $\mathbb{R}$ of the form (4.135), where we use the right side of (4.135) to generate a cardinal series in a Shannon-type sampling process (for the proof the reader again is referred to the multivariate context presented in this work).

Theorem 4.23. *Assume that Λ is an arbitrary lattice in $\mathbb{R}$. Suppose that F is a member of the class $\mathrm{C}^{(2)}_{\Lambda}(\mathbb{R})$. Then*

$$\sum_{\substack{(\mathcal{F}_\Lambda+\{g'\})\cap\mathbb{B}^1_\tau\neq\emptyset\\ g'\in\Lambda}} F^{\wedge}_{\bigcup\limits_{g\in\Lambda}(((\mathbb{B}^1_\tau\cap(\mathcal{F}_\Lambda+\{g'\}))-\{g'\})+\{g\})}(a) \tag{4.141}$$
$$= \sum_{h-a\in\Lambda^{-1}} \underbrace{\int_{\mathbb{R}^q} F(x)\, e^{-2\pi i h\cdot x}\, dx}_{=F^{\wedge}_{\mathbb{R}}(h)} \frac{1}{\|\mathcal{F}_\Lambda\|} \underbrace{\int_{\mathbb{B}^1_\tau} e^{2\pi i (h-a)\cdot y}\, dy}_{=K_{\mathbb{B}^1_\tau}(h-a)}.$$

As a consequence of (4.141) we obtain for $\Lambda=\sigma\mathbb{Z}$

$$\sum_{\substack{(\mathcal{F}_{\sigma\mathbb{Z}}+\{g'\})\cap\mathbb{B}^1_\tau\neq\emptyset\\ g'\in\sigma\mathbb{Z}}} F^{\wedge}_{\bigcup\limits_{g\in\sigma\mathbb{Z}}(((\mathbb{B}^1_\tau\cap(\mathcal{F}_{\sigma\mathbb{Z}}+\{g'\}))-\{g'\})+\{g\})}(a) - \frac{2\tau}{\sigma} \underbrace{\int_{\mathbb{R}} F(x)\, e^{-2\pi i a\cdot x}\, dx}_{=F^{\wedge}_{\mathbb{R}}(a)}$$
$$= \sum_{\substack{h-a\in\frac{1}{\sigma}\mathbb{Z}\\ h\neq a}} \underbrace{\int_{\mathbb{R}} F(x)\, e^{-2\pi i h\cdot x}\, dx}_{=F^{\wedge}_{\mathbb{R}}(h)} \frac{1}{\sigma} \underbrace{\int_{\mathbb{B}^1_\tau} e^{2\pi i (h-a)\cdot y}\, dy}_{=K_{\mathbb{B}^1_\tau}(h-a)}. \tag{4.142}$$

First we distinguish the two examples $\overline{\mathbb{B}^1_\tau}\subsetneq\overline{\mathcal{F}_{\sigma\mathbb{Z}}}$ and $\overline{\mathbb{B}^1_\tau}\supsetneq\overline{\mathcal{F}_{\sigma\mathbb{Z}}}$:

For $\overline{\mathbb{B}^1_\tau}\subsetneq\overline{\mathcal{F}_{\sigma\mathbb{Z}}}$ we obtain

$$F^{\wedge}_{\bigcup\limits_{g\in\sigma\mathbb{Z}}(\mathbb{B}^1_\tau+\{g\})}(a) - \frac{2\tau}{\sigma} \underbrace{\int_{\mathbb{R}} F(x)\, e^{-2\pi i a\cdot x}\, dx}_{=F^{\wedge}_{\mathbb{R}}(a)}$$
$$= \sum_{\substack{h-a\in\frac{1}{\sigma}\mathbb{Z}\\ h\neq a}} \underbrace{\int_{\mathbb{R}} F(x)\, e^{-2\pi i h\cdot x}\, dx}_{=F^{\wedge}_{\mathbb{R}}(h)} \frac{1}{\sigma} \underbrace{\int_{\mathbb{B}^1_\tau} e^{2\pi i (h-a)\cdot y}\, dy}_{=K_{\mathbb{B}^1_\tau}(h-a)}, \tag{4.143}$$

while for $\overline{\mathbb{B}^1_\tau} \supsetneq \overline{\mathcal{F}_{\sigma\mathbb{Z}}}$

$$\sum_{\substack{(\mathcal{F}_{\sigma\mathbb{Z}}+\{g'\})\cap\mathbb{B}^1_\tau\neq\emptyset \\ g'\in\sigma\mathbb{Z}}} F^\wedge_{\bigcup\limits_{g\in\sigma\mathbb{Z}}(((\mathbb{B}^1_\tau\cap(\mathcal{F}_{\sigma\mathbb{Z}}+\{g'\}))-\{g'\})+\{g\})}(a) - \frac{2\tau}{\sigma}\underbrace{\int_{\mathbb{R}} F(x)\, e^{-2\pi i a\cdot x}\, dx}_{=F^\wedge_{\mathbb{R}}(a)}$$

$$= \sum_{\substack{h-a\in\frac{1}{\sigma}\mathbb{Z} \\ h\neq a}} \underbrace{\int_{\mathbb{R}} F(x)\, e^{-2\pi i h\cdot x}\, dx}_{=F^\wedge_{\mathbb{R}}(h)}\ \frac{1}{\sigma}\underbrace{\int_{\mathbb{B}^1_\tau} e^{2\pi i(h-a)\cdot y}\, dx}_{=K_{\mathbb{B}^1_\tau}(h-a)}. \tag{4.144}$$

For the particular case $\overline{\mathbb{B}^1_\tau} = \overline{\mathcal{F}_{\sigma\mathbb{Z}}}$ and $a\in\mathbb{R}$ we obtain

$$\sum_{\substack{h-a\in\frac{1}{\sigma}\mathbb{Z} \\ h\neq a}} \underbrace{\int_{\mathbb{R}} F(x)\, e^{-2\pi i h\cdot x}\, dx}_{=F^\wedge_{\mathbb{R}}(h)}\ \frac{1}{\sigma}\underbrace{\int_{\mathcal{F}_{\sigma\mathbb{Z}}} e^{2\pi i(h-a)\cdot y}\, dx}_{=K_{\mathcal{F}_{\sigma\mathbb{Z}}}(h-a)} = 0. \tag{4.145}$$

Part III

Preparatory Material for Multivariate Lattice Point Summation and Shannon-Type Sampling

Chapter 5

Preparatory Tools of Vector Analysis

The preparatory material provided by the next chapters contains basic notation for our multivariate apparatus. The material is well-known, e.g., from the monograph of W. Freeden [2011], where all proofs can be found, too. Nonetheless, in the opinion of the authors, these preparatory chapters are indispensable for a consistent nomenclature and an appropriate readability of the remaining part of this work.

5.1 Cartesian Notation and Settings

As usual, we write $x, y, \ldots$ to represent the elements of the q–dimensional (real) Euclidean space $\mathbb{R}^q$ $(q \geq 1)$. In Cartesian coordinates we have the component representation (q–tuples of real numbers)

$$x = \begin{pmatrix} x_1 \\ \vdots \\ x_q \end{pmatrix}, \qquad y = \begin{pmatrix} y_1 \\ \vdots \\ y_q \end{pmatrix}. \tag{5.1}$$

Remark 5.1. *Only if necessary we write $x_{(q)}$ instead of x to point out that x is an element of $\mathbb{R}^q$.*

Cartesian Coordinates. The canonical orthonormal system in $\mathbb{R}^q$ is denoted by $\varepsilon^1, \ldots, \varepsilon^q$. More explicitly,

$$\varepsilon^1 = \begin{pmatrix} 1 \\ 0 \\ \vdots \\ 0 \end{pmatrix}, \quad \ldots, \quad \varepsilon^q = \begin{pmatrix} 0 \\ \vdots \\ 0 \\ 1 \end{pmatrix}. \tag{5.2}$$

Any element $x \in \mathbb{R}^q$ may be represented in Cartesian coordinates $x_i = x \cdot \varepsilon^i, i = 1, \ldots, q$, by

$$x = \sum_{i=1}^{q} (x \cdot \varepsilon^i)\, \varepsilon^i = \sum_{i=1}^{q} x_i\, \varepsilon^i. \tag{5.3}$$

In coordinates the *inner (scalar) product* of two elements $x, y \in \mathbb{R}^q$ is given by

$$x \cdot y = x^T y = \sum_{i=1}^{q} x_i y_i. \tag{5.4}$$

Clearly,

$$x^2 = |x|^2 = x \cdot x = x^T x, \quad x \in \mathbb{R}^q, \tag{5.5}$$

i.e., *the Euclidean norm* in $\mathbb{R}^q$ is given by

$$|x| = \sqrt{x \cdot x} = \sqrt{x^T x}, \quad x \in \mathbb{R}^q. \tag{5.6}$$

Translates. Given a vector $a \in \mathbb{R}^q$ and a set $\mathcal{M} \subset \mathbb{R}^q$. Let $\mathcal{M} + \{a\}$ denote the set of all points $y = x + a$, as x runs through the points of $\mathcal{M}$, i.e.,

$$\mathcal{M} + \{a\} = \{y = x + a : a \in \mathcal{M}\}. \tag{5.7}$$

$\mathcal{M} + \{a\}$ is the *translate* of the set $\mathcal{M}$ by a. More generally, if $\mathcal{N}$ denotes some set of vectors from $\mathbb{R}^q$ then by $\mathcal{M} + \mathcal{N}$ we understand the set of all points $y = x + a$ for an arbitrary x from $\mathcal{M}$ and an arbitrary a from $\mathcal{N}$.

Regions and Functions. If $\mathcal{G}$ is a set of points in $\mathbb{R}^q$, $\partial\mathcal{G}$ denotes its *boundary*. The set $\overline{\mathcal{G}} = \mathcal{G} \cup \partial\mathcal{G}$ is called the *closure* of $\mathcal{G}$. A set $\mathcal{G} \subset \mathbb{R}^q$ is called a *region* if and only if it is open and connected.
By a scalar or vector function (field) on a region $\mathcal{G} \subset \mathbb{R}^3$, we mean a function that assigns to each point of $\mathcal{G}$, a scalar or vectorial function value, respectively. It will be advantageous to use the following general scheme of notation:

capital letters F, G	:	scalar functions,
lower-case letters f, g	:	vector fields.

The restriction of a scalar-valued function F or a vector–valued function f to a subset M of its domain is denoted by $F|M$ or $f|M$, respectively. For a set S of functions, we set $S|M = \{F|M : F \in S\}$.

Differential Operators. Let $\mathcal{G} \subset \mathbb{R}^q$ be a region. Suppose that $F : \mathcal{G} \to \mathbb{R}$ is differentiable. $\nabla F : x \mapsto (\nabla F)(x)$, $x \in \mathcal{G}$, denotes the *gradient* of F on $\mathcal{G}$. The *partial derivatives* of F at $x \in \mathcal{G}$, are given by

$$\frac{\partial F}{\partial x_i}(x) = (\nabla_x F)(x) \cdot \varepsilon^i = ((\nabla_x F)(x))_i \quad . \tag{5.8}$$

Only if necessary we add subscripts to operators to point out the dependence on a certain variable.

Let $u : \mathcal{G} \to \mathbb{R}^q$ be a vector field, and suppose, in addition, that u is differentiable at a point $x \in \mathcal{G}$. The partial derivatives of u at $x \in \mathcal{G}$ are given by

$$\frac{\partial u_i}{\partial x_j}(x) = \varepsilon^i \cdot (\nabla u)(x) \varepsilon^j \quad . \tag{5.9}$$

The *divergence* of u at $x \in \mathcal{G}$ is the scalar value

$$\nabla_x \cdot u(x) = \mathrm{tr}\,(\nabla u)(x). \tag{5.10}$$

Let F be a differentiable scalar field on $\mathcal{G}$, and suppose, in addition, that ∇F is differentiable at $x \in \mathcal{G}$. Then we introduce the *Laplace operator (Laplacian)* of F at $x \in \mathcal{G}$ by

$$\Delta_x F(x) = \nabla_x \cdot \left(\left(\nabla_x F\right)(x)\right). \tag{5.11}$$

Multi-Indices. Let $\alpha = (\alpha_1, \ldots, \alpha_q)^{\mathrm{T}}$ be a q-tuple of non-negative integers $\alpha_1, \ldots, \alpha_q$, i.e., $\alpha \in \mathbb{N}_0{}^q$. We set

$$\alpha! = \alpha_1! \cdot \ldots \cdot \alpha_q!, \tag{5.12}$$

$$[\alpha] = \alpha_1 + \ldots + \alpha_q, \tag{5.13}$$

$$|\alpha| = \sqrt{\alpha_1^2 + \ldots + \alpha_q^2}. \tag{5.14}$$

We say $\alpha = (\alpha_1, \ldots, \alpha_q)^{\mathrm{T}}$ is a q-dimensional *multi-index of degree* n if $[\alpha] = n$. As usual, we set

$$x^\alpha = x_1^{\alpha_1} \cdot \ldots \cdot x_q^{\alpha_q}, \; x \in \mathbb{R}^q, \; \alpha \in \mathbb{N}_0^q, \tag{5.15}$$

$$(\nabla_x)^\alpha = \left(\frac{\partial}{\partial x_1}\right)^{\alpha_1} \cdots \left(\frac{\partial}{\partial x_q}\right)^{\alpha_q} = \frac{\partial^{[\alpha]}}{(\partial x_1)^{\alpha_1} \ldots (\partial x_q)^{\alpha_q}}. \tag{5.16}$$

Clearly,

$$(\nabla_x)^\alpha x^\beta = \begin{cases} 0 & , \quad \alpha \neq \beta, \; [\alpha] = [\beta] \\ \alpha! & , \quad \alpha = \beta. \end{cases} \tag{5.17}$$

In this notation of multi-indices we have

$$\left(\sum_{i=1}^{q} x_i\right)^n = \sum_{[\alpha]=n} \frac{n!}{\alpha!} x^\alpha. \tag{5.18}$$

5.2 Spherical Notation and Settings

The unit sphere in $\mathbb{R}^q$ is denoted by $\mathbb{S}^{q-1}$:

$$\mathbb{S}^{q-1} = \left\{x \in \mathbb{R}^q : \; |x| = 1\right\}. \tag{5.19}$$

Each $x \in \mathbb{R}^q, \; x = (x_1, \ldots . x_q)^{\mathrm{T}}, \; |x| \neq 0$, admits a representation of the form

$$x = r\xi, \; r = |x|, \; \xi = (\xi_1, \ldots, \xi_q)^{\mathrm{T}}, \tag{5.20}$$

where $\xi \in \mathbb{S}^{q-1}$ is the uniquely determined (unit) vector of x.

Using the canonical orthonormal basis $\varepsilon^1, \ldots, \varepsilon^q$ in $\mathbb{R}^q$ (more accurately, $\varepsilon^1_{(q)}, \ldots, \varepsilon^q_{(q)}$ in $\mathbb{R}^q$) we are able to write $\xi_{(q)} \in \mathbb{S}^{q-1}$, $q \geq 3$, in the form

$$\xi_{(q)} = t\varepsilon^q + \sqrt{1-t^2}\,\xi_{(q-1)},\ t \in [-1,1],\ \xi_{(q-1)} \in \mathbb{S}^{q-2}, \tag{5.21}$$

$$\xi_{(2)} = (\cos\varphi, \sin\varphi)^T,\ \xi_{(2)} \in \mathbb{S}^1,\ \varphi \in [0, 2\pi). \tag{5.22}$$

The *sphere in* $\mathbb{R}^q$ *with radius* R *around* $y \in \mathbb{R}^q$ is denoted by $\mathbb{S}^{q-1}_R(y)$

$$\mathbb{S}^{q-1}_R(y) = \{x \in \mathbb{R}^q : |x-y| = R\}, \tag{5.23}$$

and $\mathbb{S}^{q-1}_R$ is the sphere with radius R around 0 (i.e., $\mathbb{S}^{q-1}_R = \mathbb{S}^{q-1}_R(0)$). Furthermore, $\mathbb{B}^q_R(y)$ denotes *the ball in* $\mathbb{R}^q$ *with center* $y \in \mathbb{R}^q$ *and radius* R:

$$\mathbb{B}^q_R(y) = \{x \in \mathbb{R}^q : |x-y| < R\}. \tag{5.24}$$

We simply write $\mathbb{B}^q_R$ for the ball with radius R around the origin 0. Moreover, we let $\mathbb{B}^q_1 = \mathbb{B}^q$.

By $\mathbb{B}^q_{\rho,R}(y)$, $0 \leq \rho < R$, we denote the *ball ring with center* $y \in \mathbb{R}^q$ *and radii* ρ *and* R given by

$$\mathbb{B}^q_{\rho,R}(y) = \{x \in \mathbb{R}^q : \rho < |x-y| < R\}, \tag{5.25}$$

i.e., $\mathbb{B}^q_{\rho,R}(y) = \mathbb{B}^q_R(y)\backslash\overline{\mathbb{B}^q_\rho(y)}$.
A function G is called *radial* in $\overline{\mathbb{B}^2_{\rho,N}}$, $0 \leq \rho \leq N$, if for all $x \in \overline{\mathbb{B}^2_{\rho,N}}$

$$G(x) = G(r\xi) = G(r), \quad x = r\xi,\ r = |x|. \tag{5.26}$$

A function H is called *angular* in $\overline{\mathbb{B}^2_{\rho,N}}$, $0 < \rho \leq N$, if for all $x \in \overline{\mathbb{B}^2_{\rho,N}}$

$$H(x) = H(r\xi) = H(\xi), \quad x = r\xi,\ r = |x|. \tag{5.27}$$

By means of polar coordinates $x_{(q)} = r\xi_{(q)}$, $r = |x_{(q)}|$, $\xi_{(q)} \in \mathbb{S}^{q-1}$, the *gradient* ∇ in $\mathbb{R}^q$ can be represented in the form

$$\nabla_{x_{(q)}} = \xi_{(q)}\frac{\partial}{\partial r} + \frac{1}{r}\nabla^*_{\xi_{(q)}}, \tag{5.28}$$

where ∇^* is the *surface gradient* on $\mathbb{S}^{q-1}$. Moreover, in terms of spherical coordinates the *Laplace operator (Laplacian)* $\Delta = \nabla \cdot \nabla$ in $\mathbb{R}^q$ has the representation

$$\Delta_{x_{(q)}} = r^{1-q}\frac{\partial}{\partial r}r^{q-1}\frac{\partial}{\partial r} + \frac{1}{r^2}\Delta^*_{\xi_{(q)}}, \tag{5.29}$$

where Δ^* describes the *(Laplace–)Beltrami-operator* of the unit sphere $\mathbb{S}^{q-1}$ recursively given by

$$\Delta^*_{\xi_{(q)}} = (1-t^2)\left(\frac{\partial}{\partial t}\right)^2 - (q-1)t\frac{\partial}{\partial t} + \frac{1}{1-t^2}\Delta^*_{\xi_{(q-1)}}, \tag{5.30}$$

$$\Delta^*_{\xi_{(2)}} = \left(\frac{\partial}{\partial\varphi}\right)^2. \tag{5.31}$$

Clearly,

$$\Delta^* = \nabla^* \cdot \nabla^*, \tag{5.32}$$

where $\nabla^*\cdot$ is the *surface divergence* on $\mathbb{S}^{q-1}$ (for more details concerning the differential operators in the 3D case see, e.g., W. Freeden, M. Schreiner [2009]).

5.3 Regular Regions and Integral Theorems

We begin with some basics of vector analysis.

Regular Regions. A bounded region $\mathcal{G} \subset \mathbb{R}^q$ is called *regular*, if its boundary $\partial\mathcal{G}$ is an orientable piecewise smooth Lipschitzian manifold of dimension $q-1$ (for more details about regular regions the reader is referred to textbooks on vector analysis). Examples are ball, ellipsoid, "potato", but also cube and other polyhedra, "potato halfs", etc.
The *volume of a regular region* $\mathcal{G} \subset \mathbb{R}^q$ is written by

$$||\mathcal{G}|| = \int_{\mathcal{G}} dV, \tag{5.33}$$

where dV (or more accurately, $dV_{(q)}$) is the volume element.
The *area of the boundary* $\partial\mathcal{G}$ *of a regular region* $\mathcal{G} \subset \mathbb{R}^q$ is given by

$$||\partial\mathcal{G}|| = \int_{\partial\mathcal{G}} dS, \tag{5.34}$$

where dS (or more accurately, $dS_{(q-1)}$) is the surface element.

Green's Theorems (Laplace Operator). A regular region $\mathcal{G} \subset \mathbb{R}^q$ (cf. Section 5.1) allows the *Gauss theorem*

$$\int_{\mathcal{G}} \nabla \cdot f(x)\, dV(x) = \int_{\partial\mathcal{G}} f(x) \cdot \nu(x)\, dS(x) \tag{5.35}$$

for all continuously differentiable vector fields f on $\overline{\mathcal{G}}$, $\overline{\mathcal{G}} = \mathcal{G} \cup \partial\mathcal{G}$ (throughout this work, ν is the (unit) normal field on $\partial\mathcal{G}$ directed to the exterior of $\overline{\mathcal{G}}$). By letting $f = \nabla F, F \in \mathrm{C}^{(2)}(\overline{\mathcal{G}})$, $\mathcal{G} \subset \mathbb{R}^q$ regular, we obtain from (5.35)

$$\int_{\mathcal{G}} \Delta F(x)\, dV(x) = \int_{\partial\mathcal{G}} \frac{\partial F}{\partial \nu}(x)\, dS(x), \tag{5.36}$$

where $\frac{\partial}{\partial \nu}$ denotes the derivative in the direction of the outer (unit) normal field ν. Consequently, for all functions $F \in \mathrm{C}^{(2)}(\overline{\mathcal{G}})$ satisfying the Laplace equation $\Delta F = 0$ in $\mathcal{G}$, we have

$$\int_{\partial\mathcal{G}} \frac{\partial F}{\partial \nu}(x)\, dS(x) = 0. \tag{5.37}$$

For all vector fields $f = F\nabla G$, $F \in \mathrm{C}^{(1)}(\overline{\mathcal{G}})$, $G \in \mathrm{C}^{(2)}(\overline{\mathcal{G}})$, we obtain from the Gauss theorem the *first Green theorem.*

Theorem 5.2. *Suppose that* $\mathcal{G} \subset \mathbb{R}^q$ *is a regular region. For* $F \in \mathrm{C}^{(1)}(\overline{\mathcal{G}}), G \in \mathrm{C}^{(2)}(\overline{\mathcal{G}})$ *we have*

$$\int_{\mathcal{G}} \{F(x)\Delta G(x) + \nabla F(x) \cdot \nabla G(x)\}\, dV(x) = \int_{\partial\mathcal{G}} F(x)\frac{\partial G}{\partial \nu}(x)\, dS(x).$$

Taking $f = F\nabla G - G\nabla F$ with $F, G \in \mathrm{C}^{(2)}(\overline{\mathcal{G}})$ we obtain the *second Green theorem.*

Theorem 5.3. *Suppose that* $\mathcal{G} \subset \mathbb{R}^q$ *is a regular region. For* $F, G \in \mathrm{C}^{(2)}(\overline{\mathcal{G}})$ *we have*

$$\begin{aligned}&\int_{\mathcal{G}} \{G(x)\Delta F(x) - F(x)\Delta G(x)\}\ dV(x)\\ &\quad = \int_{\partial\mathcal{G}} \left\{G(x)\frac{\partial F}{\partial \nu}(x) - F(x)\frac{\partial G}{\partial \nu}(x)\right\}\ dS(x).\end{aligned}$$

Next we come to the well-known definition of harmonic functions.

Definition 5.4. $U \in \mathrm{C}^{(2)}(\mathcal{G})$ *is called a harmonic function (sometimes also called a harmonic) in a region* $\mathcal{G} \subset \mathbb{R}^q$ *if it satisfies the Laplace equation*

$$\Delta U(x) = 0\ ,\ x \in \mathcal{G}. \tag{5.38}$$

$U \in \mathrm{C}^{(2m)}(\mathcal{G})$, $m \in \mathbb{N}$, *is called a polyharmonic function of degree* m *in* $\mathcal{G} \subset \mathbb{R}^q$ *if*

$$\Delta^m U(x) = 0, \quad x \in \mathcal{G}. \tag{5.39}$$

Fundamental Solution (Harmonic Case). The function

$$x \mapsto F_q(|x-y|) = \begin{cases} -\frac{1}{2\pi}\ln(|x-y|) & ,\quad q = 2, \\ \frac{|x-y|^{2-q}}{(q-2)\|\mathbb{S}^{q-1}\|} & ,\quad q \geq 3 \end{cases} \tag{5.40}$$

satisfies

$$\Delta_x F_q(|x-y|),\ x \neq y. \tag{5.41}$$

By convention, (5.40) is called the *fundamental solution in* $\mathbb{R}^q$ *for the Laplace operator* Δ.

Remark 5.5. *As already mentioned, the 1D fundamental solution for the Laplacian, i.e., the operator of the second-order derivative, is given by the continuous function*

$$x \mapsto F_1(|x-y|) = -\frac{1}{2}|x-y| = -\frac{1}{2}(x-y)\ \mathrm{sign}(x-y), \tag{5.42}$$

$x, y \in \mathbb{R}$. *The finite jump of its derivative is a feature with nice applications – unfortunately only in the 1D lattice point and sampling theory.*

The fundamental solution of the Laplace operator possesses the following property (see, e.g., O.D. Kellogg [1929], N. Aronszaijn et al. [1983]).

Lemma 5.6. *For continuous functions G, H in the ball with radius $R > 0$ and center $y \in \mathbb{R}^q$, $R > r > 0$, we have*

$$\lim_{\substack{r\to 0\\ r>0}} \int_{|x-y|=r} G(x) \frac{\partial}{\partial \nu_x} F_q(|x-y|)\ dS(x) \quad = \quad -\ G(y), \tag{5.43}$$

$$\lim_{\substack{r\to 0\\ r>0}} \int_{|x-y|=r} H(x)\ F_q(|x-y|)\ dS(x) \quad = \quad 0, \tag{5.44}$$

where the (unit) normal field ν is directed to the exterior.

Solid Angle. Next we introduce an integral to represent a measure for the "solid angle" $\alpha(x)$ subtended by the boundary $\partial\mathcal{G}$ at the point $x \in \mathbb{R}^q$ (cf. Figure 5.1 and 5.2).

Definition 5.7. *For a regular region $\mathcal{G}$, the solid angle $\alpha(y)$ (more accurately, $\alpha_{\mathcal{G}}(y)$) at $y \in \mathbb{R}^q$ subtended by the surface $\partial\mathcal{G}$ is defined by*

$$\alpha(y) \quad = \quad - \int_{\partial\mathcal{G}} \frac{\partial}{\partial \nu_x} F_q(|x-y|)\ dS(x), \quad y \in \mathbb{R}^q. \tag{5.45}$$

Lemma 5.8. *Let $\mathcal{G} \subset \mathbb{R}^q$ be a regular region with continuously differentiable boundary $\partial\mathcal{G}$. Then*

$$-\int_{\partial\mathcal{G}} \frac{\partial}{\partial \nu_x} F_q(|x-y|)\ dS(x) = \begin{cases} 1 & , \quad y \in \mathcal{G} \\ \frac{1}{2} & , \quad y \in \partial\mathcal{G} \\ 0 & , \quad y \notin \overline{\mathcal{G}}. \end{cases} \tag{5.46}$$

Example 5.9. *For the cube (cf. Figure 5.2)*

$$\mathcal{G} \ = \ (-1,1)^3 \subset \mathbb{R}^3, \tag{5.47}$$

we have

(i) $\alpha(y) \ = \ 0$ *if y is located in the complement $\mathbb{R}^3 \backslash \overline{\mathcal{G}}$ of the closed cube $\overline{\mathcal{G}}$,*

(ii) $\alpha(y) \;=\; 1$ *if* y *is located in the open cube* $\mathcal{G}$,

(iii) $\alpha(y) \;=\; \frac{1}{2}$ *if* y *is located on one of the six faces of the boundary* $\partial\mathcal{G}$ *of the cube* $\mathcal{G}$ *but not on an edge or in a vertex,*

(iv) $\alpha(y) \;=\; \frac{1}{4}$ *if* y *is located on one of the eight edges of* $\partial\mathcal{G}$ *but not in a vertex,*

(v) $\alpha(y) \;=\; \frac{1}{8}$ *if* y *is located in one of the eight vertices of* $\partial\mathcal{G}$.

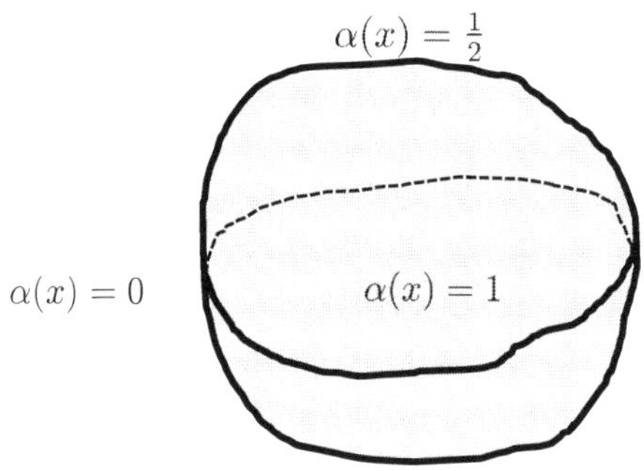

FIGURE 5.1: Solid angle subtended at $x \in \mathbb{R}^3$ by the surface $\partial\mathcal{G}$ of a regular region $\mathcal{G}$ with "smooth boundary".

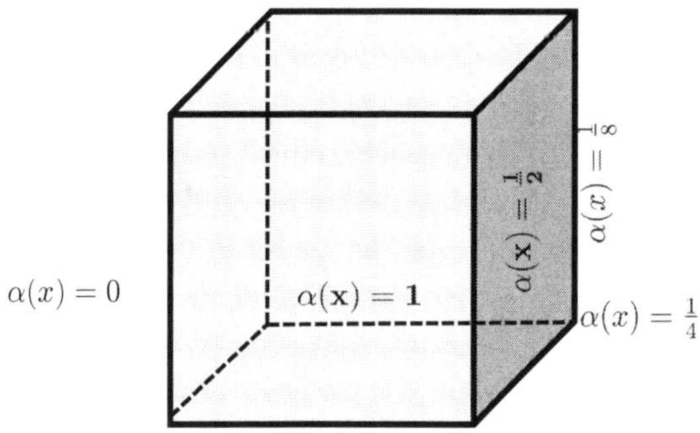

FIGURE 5.2: Solid angle subtended at $x \in \mathbb{R}^3$ by the surface $\partial\mathcal{G}$ of the "non-smooth" cube $\mathcal{G} = (-1, 1)^3$.

Lemma 5.8 is a special case of the third Green theorem in $\mathbb{R}^q$ (see, e.g., O.D. Kellogg [1929], N. Aronszaijn et al. [1983]).

Theorem 5.10. *Let $\mathcal{G}$ be a regular region. Suppose that $U : \overline{\mathcal{G}} \to \mathbb{R}$ is twice continuously differentiable, i.e., $U \in \mathrm{C}^{(2)}\left(\overline{\mathcal{G}}\right)$. Then we have*

$$\int_{\partial\mathcal{G}} \left\{ F_q(|x-y|) \left(\frac{\partial U}{\partial \nu}(x) \right) - U(x) \left(\frac{\partial}{\partial \nu_x} F_q(|x-y|) \right) \right\} dS(x) \\ - \int_{\mathcal{G}} F_q(|x-y|)\, \Delta U(x)\, dV(x) = \alpha(y)\, U(y), \tag{5.48}$$

where $\alpha(y)$, $y \in \mathbb{R}^q$, is the solid angle at y subtended by the surface $\partial\mathcal{G}$.

Poisson Differential Equation. Finally, we mention the *Poisson differential equation*, which is a classical result in potential theory (see, for example, S.G. Michlin [1970], W. Freeden, C. Gerhards [2013] for an alternative proof in $\mathbb{R}^3$).
We restrict our formulation to the case $q \geq 3$. The case $q = 2$ follows by obvious modifications.

Theorem 5.11. *(Poisson Differential Equation). Let F be of class $\mathrm{C}^{(0)}\left(\overline{\mathcal{G}}\right)$, $\overline{\mathcal{G}} = \mathcal{G} \cup \partial\mathcal{G} \subset \mathbb{R}^q$, $q \geq 3$. Then $U : \overline{\mathcal{G}} \to \mathbb{R}$ given by*

$$U(x) = \int_{\mathcal{G}} \frac{F(y)}{|x-y|^{q-2}}\, dV(y) \tag{5.49}$$

is of class $\mathrm{C}^{(1)}\left(\overline{\mathcal{G}}\right)$, and we have

$$\nabla U(x) = \int_{\mathcal{G}} F(y)\, \nabla_x \frac{1}{|x-y|^{q-2}}\, dV(y). \tag{5.50}$$

If F is bounded in $\overline{\mathcal{G}} = \mathcal{G} \cup \partial\mathcal{G}$ satisfying a Lipschitz–condition in the neighborhood of the point $x \in \mathcal{G}$ (more precisely, $|F(y) - F(z)| \leq C_F |y-z|$ for all y, z in the neighborhood of the point $x \in \mathcal{G}$), then U as given by (5.49) is twice continuously differentiable in $x \in \mathcal{G}$, and we have

$$\Delta U(x) = -(q-2) \|\mathbb{S}^{q-1}\|\; F(x). \tag{5.51}$$

Fundamental Solution (Polyharmonic Case). The *fundamental solution* $F_q^{(l)} : \mathbb{R}^q \backslash \{0\} \to \mathbb{R}$ *for the equation* $\Delta^l F_q^{(l)} = 0$ is defined by

$$F_q^{(l)}(|x|) = \begin{cases} \mu_q \frac{|x|^{2l-q}}{\gamma_{l-1}}, & q \text{ odd}, \quad l = 1, 2, \ldots, \\ \mu_q \frac{|x|^{2l-q}}{\tilde{\gamma}_{l-1}}, & q = 2r, \quad l = 1, \ldots, r-1 \\ \mu_q \left(\frac{-|x|^{2l-q} \ln(|x|)}{\tilde{\gamma}_{l-1}} + \frac{C_{l-\frac{q}{2}}}{\tilde{\gamma}_{l-1}} |x|^{2l-q} \right), & q = 2r, \quad l = r, r+1, \ldots, \end{cases} \tag{5.52}$$

where

$$\mu_q = \begin{cases} \frac{1}{(q-2)\|\mathbb{S}^{q-1}\|}, & q \neq 2, \\ \frac{1}{\|\mathbb{S}^{q-1}\|}, & q = 2. \end{cases} \tag{5.53}$$

and the constants C_l, $l = 0, 1, \ldots$, are given by

$$C_0 = 0, \tag{5.54}$$

$$C_l = \sum_{k=1}^{l} \frac{1}{2k} + \sum_{k=r}^{l+r-1} \frac{1}{2k}, \quad l > 0. \tag{5.55}$$

With this choice of the constants C_l it immediately follows (cf. N. Aronszaijn et al. [1983]) that for all dimensions $q = 2, 3, \ldots$ and all degrees $l = 1, 2, \ldots$ we have

$$\Delta^m F_q^{(l)} = F_q^{(l-m)}, \quad m < l. \tag{5.56}$$

Green's Theorems for Iterated Laplace Operators. We mention an extension of the second Green theorem.

Theorem 5.12. *For a number $\lambda \in \mathbb{R}$ and a regular region $\mathcal{G} \subset \mathbb{R}^q$, $q \geq 2$, and for a function $F \in \mathrm{C}^{(2m)}(\overline{\mathcal{G}})$, $m \in \mathbb{N}$, we have*

$$\begin{aligned} \int_{\mathcal{G}} G(x)(\Delta + \lambda)^m F(x)\, dV(x) &= \int_{\mathcal{G}} F(x)(\Delta + \lambda)^m G(x)\, dV(x) \\ &+ \sum_{r=0}^{m-1} \int_{\partial\mathcal{G}} \left(\frac{\partial}{\partial\nu}(\Delta + \lambda)^r F(x)\right)\left((\Delta + \lambda)^{m-(r+1)} G(x)\right) dS(x) \\ &- \sum_{r=0}^{m-1} \int_{\partial\mathcal{G}} \left((\Delta + \lambda)^r F(x)\right)\left(\frac{\partial}{\partial\nu}(\Delta + \lambda)^{m-(r+1)} G(x)\right) dS(x). \end{aligned} \tag{5.57}$$

Chapter 6

Preparatory Tools of the Theory of Special Functions

Next we give a brief survey on the theory of spherical harmonics of dimension q including the addition theorem, the Funk–Hecke formula, the closure and completeness properties and the characterization of spherical harmonics as eigensolutions of the Beltrami operator. Furthermore, for notational reasons, we give a short approach to the theory of Bessel functions. In particular, asymptotic estimates for Bessel, Hankel, and Kelvin functions are developed, as far as they are needed for our lattice point and sampling concerns.

6.1 Homogeneous Harmonic Polynomials

We begin with the theory on homogeneous harmonic polynomials in Euclidean space $\mathbb{R}^q$ (for a more detailed study the reader is referred to C. Müller [1952, 1966, 1998], W. Freeden [1979], and W. Freeden, M. Gutting [2013]).

Homogeneous Polynomials. Let $\mathrm{Hom}_n(\mathbb{R}^q)$ consist of all polynomials $H_n(q;\cdot)$ in q variables which are homogeneous of degree n (i.e., $H_n(q;\tau x) = \tau^n H_n(q;x)$ for all $\tau \in \mathbb{R}$ and all $x \in \mathbb{R}^q$). If $H_n(q;\cdot)$ is a member of the class $\mathrm{Hom}_n(\mathbb{R}^q)$, then there exist complex numbers $C_\alpha = C_{\alpha_1,\ldots,\alpha_q}$, such that $H_n(q;x) = \sum_{[\alpha]=n} C_\alpha x^\alpha$. In Cartesian coordinates $x = (x_1,\ldots,x_q)^T$,

$$H_n(q;x_1,\ldots,x_q) = \sum_{\alpha_1+\ldots+\alpha_q=n} C_{\alpha_1,\ldots,\alpha_q}\, x_1^{\alpha_1}\ldots x_q^{\alpha_q}. \tag{6.1}$$

It is obvious that the set of monomials $x \mapsto x^\alpha$, $[\alpha]=n$, $x \in \mathbb{R}^q$, is a basis of the space $\mathrm{Hom}_n(\mathbb{R}^q)$. The number of such monomials is known to be precisely the number of ways selecting $q-1$ elements out of a collection of $n+q-1$. This means that the dimension $\dim(\mathrm{Hom}_n(\mathbb{R}^q))$ of $\mathrm{Hom}_n(\mathbb{R}^q)$ is equal to

$$M(q;n) = \dim(\mathrm{Hom}_n(\mathbb{R}^q)) = \binom{n+q-1}{q-1} = \binom{n+q-1}{n} = \frac{\Gamma(n+q)}{\Gamma(q)\Gamma(n+1)}. \tag{6.2}$$

Homogeneous Harmonic Polynomials. Let $\mathrm{Harm}_n(\mathbb{R}^q)$ be the class of all polynomials in $\mathrm{Hom}_n(\mathbb{R}^q)$ that are harmonic:

$$\mathrm{Harm}_n(\mathbb{R}^q) = \{H_n(q;\cdot) \in \mathrm{Hom}_n(\mathbb{R}^q) : \ \Delta_x H_n(q;x) = 0, \ x \in \mathbb{R}^q\}. \tag{6.3}$$

The dimension $N(q;n)$ of $\mathrm{Harm}_n(\mathbb{R}^q)$ is given by

$$N(q;n) = \dim(\mathrm{Hom}_n(\mathbb{R}^q)) - \dim(\mathrm{Hom}_{n-2}(\mathbb{R}^q)) \tag{6.4}$$

such that

$$N(q;0) = 1, \tag{6.5}$$

$$N(q;n) = \frac{(2n+q-2)\Gamma(n+q-2)}{\Gamma(n+1)\Gamma(q-1)}, \tag{6.6}$$

so that

$$N(q;n) = O\left(n^{q-2}\right). \tag{6.7}$$

A central result in the theory of spherical harmonics is the *addition theorem.*

Theorem 6.1. *Let $K_{n,j}(q;\cdot) \in \mathrm{Harm}_n(\mathbb{R}^q)$, $j = 1,\ldots,N(q;n)$, be an $\mathrm{L}^2(\mathbb{S}^{q-1})$-orthonormal system, i.e.,*

$$\int_{\mathbb{S}^{q-1}} K_{n,j}(q;x)\,\overline{K_{r,s}(q;x)}\,dS_{(q-1)}(x) = \delta_{n,r}\delta_{j,s}. \tag{6.8}$$

Then, for $x,y \in \mathbb{R}^q, x = |x|\xi, \ y = |y|\eta, \ \xi,\eta \in \mathbb{S}^{q-1}$, we have

$$\sum_{j=1}^{N(q;n)} K_{n,j}(q;x)\,\overline{K_{n,j}(q,y)} = \frac{N(q;n)}{\|\mathbb{S}^{q-1}\|}|x|^n|y|^n P_n(q,\xi\cdot\eta), \tag{6.9}$$

where $P_n(q;\cdot)$ is defined

$$P_n(q;t) = \frac{\Gamma(n+1)\Gamma(q-2)}{\Gamma(n+q-2)\Gamma\left(\frac{q-2}{2}\right)} \sum_{m=0}^{\lfloor \frac{n}{2} \rfloor} \frac{(-1)^m \Gamma\left(\frac{q-2}{2}+n-m\right)}{\Gamma(m+1)\Gamma(n-2m+1)}(2t)^{n-2m} \tag{6.10}$$

Legendre Polynomials. The function $P_n(q;\cdot) : t \mapsto P_n(q;t)$, $t \in [-1,+1]$, $n = 0,1,\ldots$, occuring in the addition theorem is called the *Legendre polynomial of degree n and dimension q*. It is easily seen that $P_n(q;\cdot)$ satisfies the following properties:

(i) $P_n(q;\cdot)$ is a polynomial of degree n, i.e.,

$$P_n(q;t) = \alpha_n(q)\,t^n + \ldots, \tag{6.11}$$

where

$$\alpha_n(q) = \frac{2^n\Gamma(n+\frac{q}{2})}{n!\Gamma(\frac{q}{2})N(q;n))} = \frac{n!}{2^n}\frac{\Gamma(\frac{q-1}{2})}{\Gamma(\frac{2n+q-1}{2})}\binom{2n+q-3}{n}. \tag{6.12}$$

(ii) $P_n(q;1) = 1$,

(iii) $\int_{-1}^{+1} P_n(q;t)P_m(q;t)(1-t^2)^{\frac{q-3}{2}}\, dt = 0, \quad n \neq m.$

$P_n(q;\cdot)$ is uniquely determined by the properties *(i)*, *(ii)*, *(iii)*.

It is not hard to see that

$$\int_{-1}^{+1} t^k P_n(q;t)(1-t^2)^{\frac{q-3}{2}}\, dt = \begin{cases} 0 & , \quad k-n<0, \\ 0 & , \quad k-n>0, \text{ odd} \\ \frac{n!}{2^n}\binom{k}{n}\frac{\Gamma\left(\frac{q-1}{2}\right)\Gamma\left(\frac{k-n+1}{2}\right)}{\Gamma\left(\frac{k+n+q}{2}\right)} & , \quad k-n>0, \text{ even.} \end{cases} \tag{6.13}$$

Moreover, we have

$$\int_{-1}^{+1} P_n^2(q;t)(1-t^2)^{\frac{q-3}{2}}\, dt = \frac{\sqrt{\pi}\,\Gamma\left(\frac{q-1}{2}\right)}{\Gamma\left(\frac{q}{2}\right)}\frac{1}{N(q;n)} = \frac{\|\mathbb{S}^{q-1}\|}{\|\mathbb{S}^{q-2}\|\; N(q;n)}. \tag{6.14}$$

Applying the Cauchy–Schwarz inequality to the addition theorem, we obtain for $x, y \in \mathbb{R}^q$

$$\begin{aligned} \frac{N(q;n)}{\|\mathbb{S}^{q-1}\|}|x|^n|y|^n\,|P_n(q;\xi\cdot\eta)| &= \left|\sum_{j=1}^{N(q;n)} K_{n,j}(q;x)K_{n,j}(q;y)\right| \\ &\leq \sqrt{\sum_{j=1}^{N(q;n)} |K_{n,j}(q;x)|^2}\sqrt{\sum_{j=1}^{N(q;n)} |K_{n,j}(q;y)|^2} \\ &= \frac{N(q;n)}{\|\mathbb{S}^{(q-1)}\|}|x|^n|y|^n P_n(q;1). \end{aligned} \tag{6.15}$$

Consequently, we have

$$|P_n(q;t)| \leq P_n(q;1) = 1 \tag{6.16}$$

for all $t \in [-1,+1]$.

By elementary calculations we find

$$\left((1-t^2)\left(\frac{d}{dt}\right)^2 - (q-1)t\frac{d}{dt} + n(n+q-2)\right) P_n(q;t) = 0. \tag{6.17}$$

More concretely, $P_n(q;\cdot)$ is the only twice continuously differentiable eigenfunction of the “Legendre operator”

$$\mathrm{L}_t = (1-t^2)\left(\frac{d}{dt}\right)^2 - (q-1)t\frac{d}{dt} \tag{6.18}$$

on the (1D) interval $[-1,+1]$, corresponding to the eigenvalues $\mathrm{L}^\wedge(n) = n(n+q-2), n = 0, 1, \ldots,$ that is bounded on $[-1,+1]$ with $P_n(q;1) = 1$.

The Legendre polynomial can be expressed by the *Rodrigues formula*

$$P_n(q;t) = (-1)^n \frac{\Gamma\left(\frac{q-1}{2}\right)}{2^n \Gamma\left(\frac{q-1}{2}+n\right)} (1-t^2)^{\frac{3-q}{2}} \left(\frac{d}{dt}\right)^n (1-t^2)^{n+\frac{q-3}{2}}. \tag{6.19}$$

Legendre polynomials satisfy a number of recursion formulas. We mention only the following three relations:

(i)

$$(n+q-2)P_{n+1}(q;t) - (2n+q-2)tP_n(q;t) + nP_{n-1}(q;t) = 0. \tag{6.20}$$

(ii)

$$(1-t^2)P_n'(q;t) = -(n+q-2)\left(P_{n+1}(q;t) - tP_n(q;t)\right). \tag{6.21}$$

(iii)

$$(1-t^2)P_n'(q;t) = -(q-3)\left(P_{n+1}(q-2;t) - tP_n(q;t)\right). \tag{6.22}$$

The polynomial $P_n(q;\cdot)$ has exactly n different zeros in $(-1,+1)$.
It is not difficult to verify that

$$\left(\frac{d}{dt}\right)^l P_n(q;t) = \frac{2^l \Gamma\left(l+\frac{q}{2}\right)}{\Gamma\left(\frac{q}{2}\right)} \frac{N(q+2l;n-l)}{N(q;n)} P_{n-l}(q+2l;t). \tag{6.23}$$

In connection with the Rodrigues formula (6.19) we obtain by partial integration the so-called Rodrigues rule.

Lemma 6.2. *Let F be n times continuously differentiable on $[-1,+1]$. Then we have*

$$\int_{-1}^{+1} F(t)P_n(q;t)(1-t^2)^{\frac{q-3}{2}}\, dt \tag{6.24}$$

$$= \frac{\Gamma\left(\frac{q-1}{2}\right)}{2^n \Gamma\left(\frac{q-1}{2}+n\right)} \int_{-1}^{+1} (1-t^2)^{n+\frac{q-3}{2}} F^{(n)}(t)\, dt.$$

The power series

$$\phi(r) = \sum_{n=0}^{\infty} c_n(q) P_n(q;t) r^n, \quad t \in [-1,+1], \tag{6.25}$$

with

$$c_n(q) = \frac{\Gamma(n+q-2)}{\Gamma(n+1)\Gamma(q-2)}, \quad q \geq 3, \tag{6.26}$$

is absolutely and uniformly convergent for all r with $|r| \leq r_0, r_0 \in [0,1)$. By virtue of the recurrence relation (6.20) we obtain the differential equation

$$(1+r^2-2tr)\phi'(r) = -(q-2)(r-t)\phi(r). \tag{6.27}$$

Under the initial condition $\phi(0) = 1$ it is uniquely solvable. Since it is not hard to verify that $r \mapsto \phi(r) = (1 + r^2 - 2rt)^{\frac{2-q}{2}}$ solves the initial value problem, we are led to the following generating series expansion of $P_n(q;t)$.

Lemma 6.3. *For $t \in [-1,1]$ and $|r| < 1$, $q \geq 3$*

$$\sum_{n=0}^{\infty} \frac{\Gamma(n+q-2)}{\Gamma(n+1)\Gamma(q-2)} P_n(q;t) r^n = \frac{1}{(1+r^2-2rt)^{\frac{q-2}{2}}}. \tag{6.28}$$

The two-dimensional analogue of (6.28) is as follows:

$$\frac{1}{2} \ln\left(1 - 2rt + r^2\right) = -\sum_{n=1}^{\infty} \frac{1}{n} P_n(2;t)\, r^n. \tag{6.29}$$

Remark 6.4. *The Legendre polynomials $P_n(q;\cdot)$, $q \geq 3$ coincide (apart from a multiplicative constant) with the Gegenbauer polynomials. More precisely, we have (see, e.g., W. Magnus et al. [1966])*

$$\Gamma(q-2+n) P_n(q;t) = \Gamma(q-2)\Gamma(n+1)\; C_n^{\left(\frac{q-2}{2}\right)}(t), \;\; t \in [-1,1], \tag{6.30}$$

where

$$C_n^{(\alpha)}(t) \;=\; \frac{1}{\Gamma(\alpha)} \sum_{k=0}^{\lfloor \frac{n}{2} \rfloor} (-1)^k \frac{\Gamma(\alpha+n-k)}{k!\;(n-2k)!}\,(2t)^{n-k}, \quad t \in [-1,1]. \tag{6.31}$$

Spherical Harmonics. Let H_n be a homogeneous harmonic polynomial of degree n in $\mathbb{R}^q$, i.e., $H_n \in \mathrm{Harm}_n(\mathbb{R}^q)$. Its decomposition into radial and angular parts is straightforward: $H_n(x) = r^n\, Y_n(\xi)$, $x = r\xi$, $r = |x|, \xi \in \mathbb{S}^{q-1}$. The restriction $Y_n = H_n|\mathbb{S}^{q-1}$ is called a *spherical harmonic of degree n and dimension q*. The set of all spherical harmonics of degree n and dimension q, i.e., the set of all restrictions $Y_n = H_n|\mathbb{S}^{q-1}$ with $H_n \in \mathrm{Harm}_n(\mathbb{R}^q)$, is denoted by $\mathrm{Harm}_n(\mathbb{S}^{q-1})$.

Remark 6.5. *In what follows we simply write* Harm_n *instead of* $\mathrm{Harm}_n(\mathbb{R}^q)$ *(or* $\mathrm{Harm}_n(\mathbb{S}^{q-1})$*) if no confusion is likely to arise. Our purpose is to list the essential properties of functions in* Harm_n*, i.e., spherical harmonics of degree n and dimension q. For more details on spherical harmonics the reader is referred to C. Müller [1952, 1966, 1998], W. Freeden [1979], W. Freeden et al. [1998], and W. Freeden, M. Gutting [2013]. The book W. Freeden, M. Schreiner [2009] contains a vectorial as well as tensorial approach in Euclidean space $\mathbb{R}^3$. The monograph V. Michel [2013] also deals with approximation theory for the ball $\mathbb{B}^q$.*

Spherical harmonics of different degrees are orthogonal in the sense of the $\mathrm{L}^2(\mathbb{S}^{q-1})$–inner product

$$\left(Y_n(q;\cdot), Y_m(q;\cdot)\right)_{\mathrm{L}^2(\mathbb{S}^{q-1})} = \int_{\mathbb{S}^{q-1}} Y_n(q;\xi)\; \overline{Y_m(q;\xi)}\; dS_{(q-1)}(\xi) = 0 \tag{6.32}$$

if $n \neq m$. From the addition theorem of homogeneous harmonic polynomials (Theorem 3.11) it is an easy task to obtain the *addition theorem for spherical harmonics.*

Theorem 6.6. *Let* $\{Y_{n,j}(q;\cdot)\}_{j=1,\ldots,N(q;n)}$ *be an* $\mathrm{L}^2(\mathbb{S}^{q-1})$*–orthonormal system in* Harm_n*. Then, for any pair* $(\xi,\eta)\in\mathbb{S}^{q-1}\times\mathbb{S}^{q-1}$*, we have*

$$\sum_{j=1}^{N(q;n)} Y_{n,j}(q;\xi)\,\overline{Y_{n,j}(q;\eta)} = \frac{N(q;n)}{\|\mathbb{S}^{q-1}\|} P_n(q;\xi\cdot\eta). \tag{6.33}$$

An immediate consequence is that

$$Y_n(q;\xi) = \frac{N(q;n)}{\|\mathbb{S}^{q-1}\|}\int_{\mathbb{S}^{q-1}} P_n(q;\xi\cdot\eta)Y_n(q;\eta)\,dS_{(q-1)}(\eta) \tag{6.34}$$

holds for all $\xi\in\mathbb{S}^{q-1}$ and all $Y_n(q;\cdot)\in\mathrm{Harm}_n$; i.e., $K_{\mathrm{Harm}_n}(q;\cdot)$ defined by

$$K_{\mathrm{Harm}_n}(q;\xi\cdot\eta) = \frac{N(q;n)}{\|\mathbb{S}^{q-1}\|}\,P_n(q;\xi\cdot\eta),\quad \xi,\eta\in\mathbb{S}^{q-1}, \tag{6.35}$$

is the reproducing kernel of the $N(q;n)$–dimensional space Harm_n.

Lemma 6.7. *For* $q\geq 2$ *the Legendre polynomials* $P_n(q;\cdot)$ *satisfy the inequalities*

$$|P_n(q;t)| \leq \frac{\Gamma(\frac{q-1}{2})}{\sqrt{\pi}}\left(\frac{4}{n(1-t^2)}\right)^{\frac{q-2}{2}},\ n\geq 1,\ t\in(-1,1). \tag{6.36}$$

The *Funk–Hecke formula* (cf. H. Funk [1916]) plays an important role, whenever zonal functions, i.e., spherical functions depending only on the inner product of two unit vectors, should be discussed.

Theorem 6.8. *Suppose that* G *is a function of class* $\mathrm{L}^1([-1,1])$*. Then, for all* $(\xi,\eta)\in\mathbb{S}^{q-1}\times\mathbb{S}^{q-1}$ *and all* $n\in\mathbb{N}_0$*,*

$$\int_{\mathbb{S}^{q-1}} G(\xi\cdot\zeta)P_n(q;\zeta\cdot\eta)\,dS_{(q-1)}(\zeta) = G^\wedge(n)\,P_n(q;\xi\cdot\eta), \tag{6.37}$$

where

$$G^\wedge(n) = G^\wedge_{\mathbb{S}^{q-1}}(n) = \|\mathbb{S}^{q-2}\|\int_{-1}^{+1} G(s)P_n(q;s)(1-s^2)^{\frac{q-3}{2}}\,ds. \tag{6.38}$$

From Theorem 6.8 we find by multiplication with $Y_n(q;\eta)$ and integration with regard to the variable η the following *variant of the Funk–Hecke formula.*

Theorem 6.9. *Let* G *be of class* $\mathrm{L}^1([-1,+1])$*. Then, for* $Y_n(q;\cdot)\in\mathrm{Harm}_n$*,*

$$\int_{\mathbb{S}^{q-1}} G(\xi\cdot\eta)Y_n(q;\eta)\,dS_{(q-1)}(\eta) = G^\wedge(n)\,Y_n(q;\xi) \tag{6.39}$$

with

$$G^\wedge(n) = \|\mathbb{S}^{q-2}\|\int_{-1}^{+1} G(s)\,P_n(q;s)(1-s^2)^{\frac{q-3}{2}}\,ds. \tag{6.40}$$

Next we discuss the closure and completeness of the spherical harmonics in $L^2(\mathbb{S}^{q-1})$. To this end we first consider the so–called *Bernstein kernel of degree* n (cf. W. Freeden, M. Gutting [2008] for the vectorial and tensorial context in $\mathbb{R}^3$)

$$B_n(\xi \cdot \eta) = \left(\frac{1+\xi \cdot \eta}{2}\right)^n, \ \xi, \eta \in \mathbb{S}^{q-1}, \quad n \in \mathbb{N}_0. \tag{6.41}$$

Lemma 6.10. *For $q \geq 2$, the following relations hold uniformly:*

(i) For all $\xi \in \mathbb{S}^{q-1}$,

$$\left(\frac{1}{4\pi}\right)^{\frac{q-1}{2}} \frac{\Gamma(n+q-1)}{\Gamma(n+\frac{q-1}{2})} \int_{\mathbb{S}^{q-1}} \left(\frac{1+\xi \cdot \eta}{2}\right)^n dS_{(q-1)}(\eta) = 1. \tag{6.42}$$

(ii) For $\xi, \eta \in \mathbb{S}^{q-1}$ with $\xi \cdot \eta < 1$,

$$\lim_{n\to\infty} \left(\frac{1}{4\pi}\right)^{q-1} \frac{\Gamma(n+q-1)}{\Gamma(n+\frac{q-1}{2})} \left(\frac{1+\xi \cdot \eta}{2}\right)^n = 0. \tag{6.43}$$

These formulas enable us to formulate the following *approximate identity on* $C^{(0)}(\mathbb{S}^{q-1})$.

Theorem 6.11. *Let F be of class $C^{(0)}(\mathbb{S}^{q-1})$. Then*

$$\lim_{n\to\infty} \sup_{\xi \in \mathbb{S}^{q-1}} \left| \left(\frac{1}{4\pi}\right)^{\frac{q-1}{2}} \frac{\Gamma(n+q-1)}{\Gamma\left(n+\frac{q-1}{2}\right)} \int\limits_{\mathbb{S}^{q-1}} \left(\frac{1+\xi \cdot \eta}{2}\right)^n F(\eta)\, dS_{(q-1)}(\eta) - F(\xi) \right| = 0. \tag{6.44}$$

It is not difficult to see that

$$\left(\frac{1}{4\pi}\right)^{\frac{q-1}{2}} \frac{\Gamma(n+q-1)}{\Gamma\left(n+\frac{q-1}{2}\right)} \int_{-1}^{+1} \left(\frac{1+t}{2}\right)^n P_k(q;t)(1-t^2)^{\frac{q-3}{2}}\, dt = \frac{\beta_n^k(q)}{\|\mathbb{S}^{q-2}\|}, \tag{6.45}$$

where we have used the abbreviation

$$\beta_n^k(q) = \frac{n!}{(n-k)!} \frac{(n+q-2)!}{(n+k+q-2)!}. \tag{6.46}$$

Note that $\beta_n^k(q) < \beta_{n+1}^k(q)$ and $\lim_{n\to\infty} \beta_n^k(q) = 1$. Hence,

$$\left(\frac{1}{4\pi}\right)^{\frac{q-1}{2}} \frac{\Gamma(n+q-1)}{\Gamma\left(n+\frac{q-1}{2}\right)} \left(\frac{1+\xi \cdot \eta}{2}\right)^n = \sum_{k=0}^{n} \beta_n^k(q) \frac{N(q;k)}{\|\mathbb{S}^{q-1}\|} P_k(q;\xi \cdot \eta). \tag{6.47}$$

Observing these results we finally obtain in connection with the addition theorem for spherical harmonics the following theorem.

Theorem 6.12. *For all $F \in \mathrm{C}^{(0)}(\mathbb{S}^{q-1})$*

$$\lim_{n\to\infty} \sup_{\xi\in\mathbb{S}^{q-1}} \left| F(\xi) - \sum_{k=0}^{n} \sum_{j=1}^{N(q;k)} \beta_n^k(q) \int_{\mathbb{S}^{q-1}} F(\eta)\overline{Y_{k,j}(q;\eta)}\, dS_{(q-1)}(\eta) Y_{k,j}(q;\xi) \right| = 0. \tag{6.48}$$

From Theorem 6.12 we are able to prove the closure of the system of spherical harmonics.

Corollary 6.13. *The system $\{Y_{n,j}(q;\cdot)\}_{\substack{n=0,1,\ldots\\ j=1,\ldots,N(q;n)}}$ is closed in $\mathrm{C}^{(0)}(\mathbb{S}^{q-1})$; that is, for any given $\varepsilon > 0$ and each $F \in \mathrm{C}^{(0)}(\mathbb{S}^{q-1})$ there exists a linear combination $\sum_{k=0}^{n} \sum_{j=1}^{N(q;k)} d_{k,j} Y_{k,j}(q;\cdot)$ such that*

$$\left\| F - \sum_{k=0}^{n} \sum_{j=1}^{N(q;k)} d_{k,j} Y_{k,j}(q;\cdot) \right\|_{\mathrm{C}^{(0)}(\mathbb{S}^{q-1})} \leq \varepsilon. \tag{6.49}$$

Next we are interested in closure and completeness in $\mathrm{L}^2(\mathbb{S}^{q-1})$.

Lemma 6.14. *The system $\{Y_{n,j}(q;\cdot)\}_{\substack{n=0,1,\ldots\\ j=1,\ldots,N(q;n)}}$ is closed in $\mathrm{C}^{(0)}(\mathbb{S}^{q-1})$ with respect to $\|\cdot\|_{\mathrm{L}^2(\mathbb{S}^{q-1})}$.*

Indeed, Lemma 6.14 immediately follows from Corollary 6.13 by use of the norm estimate $\|F\|_{\mathrm{L}^2(\mathbb{S}^{q-1})} \leq \sqrt{\|\mathbb{S}^{q-1}\|}\, \|F\|_{\mathrm{C}^{(0)}(\mathbb{S}^{q-1})}$ for $F \in \mathrm{C}^{(0)}(\mathbb{S}^{q-1})$, so that we fnally arrive at the following result.

Theorem 6.15. *The system $\{Y_{n,j}(q;\cdot)\}_{\substack{n=0,1,\ldots\\ j=1,\ldots,N(q,n)}}$ is closed in the space $\mathrm{L}^2(\mathbb{S}^{q-1})$ with respect to $\|\cdot\|_{\mathrm{L}^2(\mathbb{S}^{q-1})}$.*

Note that $\mathrm{C}^{(0)}(\mathbb{S}^{q-1})$ is dense in $\mathrm{L}^2(\mathbb{S}^{q-1})$, that is, for every $\varepsilon > 0$ and every $F \in \mathrm{L}^2(\mathbb{S}^{q-1})$, there exists a function $G \in \mathrm{C}^{(0)}(\mathbb{S}^{q-1})$ with $\|F-G\|_{\mathrm{L}^2(\mathbb{S}^{q-1})} \leq \varepsilon$. The function $G \in \mathrm{C}^{(0)}(\mathbb{S}^{q-1})$ admits an arbitrarily close approximation by finite linear combinations of spherical harmonics. Therefore, the closure in Theorem 6.15 is clear.

Theorem 6.16. *(Expansion Theorem). Suppose that F is of class $\mathrm{C}^{(0)}(\mathbb{S}^{q-1})$. Then*

$$F(\xi) = \lim_{\substack{r\to 1\\ r<1}} \sum_{n=0}^{\infty} r^n \sum_{j=1}^{N(q,n)} \int_{\mathbb{S}^{q-1}} F(\eta)\overline{Y_{n,j}(q;\eta)}\, dS_{(q-1)}(\eta)\, Y_{nj}(q;\xi) \tag{6.50}$$

holds uniformly with respect to $\xi \in \mathbb{S}^{q-1}$. For a Lipschitz-continuous function, the limits may be interchanged.

6.2 Bessel Functions

The point of departure is the entire solution $U_n : \mathbb{R}^q \to \mathbb{C}$ of the Helmholtz equation $\Delta U_n + U_n = 0$ in $\mathbb{R}^q$ of the form

$$U_n(x) = \frac{i^{-n}}{\|\mathbb{S}^{q-1}\|} \int_{\mathbb{S}^{q-1}} e^{ix\cdot\eta} \; Y_n(q;\eta) \; dS_{(q-1)}(\eta), \tag{6.51}$$

where $Y_n(q;\cdot)$ is a member of Harm_n. In terms of (standard) polar coordinates $x = r\xi, r = |x|, \xi \in \mathbb{S}^{q-1}$, the Funk–Hecke formula (cf. Theorem 6.8) yields the decomposition

$$U_n(r\xi) = i^{-n} \frac{\|\mathbb{S}^{q-2}\|}{\|\mathbb{S}^{q-1}\|} \int_{-1}^{1} e^{irt} P_n(q;t)(1-t^2)^{\frac{q-3}{2}} \; dt \;\; Y_n(q;\xi). \tag{6.52}$$

In other words, a separation of the variables into a radial and an angular part is achieved in the form

$$U_n(x) = J_n(q;r) \; Y_n(q;\xi), \; x = r\xi, \; \xi \in \mathbb{S}^{q-1}. \tag{6.53}$$

Definition 6.17. *The function $J_n(q;\cdot)$ given by*

$$J_n(q;r) = i^{-n} \frac{\|\mathbb{S}^{q-2}\|}{\|\mathbb{S}^{q-1}\|} \int_{-1}^{1} e^{irt} P_n(q;t)(1-t^2)^{\frac{q-3}{2}} \; dt, \quad r \geq 0, \tag{6.54}$$

is called the Bessel function of order n and dimension q (more accurately, regular Bessel function of order n and dimension q).

Applying the Rodrigues rule (i.e., Lemma 6.2) we get

$$J_n(q;r) = \frac{\Gamma\left(\frac{q}{2}\right)\left(\frac{r}{2}\right)^n}{\Gamma\left(n+\frac{q-1}{2}\right)\Gamma\left(\frac{1}{2}\right)} \int_{-1}^{1} e^{irt}(1-t^2)^{n+\frac{q-3}{2}} \; dt \tag{6.55}$$

such that

$$J_n(q;r) = \frac{\Gamma\left(\frac{q}{2}\right)\left(\frac{r}{2}\right)^n}{\Gamma\left(n+\frac{q-1}{2}\right)\Gamma\left(\frac{1}{2}\right)} \int_{-1}^{1} (1-t^2)^{n+\frac{q-3}{2}} \cos(rt) \; dt. \tag{6.56}$$

In connection with the well known results of the theory of the gamma function we have

$$\left| \int_{-1}^{1} (1-t^2)^{n+\frac{q-3}{2}} \cos(rt) \; dt \right| \leq 2 \int_{0}^{1} (1-t^2)^{n+\frac{q-3}{2}} \; dt = \frac{\Gamma\left(\frac{1}{2}\right)\Gamma\left(n+\frac{q-1}{2}\right)}{\Gamma\left(n+\frac{q}{2}\right)}. \tag{6.57}$$

Lemma 6.18. *For fixed $r \in [0,\infty)$ and $n \to \infty$,*

$$|J_n(q;r)| \leq \frac{\Gamma\left(\frac{q}{2}\right)}{\Gamma\left(n+\frac{q}{2}\right)} \left(\frac{r}{2}\right)^n. \tag{6.58}$$

The Bessel functions can be seen as orthogonal (i.e., Fourier) coefficients of an orthogonal series expansion in terms of Legendre polynomials.

Lemma 6.19. *For $x \in \mathbb{R}^q$, $x = r\xi$, $r = |x|$, $\xi \in \mathbb{S}^{q-1}$, and $\eta \in \mathbb{S}^{q-1}$ we have*

$$e^{ix\cdot\eta} = e^{ir\xi\cdot\eta} = \sum_{n=0}^{\infty} i^n N(q;n)\ J_n(q;r)\ P_n(q;\xi\cdot\eta), \tag{6.59}$$

where, for each $r \in [0,\infty)$, the series converges absolutely and uniformly with respect to $\xi \cdot \eta$.

Indeed, in connection with the Funk–Hecke formula of the theory of spherical harmonics we obtain the following decomposition into radial and angular components

$$\frac{N(q;n)}{\|\mathbb{S}^{q-1}\|} \int_{\mathbb{S}^{q-1}} e^{ir\xi\cdot\zeta} P_n(q;\eta\cdot\zeta)\ dS_{(q-1)}(\zeta) = i^n\ N(q;n)\ J_n(q;r)\ P_n(q;\xi\cdot\eta). \tag{6.60}$$

This explains the identity (6.59) of Lemma 6.19.

The series representation (Lemma 6.19) has an immediate consequence known as the *addition theorem of Bessel functions.* To this end we consider, for $x, y \in \mathbb{R}^q$, the integral

$$J_0(q;|x-y|) = \frac{1}{\|\mathbb{S}^{q-1}\|} \int_{\mathbb{S}^{q-1}} e^{ix\cdot\zeta} e^{-iy\cdot\zeta}\ dS_{(q-1)}(\zeta). \tag{6.61}$$

The Parseval identity for the (orthogonal system of) Legendre polynomials shows that

$$\begin{aligned} &\sum_{n=0}^{\infty} (N(q;n))^2 J_n(q;|x|) J_n(q;|y|) \int_{\mathbb{S}^{q-1}} P_n(q;\xi\cdot\zeta) P_n(q;\eta\cdot\zeta)\ dS_{(q-1)}(\zeta) \\ &\quad = \|\mathbb{S}^{q-1}\| \sum_{n=0}^{\infty} N(q;n) J_n(q;|x|) J_n(q;|y|) P_n(q;\xi\cdot\eta). \end{aligned} \tag{6.62}$$

Lemma 6.20. *(Addition Theorem of Bessel Functions). For $x, y \in \mathbb{R}^q$ with $x = |x|\xi, y = |y|\eta$, $\xi, \eta \in \mathbb{S}^{q-1}$, we have*

$$J_0(q;|x-y|) = \sum_{n=0}^{\infty} N(q;n) J_n(q;|x|) J_n(q;|y|) P_n(q;\xi\cdot\eta). \tag{6.63}$$

It is not hard to see that

Lemma 6.21. *For $r \in [0,\infty)$, the Bessel function $J_n(q;\cdot)$ permits the representation by the power series*

$$J_n(q;r) = \Gamma\left(\frac{q}{2}\right) \left(\frac{r}{2}\right)^n \sum_{k=0}^{\infty} \frac{(-1)^k}{k!} \frac{\left(\frac{r}{2}\right)^{2k}}{\Gamma\left(n+k+\frac{q}{2}\right)}. \tag{6.64}$$

Notational Conventions for Bessel Functions. The traditional notation, which is mostly used in the literature, is based on the observation that the Bessel functions of integral order n and dimension q are expressible as Bessel functions of order $n + \frac{q-2}{2}$ and dimension 2. We discuss this aspect in more detail: Letting, as usual (see G.N. Watson [1944]),

$$J_\nu(r) = J_\nu(2; r) = \left(\frac{r}{2}\right)^\nu \sum_{k=0}^{\infty} \frac{(-1)^k}{k!} \frac{\left(\frac{r}{2}\right)^{2k}}{\Gamma(\nu + k + 1)}, \quad \nu \geq 0, \tag{6.65}$$

and observing

$$J_n(q; r) = \left(\frac{r}{2}\right)^n \Gamma\left(\frac{q}{2}\right) \sum_{k=0}^{\infty} \frac{(-1)^k}{k!} \frac{\left(\frac{r}{2}\right)^{2k}}{\Gamma(n + k + \frac{q}{2})}, \quad n \in \mathbb{N}_0, \tag{6.66}$$

we are led by comparison to the relation (see C. Müller [1998])

$$J_n(q; r) = \Gamma\left(\frac{q}{2}\right) \left(\frac{r}{2}\right)^{\frac{2-q}{2}} J_{n+\frac{q-2}{2}}(r), \quad n \in \mathbb{N}_0. \tag{6.67}$$

In particular, we have

$$J_0(3; r) = \sqrt{\frac{\pi}{2}} \frac{J_{\frac{1}{2}}(2; r)}{\sqrt{r}} = \sqrt{\frac{\pi}{2}} \frac{J_{\frac{1}{2}}(r)}{\sqrt{r}}. \tag{6.68}$$

Furthermore we see that $J_0(3; \cdot)$ is the sinc-function

$$\begin{aligned} J_0(3; r) &= \frac{1}{2}\sqrt{\pi} \sum_{k=0}^{\infty} \frac{(-1)^k r^{2k}}{2^{2k}\Gamma(k+1)\Gamma\left(k + \frac{3}{2}\right)} \\ &= \sum_{k=0}^{\infty} \frac{(-1)^k r^{2k}}{(2k+1)!} \\ &= \frac{\sin(r)}{r}. \end{aligned} \tag{6.69}$$

This yields

$$J_0(3; r) = \frac{\sin(r)}{r} = \operatorname{sinc}(r). \tag{6.70}$$

Finally it should be remarked that the relations (6.65), (6.66), and (6.67) also allow us to understand $J_\nu(q; \cdot)$ for non-negative real values ν by

$$J_\nu(q; r) = \Gamma\left(\frac{q}{2}\right) \left(\frac{r}{2}\right)^{\frac{2-q}{2}} J_{\nu+\frac{q-2}{2}}(r). \tag{6.71}$$

Since the spherical harmonics of degree n and dimension q are the eigenfunctions of the Beltrami operator corresponding to the eigenvalues $(\Delta^*)^\wedge(n) =$

$n(n+q-2)$, $n = 0, 1, \ldots$, i.e., $(\Delta^* + (\Delta^*)^\wedge(n))Y_n(q;\cdot) = 0$, $Y_n(q;\cdot) \in \mathrm{Harm}_n$, we get from the Helmholtz equation

$$\left(\underbrace{\frac{1}{r^{q-1}} \frac{\partial}{\partial r} r^{q-1} \frac{\partial}{\partial r} + \frac{1}{r^2} \Delta^*_{(q)}}_{=\Delta_{(q)}} + 1 \right) J_n(q;r) Y_n(q;\xi) = 0 \tag{6.72}$$

the following differential equation for the Bessel functions

$$J_n''(q;r) + \frac{q-1}{r} J_n'(q;r) + \left(1 - \frac{n(n+q-2)}{r^2}\right) J_n(q;r) = 0. \tag{6.73}$$

In connection with the recursion relation (6.20) for the Legendre polynomial we are able to deduce the recursion relation

$$(n+q-2) J_{n+1}(q;r) + (2n+q-2) J_n'(q;r) = n J_{n-1}(q;r). \tag{6.74}$$

Even more, we are able to verify the following recursion relations.

Lemma 6.22. *The following recurrence relations for Bessel functions hold true:*

(i)

$$J_{n-1}(q;r) + J_{n+1}(q;r) = \frac{2n+q-2}{r} J_n(q;r), \tag{6.75}$$

(ii)

$$J_{n-1}(q;r) - J_{n+1}(q;r) = 2J_n'(q;r) + \frac{q-2}{r} J_n(q;r), \tag{6.76}$$

(iii)

$$J_{n+1}(q;r) = -r^n \frac{d}{dr} \left(r^{-n} J_n(q;r) \right), \tag{6.77}$$

(iv)

$$J_{n-1}(q;r) = r^{2-q-n} \frac{d}{dr} \left(r^{n+q-2} J_n(q;r) \right). \tag{6.78}$$

Let $H_n(q;\cdot)$ be a homogeneous harmonic polynomial of the form $H_n(q;x) = r^n Y_n(q;\xi)$, $x \in \mathbb{R}^q$, $x = r\xi$, $r = |x|$, $x \in \mathbb{S}^{q-1}$. Then it is not hard to see that

$$H_n(q;\nabla_x)\, e^{ix\cdot\eta} = i^n e^{ix\cdot\eta}\, H_n(q;\eta), \quad \eta \in \mathbb{S}^{q-1}. \tag{6.79}$$

In connection with

$$J_0(q;|x|) = \frac{1}{\|\mathbb{S}^{q-1}\|} \int_{\mathbb{S}^{q-1}} e^{ix\cdot\eta}\, dS_{(q-1)}(\eta) \tag{6.80}$$

this yields

$$H_n(q;\nabla_x)J_0(q;|x|) = \frac{i^n}{\|\mathbb{S}^{q-1}\|}\int_{\mathbb{S}^{q-1}} e^{ix\cdot\eta}Y_n(q;\eta)\, dS_{(q-1)}(\eta), \tag{6.81}$$

such that

$$H_n(q;\nabla_x)J_0(q;|x|) = (-1)^n J_n(q;r)H_n(q;\xi). \tag{6.82}$$

In other words, Bessel functions of any order may be obtained by differentiation of the Bessel function $J_0(q;\cdot)$ of order 0.

After these preliminaries we are prepared to formulate the following result.

Lemma 6.23. *For fixed $r > 0$,*

$$\lim_{n\to\infty} \frac{J_n(q;r)}{\frac{\Gamma(\frac{q}{2})(\frac{r}{2})^n}{\Gamma(n+\frac{q}{2})}} = 1. \tag{6.83}$$

From the power series of the Bessel functions we borrow the following asymptotic relation.

Lemma 6.24. *For fixed n we have*

$$J_n(q;r) = \Gamma\left(\frac{q}{2}\right)\left(\frac{r}{2}\right)^n + O(r^{n+2}), \quad r \to 0. \tag{6.84}$$

The relevant properties of Bessel functions for our later work can be summarized as follows:

Lemma 6.25. *The Bessel function $J_n(q;\cdot)$ satisfies the following relations:*

(i) For $r > 0,\ \xi,\eta \in \mathbb{S}^{q-1}$,

$$e^{r\xi\cdot\eta} = \sum_{n=0}^{\infty} N(q;n)J_n(q;r)P_n(q,\xi\cdot\eta),$$

(ii) For $x = |x|\xi,\ y = |y|\eta$

$$J_0(q;|x-y|) = \sum_{n=0}^{\infty} N(q;n)J_n(q;|x|)J_n(q;|y|)P_n(q;\xi\cdot\eta),$$

(iii) $J_n(q;\cdot)$ *is a solution of the differential equation*

$$J_n''(q;r) + \frac{q-1}{r}J_n'(q;r) + \left(1 - \frac{n(n+q-2)}{r^2}\right)J_n(q;r) = 0,$$

(iv) $J_n(q;\cdot)$ *satisfies the recursion relation*

$$(n+q-2)J_{n+1}(q;r) + (2n+q-2)J_n'(q;r) - nJ_{n-1}(q;r) = 0,$$

(v)

$$J_n(q;r) = (-1)^n \left(\frac{1}{r}\frac{d}{dr}\right)^n J_n(q;r),$$

$$J_{n-1}(q;r) = r^{2-q-n}\frac{d}{dr}\left(r^{n+q-2}J_n(q;r)\right),$$

(vi)

$$\lim_{n\to\infty} \frac{J_n(q;r)}{\frac{\Gamma(\frac{q}{2})(\frac{r}{2})^n}{\Gamma(n+\frac{q}{2})}} = 1.$$

(vii) For homogeneous harmonic polynomials $H_n(q;\cdot)$ of degree n

$$H_n(q;\nabla_x)J_0(q;|x|) = (-1)^n J_n(q;|x|)H_n(q;\xi),$$

(viii) For n fixed and $r \to 0$,

$$J_n(q;r) = \Gamma\left(\frac{q}{2}\right)\left(\frac{r}{2}\right)^n + O(r^{n+2}).$$

Hankel Functions. The Bessel functions are regular at 0. We now discuss a pair of functions which turn out to be regular at infinity.

Definition 6.26. *For $r > 0$, the functions $H_n^{(1)}(q;\cdot)$, $H_n^{(2)}(q;\cdot)$ defined by*

$$H_n^{(1)}(q;r) = -2i^{-n}\frac{\|\mathbb{S}^{q-2}\|}{\|\mathbb{S}^{q-1}\|}\int_{1+0i}^{1+\infty i} e^{irt}P_n(q;t)(1 \quad t^2)^{\frac{q-3}{?}}\,dt, \tag{6.85}$$

$$H_n^{(2)}(q;r) = 2i^{-n}\frac{\|\mathbb{S}^{q-2}\|}{\|\mathbb{S}^{q-1}\|}\int_{-1+0i}^{-1+\infty i} e^{irt}P_n(q;t)(1-t^2)^{\frac{q-3}{2}}\,dt \tag{6.86}$$

are called Hankel functions of the first- and second-kind of order n and dimension q, respectively.

The Hankel functions have certain characteristic asymptotic properties.

Lemma 6.27. *For n fixed and $r \to \infty$ we have*

$$\begin{aligned} H_n^{(1)}(q;r) &= \frac{2}{\|S^{q-1}\|}\left(\frac{2\pi}{r}\right)^{\frac{q-1}{2}} e^{i(r-n\frac{\pi}{2}-(q-1)\frac{\pi}{4})} + O(r^{-\frac{q+1}{2}}) \\ &= \frac{\Gamma(\frac{q}{2})}{\sqrt{\pi}}\left(\frac{2}{r}\right)^{\frac{q-1}{2}} i^{-n-\frac{q-1}{2}}e^{ir} + O\left(r^{-\frac{q+1}{2}}\right), \\ H_n^{(2)}(q;r) &= \frac{\Gamma(\frac{q}{2})}{\sqrt{\pi}}\left(\frac{2}{r}\right)^{\frac{q-1}{2}} i^{n+\frac{q-1}{2}}e^{-ir} + O\left(r^{-\frac{q+1}{2}}\right). \end{aligned}$$

Observing the identity

$$J_n(q;r) = \frac{1}{2}\left(H_n^{(1)}(q;r) + H_n^{(2)}(q;r)\right) \tag{6.87}$$

we obtain from (6.87)

Lemma 6.28. *For n fixed and $r \to \infty$,*

$$J_n(q;r) = \frac{\Gamma(\frac{q}{2})}{\sqrt{\pi}}\left(\frac{2}{r}\right)^{\frac{q-1}{2}} \cos\left(n\frac{\pi}{2} + (q-1)\frac{\pi}{4} - r\right) + O\left(r^{-\frac{q+1}{2}}\right).$$

Based on techniques from G.N. Watson [1944] (see also W. Magnus et al. [1949]) the O–term in Lemma 6.28 can be written out in more detail.

Lemma 6.29. *For $n \in \mathbb{N}_0,\ m \in \mathbb{N}$ fixed and $r \to \infty$*

$$\begin{aligned} J_n(q;r) \;=\; & \frac{\Gamma(\frac{q}{2})}{2^{\frac{3-q}{2}}\sqrt{\pi}} r^{\frac{1-q}{2}} \\ & \times \left(e^{i(r-\frac{\pi}{2}(n+\frac{q-2}{2})-\frac{\pi}{4})} \sum_{l=0}^{m-1} \frac{(n+\frac{q-2}{2},l)}{(-2ir)^l} \right. \\ & \qquad \left. + \; e^{-i(r-\frac{\pi}{2}(n+\frac{q-2}{2})-\frac{\pi}{4})} \sum_{l=0}^{m-1} \frac{(n+\frac{q-2}{2},l)}{(2ir)^l} \right) \\ & + O\left(r^{-(m+\frac{q-1}{2})}\right), \end{aligned} \tag{6.88}$$

where

$$\left(n + \frac{q-2}{2}, 0\right) = 1 \tag{6.89}$$

and

$$\begin{aligned} \left(n + \frac{q-2}{2}, l\right) \;=\; & \frac{1}{l!}\left(n(n+q-2) + \frac{(q-3)(q-1)}{4}\right) \\ & \cdot \ldots \cdot \left(n(n+q-2) + \frac{(q+2l-3)(q-2l-1)}{4}\right). \end{aligned} \tag{6.90}$$

Next we are interested in an asymptotic relation for $n \to \infty$ and fixed r.

Lemma 6.30. *For $n \to \infty$ and fixed $r > 0$*

$$\lim_{n\to\infty} \frac{H_n^{(1)}(q;r)}{\frac{i}{\pi}\Gamma\left(\frac{q}{2}\right)\Gamma\left(n+\frac{q}{2}-1\right)\left(\frac{2}{r}\right)^{n+q-2}} = 1. \tag{6.91}$$

Kelvin Functions. Next we come to the Neumann function, which together with the Bessel function implies the Hankel functions.

Definition 6.31. *For $r > 0$, the Neumann function $N_n(q;\cdot)$ of order n and dimension q is defined by*

$$\begin{aligned} N_n(q;r) &= 2\frac{\|\mathbb{S}^{q-2}\|}{\|\mathbb{S}^{q-1}\|}\int_0^1 \sin\left(rt - n\frac{\pi}{2}\right) P_n(q;t)(1-t^2)^{\frac{q-3}{2}}\, dt \\ &\quad - 2i^{-n}\frac{\|\mathbb{S}^{q-2}\|}{\|\mathbb{S}^{q-1}\|}\int_0^\infty e^{-rt} P_n(q;it)(1-t^2)^{\frac{q-3}{2}}\, dt. \end{aligned} \tag{6.92}$$

As already announced, Hankel functions can be obtained by combination of Bessel and Neumann functions. More explicitly, we have the following identities.

Lemma 6.32. *For $r > 0$,*

$$\begin{aligned} H_n^{(1)}(q;r) &= J_n(q;r) + iN(q;r), \\ H_n^{(2)}(q;r) &= J_n(q;r) - iN(q;r). \end{aligned}$$

Let $C_n(q;\cdot)$ stand for any of the so-called "*cylinder functions*" $J_n(q;\cdot)$, $N_n(q;\cdot)$, $H_n^{(1)}(q;\cdot)$, and $H_n^{(2)}(q;\cdot)$. Then, for $r > 0$, the following recursion relation holds true

$$C_{n-1}(q;r) + C_{n+1}(q;r) = \frac{2n+q-2}{r} C_n(q;r). \tag{6.93}$$

Furthermore, we have

$$(2n+q-2)C_n'(q;r) = nC_{n-1}(q;r) - (n+q-2)C_{n+1}(q;r). \tag{6.94}$$

These identities immediately follow from the recursion relation for the Legendre polynomial.

Definition 6.33. *For $r > 0$, the function $K_n(q;\cdot)$ given by*

$$K_n(q;r) = \int_1^\infty e^{-rt} P_n(q;t)(t^2-1)^{\frac{q-3}{2}}\, dt \tag{6.95}$$

is called the Kelvin function (or modified Hankel function) of order n and dimension q.

It can be shown that

$$H_n^{(1)}(q;ir) = 2\frac{\|\mathbb{S}^{q-2}\|}{\|\mathbb{S}^{q-1}\|} i^{1-q-n} K_n(q;r), \tag{6.96}$$

$$H_n^{(2)}(q;-ir) = 2\frac{\|\mathbb{S}^{q-2}\|}{\|\mathbb{S}^{q-1}\|} i^{n+q-1} K_n(q;r). \tag{6.97}$$

Moreover, we find for $r > 0$ (see (6.93) and (6.94), respectively)

$$K_{n-1}(q;r) - K_{n+1}(q;r) = -\frac{2n+q-2}{r} K_n(q;r) \tag{6.98}$$

and

$$nK_{n-1}(q;r) + (n+q-2)K_{n+1}(q;r) = (2n+q-2)K_n'(q;r), \tag{6.99}$$

which again follows from the recursion relation for the Legendre polynomial. Keeping r fixed we obtain the following lemma by similar techniques as used for the proof of Lemma 6.30.

Lemma 6.34. *For $n \to \infty$ and fixed $r > 0$ we have*

$$\lim_{n\to\infty} \frac{K_n(q;r)}{\frac{1}{2\pi}\Gamma(\frac{q-1}{2})\Gamma\left(n+\frac{q}{2}-1\right)\left(\frac{2}{r}\right)^{n+q-2}} = 1. \tag{6.100}$$

The Hankel functions as well as the Kelvin functions are not defined at the origin, but they have characteristic singularities there.

Lemma 6.35. *For $r \to 0$ and $n+q-3 \geq 0$ we have*

$$\begin{aligned} H_n^{(1)}(q;r) &= \frac{i}{\pi}\Gamma\left(n+\frac{q-2}{2}\right)\Gamma\left(\frac{q}{2}\right)\left(\frac{2}{r}\right)^{n+q-2} \\ &\quad + O\left(r^{-n-q+4}\right), \end{aligned} \tag{6.101}$$

$$\begin{aligned} K_n(q;r) &= \frac{\Gamma(q-1)}{2^{q-1}\Gamma\left(\frac{q}{2}\right)}\Gamma\left(n+\frac{q-2}{2}\right)\left(\frac{2}{r}\right)^{n+q-2} \\ &\quad + O\left(r^{-n-q+4}\right). \end{aligned} \tag{6.102}$$

The identity

$$\begin{aligned} K_0(2;r) &= \int_1^\infty e^{-rt}(t^2-1)^{-\frac{1}{2}}\,dt = \int_r^\infty e^{-u}(u^2-r^2)^{-\frac{1}{2}}\,du \\ &= -e^{-r}\ln(r) + \int_r^\infty e^{-u}\ln\left(u+\sqrt{u^2+r^2}\right)\,du \end{aligned} \tag{6.103}$$

shows us that $K_0(2,r)$ has a logarithmic singularity. In fact, for $n=0$, $q=2$, we have

$$K_0(2;r) = -\ln(r) + O(1). \tag{6.104}$$

In the same way we obtain

$$H_0^{(1)}(2;r) = \frac{2i}{\pi}\ln(r) + O(1). \tag{6.105}$$

Remark 6.36. *The function $x \mapsto K_0(q;|x|)$, $x \in \mathbb{R}^q\backslash\{0\}$, shows the same singularity behavior at the origin as the fundamental solution $x \mapsto F_q(|x|)$, $x \in \mathbb{R}^q\backslash\{0\}$, for the Laplace operator in $\mathbb{R}^q$. This fact is of particular significance for the characterization of the singularity behavior of the multi-dimensional Euler–Green function in lattice points.*

For $r \to \infty$ and $n \geq 0$ fixed we obtain

$$\begin{aligned} K_n(q;r) &= \int_1^\infty e^{-rt} P_n(q;t)(t^2-1)^{\frac{q-3}{2}}\, dt \qquad (6.106)\\ &\overset{t=1+s}{=} e^{-r} \int_0^\infty e^{-rs} P_n(q;1+s)((2+s)s)^{\frac{q-2}{2}}\, ds \\ &= e^{-r}\left(2^{\frac{q-3}{2}} \frac{\Gamma\left(\frac{q-1}{2}\right)}{r^{\frac{q-1}{2}}} + O(r^{-\frac{q+1}{2}}) \right). \end{aligned}$$

This leads to the following asymptotic relation.

Lemma 6.37. *For $r \to \infty$ and $n \geq 0$ fixed,*

$$K_n(q;r) = \frac{e^{-r}}{2}\Gamma\left(\frac{q-1}{2}\right)\left(\left(\frac{2}{r}\right)^{\frac{q-1}{2}} + O\left(r^{-\frac{q+1}{2}}\right)\right). \qquad (6.107)$$

The properties of the Kelvin function relevant for our purposes in the analytic theory of numbers can be summarized as follows.

Lemma 6.38. *The Kelvin function $K_n(q;\cdot)$ satisfies the following relations:*

(i) For $x = |x|\xi$, $y = |y|\eta$,

$$K_0(q;|x-y|) = \sum_{n=0}^\infty N(q;n) I_n(q;|x|) K_n(q;|y|) P_n(q;\xi\cdot\eta),$$

(ii) $K_n(q;\cdot)$ is a solution of the differential equation

$$K_n''(q;r) - \frac{q-1}{r}K_n'(q;r) - \left(1 + \frac{n(n+q-2)}{r^2}\right) K_n(q;r) = 0,$$

(iii) $K_n(q;\cdot)$ satisfies the recursion relations

$$(n+q-2)K_{n+1}(q;r) + (2n+q-2)K_n'(q;r) + nK_{n-1}(q;r) = 0,$$

$$K_{n-1}(q;r) - K_{n+1}(q;r) = -\frac{2n+q-2}{r} K_n(q;r),$$

(iv)

$$(-1)^m r^{-n-m} K_{n+m}(q;r) = \left(\frac{1}{r}\frac{d}{dr}\right)^m \left(K_n(q;r)\, r^{-n}\right),$$

$$(-1)^m r^{n-m+q-2} K_{n-m}(q;r) = \left(\frac{1}{r}\frac{d}{dr}\right)^m \left(r^{n+q-2} K_n(q;r)\right),$$

(v)

$$\lim_{n\to\infty} \frac{2\pi\, K_n(q;r)}{\Gamma\left(\frac{q-1}{2}\right)\left(\frac{2}{r}\right)^{n+q-2}\Gamma\left(n+\frac{q}{2}-1\right)} = 1,$$

(vi) For $n+q-3>0$ fixed and $r \to 0$,

$$K_n(q;r) = \frac{\Gamma(q-1)}{2^{q-1}\Gamma\left(\frac{q}{2}\right)}\Gamma\left(n+\frac{q-2}{2}\right)\left(\frac{2}{r}\right)^{n+q-2} + O(r^{-n-q+4})$$

and, for $n=0$, $q=2$ and $r \to 0$,

$$K_0(q;r) \;=\; -\ln(r)+O(1).$$

(vii) For $r \to \infty$ and $n \geq 0$ fixed

$$K_n(q;r) = \frac{e^{-r}}{2}\Gamma\left(\frac{q-1}{2}\right)\left(\left(\frac{2}{r}\right)^{\frac{q-1}{2}} + O\left(r^{-\frac{q+1}{2}}\right)\right).$$

In particular,

$$K_0(3;r) = \frac{e^{-r}}{r} \tag{6.108}$$

and

$$K_n(3;r) = r^n\left(\frac{1}{r}\frac{d}{dr}\right)^n \frac{e^{-r}}{r}. \tag{6.109}$$

Notational Conventions for Kelvin Functions. It is interesting to relate the Kelvin and Hankel functions of dimension q to their counterparts of dimension 2. For real values ν we have (see G.N. Watson [1944], W. Magnus et al. [1966])

$$K_\nu(r) = K_\nu(2;r) = \frac{\sqrt{\pi}}{\Gamma\left(\nu+\frac{1}{2}\right)}\left(\frac{r}{2}\right)^\nu \int_1^\infty e^{-rt}(t^2-1)^{\nu-\frac{1}{2}}\,dt. \tag{6.110}$$

By comparison we find

$$K_n(q;r) = \frac{\Gamma\left(\frac{q-1}{2}\right)}{\sqrt{\pi}}\left(\frac{r}{2}\right)^{\frac{2-q}{2}} K_{n+\frac{q-2}{2}}(r). \tag{6.111}$$

Similarly,

$$H_n^{(1)}(q;r) = \Gamma\left(\frac{q}{2}\right)\left(\frac{r}{2}\right)^{\frac{2-q}{2}} H^{(1)}_{n+\frac{q-2}{2}}(r), \tag{6.112}$$

where

$$H_\nu^{(1)}(r) = H_\nu^{(1)}(2;r) = \frac{-2}{\sqrt{\pi}\Gamma\left(\nu+\frac{1}{2}\right)}\left(\frac{r}{2}\right)^\nu \int_{1+0i}^{1+\infty i} e^{irt}(1-t^2)^{\nu-\frac{1}{2}}\,dt. \tag{6.113}$$

6.3 Asymptotic Expansions

As is well-known, any finite sum of the form

$$U^{(N)}(x) = (2\pi)^{\frac{1-q}{2}} \left\|\mathbb{S}^{q-1}\right\| \sum_{n=0}^{N} i^n J_n(q;r) Y_n(q;\xi) \tag{6.114}$$

with $x = r\xi$, $r = |x|$, $\xi \in \mathbb{S}^{q-1}$, $Y_n(q;\cdot) \in \mathrm{Harm}_n(\mathbb{S}^{q-1})$, and $N \in \mathbb{N}_0$, satisfies the Helmholtz wave equation

$$\Delta U^{(N)}(x) + U^{(N)}(x) = 0, \quad x \in \mathbb{R}^q; \tag{6.115}$$

i.e., $U^{(N)}$ is an entire solution of the Helmholtz equation (6.115) in $\mathbb{R}^q$. Remembering the asymptotic expansion of the Bessel function (Lemma 6.28)

$$J_n(q;r) = \frac{\Gamma\left(\frac{q}{2}\right)}{\sqrt{\pi}} \left(\frac{2}{r}\right)^{\frac{q-1}{2}} \cos\left(n\frac{\pi}{2} + (q-1)\frac{\pi}{4} - r\right) + O\left(r^{-\frac{q+1}{2}}\right) \tag{6.116}$$

we get

$$\begin{aligned} r^{\frac{q-1}{2}} U^{(N)}(r\xi) &= \sum_{n=0}^{N} i^n \left(e^{i\left(n\frac{\pi}{2}+(q-1)\frac{\pi}{4}-r\right)} + e^{-i\left(n\frac{\pi}{2}+(q-1)\frac{\pi}{4}-r\right)}\right) Y_n(q;\xi) \\ &+ o(1), \end{aligned} \tag{6.117}$$

i.e.,

$$r^{\frac{q-1}{2}} U^{(N)}(r\xi) = e^{i\left(r-(q-1)\frac{\pi}{4}\right)} F^{(N)}(q;\xi) + e^{-i\left(r-(q-1)\frac{\pi}{4}\right)} F^{(N)}(q;-\xi) + o(1), \tag{6.118}$$

where

$$F^{(N)}(q;\xi) = \sum_{n=0}^{N} Y_n(q;\xi), \quad F^{(N)}(q;-\xi) = \sum_{n=0}^{N} (-1)^n Y_n(q;\xi). \tag{6.119}$$

Using the estimate (Lemma 6.29)

$$\begin{aligned} J_n(q;r) &= \frac{\Gamma\left(\frac{q}{2}\right)}{\sqrt{\pi}2^{\frac{3-q}{2}}} r^{\frac{1-q}{2}} \\ &\quad \times \left(e^{i\left(r-\frac{\pi}{2}\left(n+\frac{q-2}{2}\right)-\frac{\pi}{4}\right)} \sum_{l=0}^{m-1} \frac{\left(n+\frac{q-2}{2},l\right)}{(-2ir)^l}\right. \\ &\qquad\qquad \left. + \ e^{-i\left(r-\frac{\pi}{2}\left(n+\frac{q-2}{2}\right)-\frac{\pi}{4}\right)} \sum_{l=0}^{m-1} \frac{\left(n+\frac{q-2}{2},l\right)}{(-2ir)^l}\right) \\ &+ \quad O\left(\frac{1}{r^{m+\frac{q-1}{2}}}\right), \quad m \geq 1, \end{aligned} \tag{6.120}$$

with (see (6.89) and (6.90))

$$\left(n+\frac{q-2}{2},\ 0\right)=1 \tag{6.121}$$

and

$$\begin{aligned}\left(n+\frac{q-2}{2},\ l\right) &= \frac{1}{l!}\left(n(n+q-2)+\frac{(q-1)(q-3)}{4}\right) \\ &\quad \cdot \ldots \cdot \left(n(n+q-2)+\frac{(q+2l-3)(q-2l-1)}{4}\right),\end{aligned} \tag{6.122}$$

we are led to introduce the "operators"

$$O(q;0) = 1, \tag{6.123}$$

$$\begin{aligned}O(q;l) &= \frac{1}{l!}\left(\frac{(q-1)(q-3)}{4}-\Delta^*\right) \\ &\quad \cdot \ldots \cdot \left(\frac{(q+2l-3)(q-2l-1)}{4}-\Delta^*\right).\end{aligned} \tag{6.124}$$

In terms of (6.123) and (6.124) we obtain the asymptotic relation

$$\begin{aligned} r^{\frac{q-1}{2}} U^{(N)}(r\xi) &= (2\pi)^{\frac{1-q}{2}} \left\|\mathbb{S}^{q-1}\right\| \sum_{n=0}^{N} i^n J_n(q;r) Y_n(q;\xi) \\ &= i^{\frac{1-q}{2}} e^{ir} \sum_{l=0}^{m-1} \left(\frac{1}{-2\pi i r}\right)^l O(q;l) F^{(N)}(q;\xi) \\ &\qquad + i^{\frac{q-1}{2}} e^{-ir} \sum_{l=0}^{m-1} \left(\frac{1}{2\pi i r}\right)^l O(q;l) F^{(N)}(q;-\xi) \\ &\quad + o\left(r^{1-m}\right), \quad r \to \infty. \end{aligned} \tag{6.125}$$

Asymptotic expansions of the type (6.118) and (6.125), respectively, are characteristic for entire solutions of the Helmholtz wave equation.

Remark 6.39. *Unfortunately, when the attempt is made to formulate an asymptotic expansion of type (6.125) for $N \to \infty$, we are confronted with serious problems of the convergence for the right side of (6.125). A way out can be found in certain mean values as proposed by G. Herglotz and C. Müller (see C. Müller [1952] and, in addition, W. Magnus et al. [1949], and many others). However, this concept is not of interest for our purposes.*

Our lattice pont summations require an asymptotic expansion of the integral (cf. C. Müller [1952])

$$U(r\xi) = \left(\frac{r}{2\pi}\right)^{\frac{q-1}{2}} \int_{\mathbb{S}^{q-1}} F(\eta) e^{ir\xi\cdot\eta}\, dS_{(q-1)}(\eta), \tag{6.126}$$

where certain smoothness imposed on the integrand F defined on $\mathbb{S}^{q-1}$ is mandatory to circumvent problems of convergence. Clearly, the integral U as defined by (6.126) satisfies the equation $\Delta U(x) + U(x) = 0,\ x \in \mathbb{R}^q$, i.e., U is an entire solution. In connection with Lemma 6.29 these observations lead to the following asymptotic relation: Let F be of class $\mathrm{C}^{(k)}(\mathbb{S}^{q-1})$, $k \geq \frac{q}{2} + 1$. Then, uniformly with respect to all $\xi \in \mathbb{S}^{q-1}$, we have for $r \to \infty$

$$\left(\frac{r}{2\pi}\right)^{\frac{q-1}{2}} \int_{\mathbb{S}^{q-1}} F(\eta) e^{ir\xi\cdot\eta}\, dS_{(q-1)}(\eta) = i^{\frac{1-q}{2}} e^{ir} F(\xi) + i^{\frac{q-1}{2}} e^{-ir} F(-\xi) \;+\; o(1). \tag{6.127}$$

In an analogous way as in the theory of Bessel functions, (6.127) admits a canonical extension.

Theorem 6.40. *(Expansion Theorem). Let F be of class $\mathrm{C}^{(k)}(\mathbb{S}^{q-1})$ with $k \geq \frac{q}{2} + 2m - 1, m \in \mathbb{N}$. Then*

$$\begin{aligned}\left(\frac{r}{2\pi}\right)^{\frac{q-1}{2}} &\int_{\mathbb{S}^{q-1}} e^{ir\xi\cdot\eta} F(\eta)\, dS_{(q-1)}(\eta) \\ = &\; i^{\frac{1-q}{2}} e^{ir} \sum_{l=0}^{m-1} \left(\frac{1}{-2ir}\right)^l O(q;l) F(\xi) + i^{\frac{q-1}{2}} e^{-ir} \sum_{l=0}^{m-1} \left(\frac{1}{2ir}\right)^l O(q;l) F(-\xi) \\ &+ o\left(r^{1-m}\right),\end{aligned} \tag{6.128}$$

where

$$O(q;0) = 1, \tag{6.129}$$

$$O(q;l) = \frac{1}{l!} \prod_{j=1}^{l} \left(\frac{(q+2j-3)(q-2j-1)}{4} - \Delta^* \right),\ l \in \mathbb{N}, \tag{6.130}$$

and, as always, Δ^* is the Laplace–Beltrami operator on the unit sphere $\mathbb{S}^{q-1}$.

Obviously, for the constant function $F = 1$, Corollary 6.40 leads back to the theory of Bessel functions (cf. Lemma 6.29).

Chapter 7

Preparatory Tools of Lattice Point Theory

In this chapter, we replicate some basic results of geometric and analytic number theory (for more details the reader is referred to textbooks of number theory, e.g., G.H. Hardy, E.M. Wright [1958], J.W.S. Cassels [1968], F. Fricker [1982], E. Hlawka et al. [1991], E. Krätzel [2000], and many others). Our aim is twofold: On the one side, key information about lattices should be provided as helpful preparation. On the other side, the material should serve as an appropriate reference for our later work on lattice point summation.

The layout of this chapter is as follows: First we introduce the concept of lattices and their inverse counterparts in Euclidean spaces $\mathbb{R}^q$. Then we are devoted to some fundamentals of geometric theory of numbers. Minkowski's theorem is formulated within its standard framework. We list the asymptotic relations for the total number of lattice points inside circles starting from the original work of C.F. Gauss [1801, 1826]. The discussion of upper bounds leads to the famous Hardy's conjecture. Finally, some basic information about lattice points inside spheres in Euclidean spaces $\mathbb{R}^q$ is given for the convenience of the reader.

7.1 Lattices in Euclidean Spaces

Let $g_1, \ldots, g_q$ be linearly independent vectors in q-dimensional Euclidean space $\mathbb{R}^q$.

Definition 7.1. *The set* Λ *(more precisely,* $\Lambda_{(q)}$*) of all points*

$$g = n_1 g_1 + \ldots + n_q g_q \tag{7.1}$$

$n_i \in \mathbb{Z},\ i = 1, \ldots, q$, *is called a lattice in* $\mathbb{R}^q$ *with basis* $g_1, \ldots, g_q$.

Definition 7.2. *The half-open parallelotope consisting of the points* $x \in \mathbb{R}^q$ *with*

$$x = t_1 g_1 + \ldots + t_q g_q, \quad -\frac{1}{2} \leq t_i < \frac{1}{2}, \tag{7.2}$$

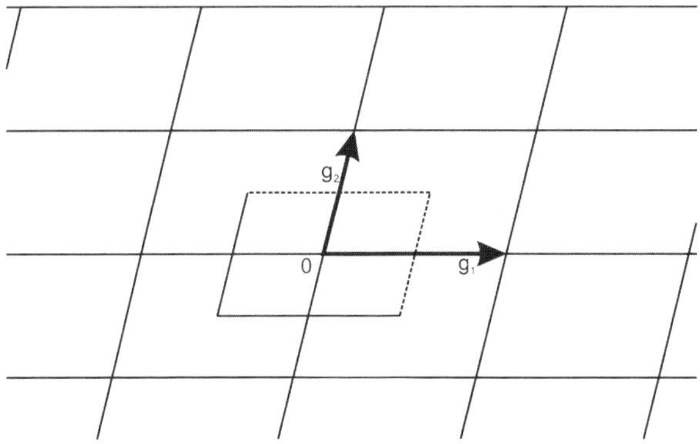

FIGURE 7.1: Two-D lattice generated by g_1, g_2.

$i = 1, \ldots, q$, is called the fundamental cell $\mathcal{F}_\Lambda$ of the lattice Λ (cf. Figure 7.1 for the 2D case).

Remark 7.3. *Obviously there are infinitely many lattice cells of Λ reflecting the Λ–periodicity. $\mathcal{F}_\Lambda$, as specified by (7.2), is both simple and appropriate for our purposes, since it includes the origin in its interior.*

From linear algebra (see, e.g., P.J. Davis [1963]) it is well-known that the volume of $\mathcal{F}_\Lambda$ is equal to the quantity

$$||\mathcal{F}_\Lambda|| = \int_{\mathcal{F}_\Lambda} dV = \sqrt{\det\left((g_i \cdot g_j)_{i,j=1,\ldots,q}\right)}. \tag{7.3}$$

For each $g \in \Lambda$ we have

$$||\mathcal{F}_\Lambda|| = ||\mathcal{F}_\Lambda + \{g\}||. \tag{7.4}$$

Clearly, because of $(\mathcal{F}_\Lambda + \{g\}) \cap (\mathcal{F}_\Lambda + \{g'\}) = \emptyset$ for $g \neq g'$, $g, g' \in \Lambda$, we have

$$\mathbb{R}^q = \bigcup_{g \in \Lambda} (\mathcal{F}_\Lambda + \{g\}) = \bigcup_{g \in \Lambda} (\mathcal{F}_\Lambda - \{g\}). \tag{7.5}$$

Since the generating vectors $g_1, \ldots, g_q$ are assumed to be linearly independent, there exists a system of vectors $h_1, \ldots, h_q$ in $\mathbb{R}^q$ such that

$$h_j \cdot g_i = \delta_{i,j} = \begin{cases} 0, & i \neq j, \\ 1, & i = j \end{cases} \tag{7.6}$$

($\delta_{i,j}$ is the Kronecker symbol). In more detail, for $i, j = 1, \ldots, q$, we let

$$g_i \cdot g_j = \gamma_{i,j}. \tag{7.7}$$

The scalars $\gamma^{i,j}$, $i,j = 1,\ldots,q$, are defined by

$$\sum_{j=1}^{q} \gamma^{i,j}\gamma_{j,k} = \delta_{i,k}. \tag{7.8}$$

The vectors h_j, $j = 1,\ldots,q$, given by

$$h_j = \sum_{k=1}^{q} \gamma^{j,k} g_k, \quad j = 1,\ldots,q, \tag{7.9}$$

satisfy the equations

$$h_j \cdot g_i = \sum_{k=1}^{q} \gamma^{j,k} g_k \cdot g_i = \sum_{k=1}^{q} \gamma^{j,k}\gamma_{k,i} = \delta_{j,i}, \tag{7.10}$$

$i,j = 1,\ldots,q$. Moreover, we find

$$\begin{aligned} h_i \cdot h_j &= \sum_{k=1}^{q} \gamma^{i,k} g_k \cdot \sum_{l=1}^{q} \gamma^{j,l} g_l \\ &= \sum_{l=1}^{q} \gamma^{j,l} \sum_{k=1}^{q} \gamma^{i,k}\gamma_{k,l} \\ &= \gamma^{j,i}, \quad i,j = 1,\ldots,q. \end{aligned} \tag{7.11}$$

Definition 7.4. *The lattice with the basis $h_1,\ldots,h_q$ given by (7.9) is called the inverse (or dual) lattice Λ^{-1} to Λ.*

The inverse lattice Λ^{-1} consists of all vectors $h \in \mathbb{R}^q$, such that the inner product $h \cdot g$ is an integer for all $g \in \Lambda$.

Obviously, we have

$$\Lambda = (\Lambda^{-1})^{-1}. \tag{7.12}$$

Moreover, for the fundamental cell $\mathcal{F}_{\Lambda^{-1}}$ of the inverse lattice Λ^{-1} we have

$$\|\mathcal{F}_{\Lambda^{-1}}\| = \|\mathcal{F}_{\Lambda}\|^{-1}. \tag{7.13}$$

Example 7.5. *Suppose that $\Lambda = \sigma\mathbb{Z}^q$ is the lattice generated by the "dilated" basis $\sigma\varepsilon^1,\ldots,\sigma\varepsilon^q$, where σ is a positive number and $\varepsilon^1,\ldots,\varepsilon^q$ forms the canonical orthonormal basis in $\mathbb{R}^q$. Then, the volume of the fundamental cell of $\sigma\mathbb{Z}^q$ is $\|\mathcal{F}_{\Lambda}\| = \sigma^q$. Generating vectors of the inverse lattice Λ^{-1} are $\sigma^{-1}\varepsilon^1,\ldots,\sigma^{-1}\varepsilon^q$. The volume of the fundamental cell of the inverse lattice is given by*

$$\|\mathcal{F}_{\Lambda^{-1}}\| = \sigma^{-q} = \|\mathcal{F}_{\Lambda}\|^{-1}. \tag{7.14}$$

In particular, for $\sigma = 1$, i.e., the lattice $\Lambda = \mathbb{Z}^q$, we have $\Lambda^{-1} = \mathbb{Z}^q = \Lambda$ such that

$$\|\mathcal{F}_{\Lambda^{-1}}\| = 1 = \|\mathcal{F}_{\Lambda}\|. \tag{7.15}$$

7.2 Figure Lattices in Euclidean Spaces

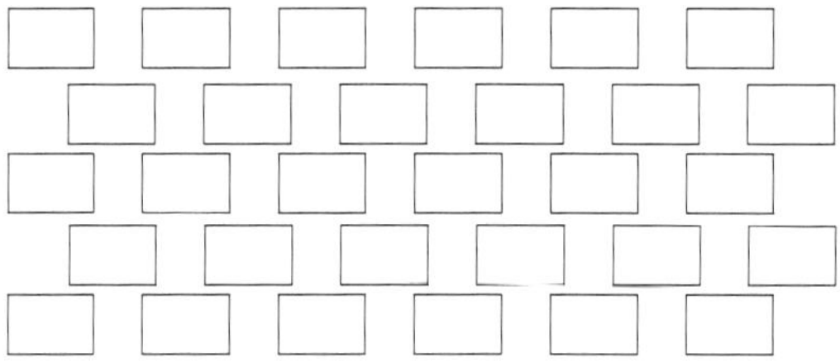

FIGURE 7.2: Figure lattice associated to a rectangle.

If $\Lambda \subset \mathbb{R}^q$ is a lattice and $x \in \mathbb{R}^q$, then $\Lambda + \{x\} \subset \mathbb{R}^q$ is called the *(point) lattice based at* x. If $\Lambda \subset \mathbb{R}^q$ is a lattice and $\mathcal{M} \subset \mathbb{R}^q$ is a set, then the sum $\mathcal{M} + \Lambda$ is called the *figure lattice* associated to $\mathcal{M}$ by Λ (see the example illustrated by Figure 7.2). If $\mathcal{M}$ is a ball around the origin 0, then the figure lattice $\mathcal{M} + \Lambda$ is called a *ball lattice* (see Figure 7.3).

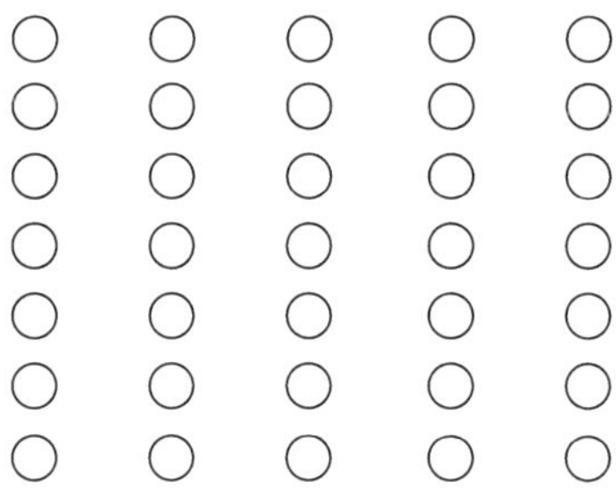

FIGURE 7.3: Example of a ball lattice.

For a set $\mathcal{M} \subset \mathbb{R}^q$ we define *the characteristic function* $\mathcal{X}_{\mathcal{M}}$ by

$$\mathcal{X}_{\mathcal{M}}(x) = \begin{cases} 1, & x \in \mathcal{M} \\ 0, & x \notin \mathcal{M}. \end{cases} \tag{7.16}$$

From the considerations above it follows that a point $x \in \mathbb{R}^q$ belongs to the figure lattice $\mathcal{M} + \Lambda$ if and only if

$$\sum_{g \in \Lambda} \mathcal{X}_{\mathcal{M}}(x + g) \geq 1. \tag{7.17}$$

The point $x \in \mathbb{R}^q$ does not belong to the figure lattice $\mathcal{M} + \Lambda$ if and only if

$$\sum_{g \in \Lambda} \mathcal{X}_{\mathcal{M}}(x+g) = 0. \tag{7.18}$$

If, for all points from $\mathbb{R}^q$, the formula

$$\sum_{g \in \Lambda} \mathcal{X}_{\mathcal{M}}(x+g) \geq 1 \tag{7.19}$$

is valid, then we call Λ a *covering lattice* for $\mathcal{M}$ (see Figure 7.4), because in this case $\mathbb{R}^q = \mathcal{M} + \Lambda$. If, on the other hand, for all points $x \in \mathbb{R}^q$ the formula

$$\sum_{g \in \Lambda} \mathcal{X}_{\mathcal{M}}(x+g) \leq 1 \tag{7.20}$$

holds true, the lattice Λ is called a *filling lattice* for $\mathcal{M}$ (see Figure 7.5). In this case, an arbitrary point $x \in \mathbb{R}^q$ can belong to at most one of the sets $\mathcal{M} + \{g\}$ with $g \in \Lambda$.

FIGURE 7.4: Example of a covering lattice.

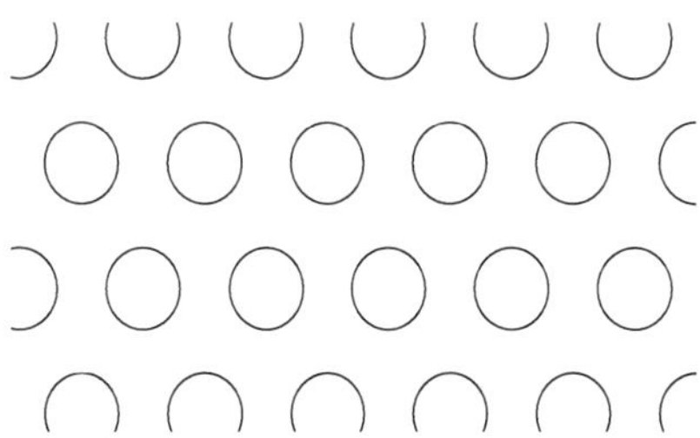

FIGURE 7.5: Example of a filling lattice.

7.3 Basic Results of the Geometry of Numbers

For the fundamental cell $\mathcal{F}_\Lambda$, the lattice Λ simultaneously is a covering and filling lattice for $\mathcal{F}$. Furthermore, it is clear that

$$\begin{aligned} \int_{\mathbb{R}^q} F(x)\, dV_{(q)}(x) &= \sum_{g\in\Lambda} \int_{\mathcal{F}_\Lambda - \{g\}} F(x)\, dV_{(q)}(x) \\ &= \sum_{g\in\Lambda} \int_{\mathcal{F}_\Lambda} F(x+g)\, dV_{(q)}(x) \\ &= \int_{\mathcal{F}_\Lambda} \left(\sum_{g\in\Lambda} F(x+g) \right) dV_{(q)}(x) \\ &= \int_{\mathcal{F}_\Lambda} \left(\sum_{g\in\Lambda} F(x-g) \right) dV_{(q)}(x) \end{aligned} \tag{7.21}$$

holds for an integrable function F in $\mathbb{R}^q$ (note that $-g$ as well as g runs through the lattice Λ). Consequently, if for some positive constant C we have

$$\int_{\mathbb{R}^q} F(x)\, dV_{(q)}(x) \geq C\ \|\mathcal{F}_\Lambda\|, \tag{7.22}$$

then for at least one point x of the fundamental cell $\mathcal{F}_\Lambda$ we must have

$$\sum_{g\in\Lambda} F(x+g) \geq C. \tag{7.23}$$

If conversely

$$\sum_{g\in\Lambda} F(x+g) < C \tag{7.24}$$

holds for all points $x \in \mathcal{F}_\Lambda$, the calculation

$$\begin{aligned} \int_{\mathbb{R}^q} F(x)\, dV_{(q)}(x) &= \int_{\mathcal{F}_\Lambda} \left(\sum_{g\in\Lambda} F(x+g) \right) dV_{(q)}(x) \\ &< C \int_{\mathcal{F}_\Lambda} dV_{(q)}(x) \\ &= C\ \|\mathcal{F}_\Lambda\| \end{aligned} \tag{7.25}$$

would give a contradiction to the assumption.
If we restrict ourselves to the function $F = \mathcal{X}_\mathcal{M}$, where $\mathcal{M}$ is a bounded measurable set, the volume $\|\mathcal{M}\|$ of $\mathcal{M}$ is given by

$$\int_{\mathbb{R}^q} \mathcal{X}_\mathcal{M}(x)\, dV_{(q)}(x) = \|\mathcal{M}\|. \tag{7.26}$$

The relation

$$\sum_{g \in \Lambda} \mathcal{X}_{\mathcal{M}}(x+g) \geq \frac{\|\mathcal{M}\|}{\|\mathcal{F}_\Lambda\|} \tag{7.27}$$

says that at least $\lceil \|\mathcal{M}\| \ / \ \|\mathcal{F}_\Lambda\| \rceil$ points of the form $x+g$, $g \in \Lambda$, lie in the set $\mathcal{M}$.

Theorem 7.6. *(Blichfeldt's Theorem). For each lattice Λ and each non-empty, bounded, measurable set $\mathcal{M}$ there exists a point $x \in \mathbb{R}^q$ such that the number N of lattice points of the lattice $\Lambda + \{x\}$ based at $x \in \mathbb{R}^q$ lying inside the set $\mathcal{M}$ satisfies the relation*

$$N \geq \frac{\|\mathcal{M}\|}{\|\mathcal{F}_\Lambda\|} \tag{7.28}$$

In Theorem 7.6 little is assumed about the set $\mathcal{M}$. In particular, the point $x \in \mathbb{R}^q$ serving as origin for the lattice $\Lambda + \{x\}$ remains completely unspecified. If one wishes to avoid this lack of information, additional assumptions on the set $\mathcal{M}$ are necessary.

A set $\mathcal{M}$ is said to be *symmetric* with respect to a point $z \in \mathbb{R}^q$ if $x = z + a$ belonging to the set implies $x' = z - a$ also does. The point z is called the center of symmetry of this set. A set $\mathcal{M}$ is said to be *star-shaped* with respect to a point z if $x = z + a$ belonging to the set implies that $x_t = z + ta$ does also for all $t \in [0,1]$. A set $\mathcal{M}$ is called *convex* if it is star-shaped with respect to each of its points. A criterion for convex sets reads: A set $\mathcal{M}$ is convex if and only if, for each two points x and z belonging to it, so does the line segment consisting of all points $y_t = x + t(z-x)$, $t \in [0,1]$.

With these preparations we are able to formulate *Minkowski's lattice point theorem*, that is a classical result in the geometry of numbers. For more details the reader is referred, e.g., to L.J. Mordell [1935], E. Hlawka [1954], E. Hlawka et al. [1991], and the references therein).

Theorem 7.7. *If x denotes the center of symmetry of the symmetric, bounded, and convex region $\mathcal{G} \subset \mathbb{R}^q$, and $\Lambda \subset \mathbb{R}^q$ is a lattice with*

$$\|\mathcal{G}\| > 2^q \ \|\mathcal{F}_\Lambda\|, \tag{7.29}$$

then $\mathcal{G}$ contains at least one lattice point $z = x+g$, distinct from $x \in \mathbb{R}^q, g \in \Lambda$, from the lattice $\Lambda + \{x\}$ based at x.

From the example of the cube

$$\mathcal{G} = \left\{ x + \sum_{i=1}^{q} t_i \varepsilon^i \in \mathbb{R}^q \ : \ t_i \in (-1,1), \ i = 1, \ldots, q \right\}, \tag{7.30}$$

we see that the condition $\|\mathcal{G}\| > 2^q \ \|\mathcal{F}\|$ cannot be weakened for Λ being the lattice $\mathbb{Z}^q$. However, we are able to formulate a variant of Minkowski's theorem (see Figure 7.6) showing the not-restrictive inequality $\|\mathcal{G}\| \geq 2^q \ \ \|\mathcal{F}_\Lambda\|$ (see, e.g., E. Hlawka et al. [1991]).

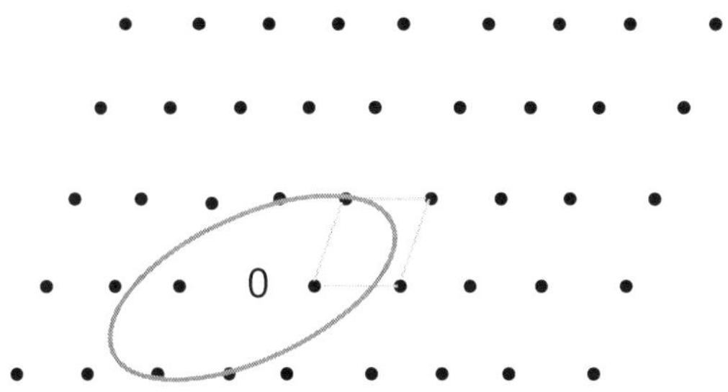

FIGURE 7.6: Two-D geometric situation of Minkowski's result.

Corollary 7.8. *If x denotes the center of symmetry of the symmetric, compact, convex set $\mathcal{K}$, and if $\Lambda \subset \mathbb{R}^q$ is a lattice with*

$$\|\mathcal{K}\| \geq 2^q \ \|\mathcal{F}_\Lambda\|, \tag{7.31}$$

then $\mathcal{K}$ contains at least one lattice point $y = x+g$ distinct from $x \in \mathbb{R}^q, g \in \Lambda$, from the lattice $\Lambda + \{x\}$ based at x.

The standard proof of Minkowski's Theorem (Theorem 7.7) is based on Mordell's approach (cf. L.J. Mordell [1935], for more details the reader is referred to L.J. Mordell [1969], G.H. Hardy, E.M. Wright [1958], E. Hlawka et al. [1991], P. Erdös et al. [1989], R. Ait–Haddou et al. [2000], and many others). Later on, we give a different proof of Minkowski's theorem based on a purely analytical proposition, namely a multi-dimensional variant of the Poisson summation formula for regular regions.

7.4 Lattice Points Inside Spheres

The branch of analytic theory of numbers concerned with lattice point summation has a long history, which reaches back to L. Euler [1736] and C.F. Gauss [1801, 1826]. Enlightening accounts of the developments within this theory are, e.g., attributed to G.H. Hardy [1915], E. Landau [1927], A. Walfisz [1960], to mention just a few.

Bivariate Theory. In what follows, we are interested in recapitulating some results on the number of lattice points inside circles $\mathbb{S}^2_N$ of radii $N > \frac{\sqrt{2}}{2}$ around the origin 0; more conctretely, we deal with closed disks $\overline{\mathbb{B}^2_N}$ of radii $N > \frac{\sqrt{2}}{2}$ (for more background material the reader is referred, e.g., to the monographs F. Fricker [1982], E. Krätzel [1988], and the survey paper A. Ivic et al. [2004]).

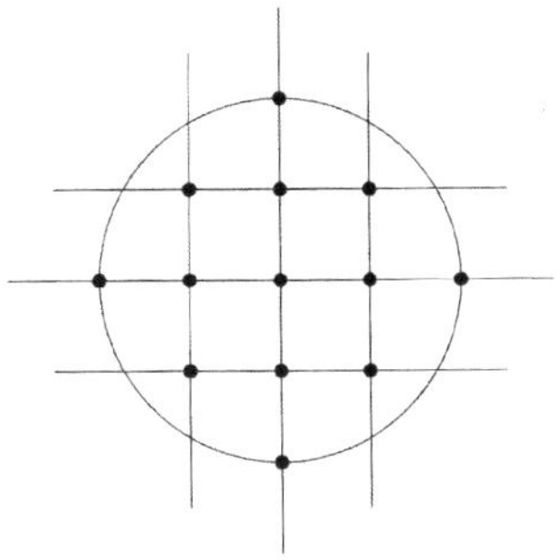

FIGURE 7.7: Two-dimensional lattice points inside a circle.

The problem of determining the total number of lattice points of $\mathbb{Z}^2$ inside and on a circle with radius N, i.e., the determination of the quantity

$$\#_{\mathbb{Z}^2}\left(\overline{\mathbb{B}^2_N}\right) = \#\left\{(n_1, n_2)^T \in \mathbb{Z}^2 : n_1^2 + n_2^2 \leq N^2\right\} \tag{7.32}$$

is very old. In fact, for the calculation of sum

$$\#_{\mathbb{Z}^2}\left(\overline{\mathbb{B}^2_N}\right) = \sum_{\substack{n_1^2+n_2^2 \leq N^2 \\ (n_1,n_2)\in\mathbb{Z}^2}} 1, \tag{7.33}$$

C.F. Gauss [1826] found a simple, but efficient method for its estimation (see Figure 7.7): Associate to every square the north-west edge as lattice point. The union of all squares with lattice points inside $\overline{\mathbb{B}^2_N}$ defines a polyhedral set $\mathbb{P}^2_N$ with area $\#_{\mathbb{Z}^2}\left(\overline{\mathbb{B}^2_N}\right)$ (cf. Figure 7.8). Since the diagonal of each square is $\sqrt{2}$, the geometry of Figure 7.8 tells us that

$$\pi\left(N - \frac{\sqrt{2}}{2}\right)^2 \leq \#_{\mathbb{Z}^2}\left(\overline{\mathbb{B}^2_N}\right) \leq \pi\left(N + \frac{\sqrt{2}}{2}\right)^2. \tag{7.34}$$

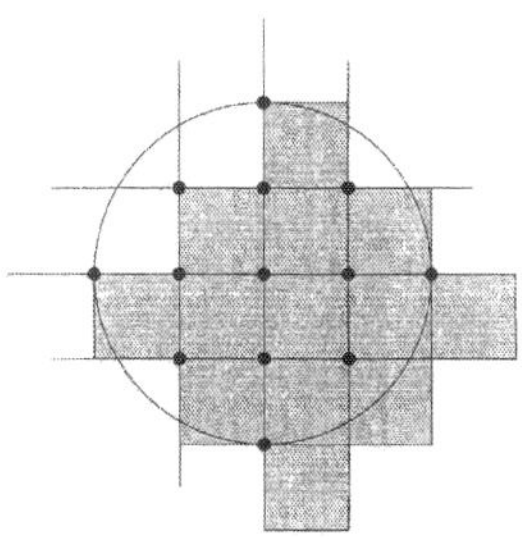

FIGURE 7.8: The polyhedral set $\mathbb{P}^2_N$.

Therefore, the "discrepancy" $\#_{\mathbb{Z}^2}\left(\overline{\mathbb{B}^2_N}\right) - \pi N^2$ after division by N is bounded for $N \to \infty$, i.e.,

$$\left| \#_{\mathbb{Z}^2}\left(\overline{\mathbb{B}^2_N}\right) - \pi N^2 \right| < \sqrt{2}\pi N + \frac{\pi}{2}, \tag{7.35}$$

which can be written as

$$\#_{\mathbb{Z}^2}\left(\overline{\mathbb{B}^2_N}\right) = \pi N^2 + O(N). \tag{7.36}$$

In other words, the number of lattice points in $\overline{\mathbb{B}^2_N}$ is equal to the area of that circle plus a remainder (i.e., discrepancy) of the order of the boundary. In particular,

$$\lim_{N\to\infty} \frac{\#_{\mathbb{Z}^2}\left(\overline{\mathbb{B}^2_N}\right)}{N^2} = \pi. \tag{7.37}$$

Remark 7.9. *C.F. Gauss [1826] illustrated his result by taking* $N^2 = 100\,000$. *In this case, he calculated*

$$\#_{\mathbb{Z}^2}\left(\overline{\mathbb{B}^2_N}\right) = 314197. \tag{7.38}$$

This determines the number π *up to three decimals.*

In our nomenclature, the formula (7.32) attributed to C.F. Gauss [1826] allows the representation

$$\#_{\mathbb{Z}^2}\left(\overline{\mathbb{B}^2_N}\right) = \sum_{\substack{|g|^2 \leq N^2 \\ g\in\mathbb{Z}^2}} 1 = \pi N^2 + O(N). \tag{7.39}$$

The so-called *circle problem* is concerned with the question of determining the bound

$$\alpha_2 = \inf\left\{\gamma : \#_{\mathbb{Z}^2}\left(\overline{\mathbb{B}^2_{\sqrt{N}}}\right) = \pi N + O(N^\gamma)\right\}. \tag{7.40}$$

Until now, we know from (7.39) that $\alpha_2 \leq \frac{1}{2}$. The improvement of the situation, however, turns out to be very laborious, in fact, requiring a great effort. A first remarkable result is attributed to W. Sierpit'nski [1906], who proved by use of a method of his teacher G. Voronoi [1903] that

$$\#_{\mathbb{Z}^2}\left(\overline{\mathbb{B}^2_{\sqrt{N}}}\right) = \pi N + O\left(N^{\frac{1}{3}}\right) \tag{7.41}$$

i.e., $\alpha_2 \leq \frac{1}{3}$. The proof of Sierpiniski is elementary, it is a link between geometry and number theory. By use of advanced methods on exponential sums (based on the work by, e.g., H. Weyl [1916], L.–K. Hua [1959], J.-R. Chen [1963], and many others) the estimate $\frac{1}{3}$ could be strengthened to some extent. G. Kolesnik [1985] had as his sharpest result with these techniques

$$\#_{\mathbb{Z}^2}\left(\overline{\mathbb{B}^2_{\sqrt{N}}}\right) - \pi N = O\left(N^{\frac{139}{429}+\varepsilon}\right), \quad \varepsilon > 0. \tag{7.42}$$

M.N. Huxley [2003] devised a substantially new approach; his strongest result was the estimate

$$\#_{\mathbb{Z}^2}\left(\overline{\mathbb{B}^2_{\sqrt{N}}}\right) - \pi N = O\left(N^{\frac{131}{416}+\varepsilon}\right), \quad \varepsilon > 0. \tag{7.43}$$

Note that $\frac{139}{429} = 0.324009...$, while $\frac{131}{416} = 0.315068...$. Nevertheless, Hardy's conjecture

$$\#_{\mathbb{Z}^2}\left(\overline{\mathbb{B}^2_{\sqrt{N}}}\right) - \pi N = O\left(N^{\frac{1}{4}+\varepsilon}\right) \tag{7.44}$$

for <u>every</u> $\varepsilon > 0$ seems to be still a challenge for future work.

The following Table 7.1 summarizes incomplete incremental improvements in the upper limit for the circle problem.

TABLE 7.1: Improvements in the Upper Limit for the Circle Problem.

$\frac{1}{3}$	G. Voronoi (1903)
$\frac{37}{112}$	J.E. Littlewood, A. Walfisz (1924)
$\frac{27}{82}$	J.G. van der Corput (1928)
$\frac{12}{37}$	J.-R. Chen (1963)
$\frac{139}{429}$	G. Kolesnik (1985)
$\frac{131}{416}$	M.N. Huxley (2003)

In all cases, the proofs became rather long and made use of some of the more heavy machinery in analysis.

G.H. Hardy [1915] also contributed significantly to the understanding of the circle problem by detecting a lower bound, i.e.,

$$\#_{\mathbb{Z}^2}\left(\overline{\mathbb{B}^2_{\sqrt{N}}}\right) - \pi N = \Omega\left(N^{\frac{1}{4}}\right), \tag{7.45}$$

i.e., there is a constant C and arbitrarily large values of N so that

$$\left|\#_{\mathbb{Z}^2}\left(\overline{\mathbb{B}^2_{\sqrt{N}}}\right) - \pi N\right| > CN^{\frac{1}{4}} \tag{7.46}$$

is satisfied.

There are many perspectives to formulate variants of lattice point problems for the circle (showing a relationship to Shannon-type sampling). It already was the merit of E. Landau [1924] to point out certain fields of interest, such as

- general lattices can be used instead of the unit lattice $\mathbb{Z}^2$ (clearly, analogous estimates as known in the case of $\mathbb{Z}^2$ for the circle problem also apply to all general lattices $\Lambda \subset \mathbb{R}^2$),
- lattice points can be affected by non-constant weights,
- multi-dimensional generalizations can be described, e.g., the lattice point problem for spheres and ellipsoids, using general lattice Λ in $\mathbb{R}^q$.

Multivariate Theory. Just as in the planar case, we may discuss the problem of counting the number of points $(n_1, \ldots, n_q)^T$ with all the n_i integers inside spheres $\mathbb{S}_N^q$ of radii N centered at the origin 0. More precisely, by

$$\#_{\mathbb{Z}^q}\left(\overline{\mathbb{B}_N^q}\right) = \sum_{\substack{|g| \leq N \\ g \in \mathbb{Z}^q}} 1 \tag{7.47}$$

we denote the *total number of lattice points* g of the lattice $\mathbb{Z}^q$, such that $|g| \leq N$, i.e., g is a member of $\overline{\mathbb{B}_N^q}$. Following the idea of C.F. Gauss [1826], we consider the polyhedron

$$\mathbb{P}_N^q = \bigcup_{\substack{g \in \overline{\mathbb{B}_N^q(0)} \\ g \in \mathbb{Z}^q}} \left(\mathcal{F}_{\mathbb{Z}^q} + \{g\}\right), \tag{7.48}$$

where, as usual, $\mathcal{F}_{\mathbb{Z}^q} + \{g\}$ is the translate of $\mathcal{F}_{\mathbb{Z}^q}$ by the lattice point $g \in \Lambda$. Obviously, the volume $\|\mathbb{P}_N^q\|$ of $\mathbb{P}_N^q$ is equal to $\#_\Lambda\left(\overline{\mathbb{B}_N^q}\right)$. Now it is clear that there exists a constant d such that the estimate $|x| \leq d$ holds true for all $x \in \mathcal{F}_{\mathbb{Z}^q}$ (for example, $d = \frac{\sqrt{q}}{2}$), so that for $\mathbb{P}_N^q$ the following inclusion is valid $(N > d)$:

$$\overline{\mathbb{B}_{N-d}^q} \subset \mathbb{P}_N^q \subset \overline{\mathbb{B}_{N+d}^q}. \tag{7.49}$$

This leads us to the following inequality:

$$\left\|\overline{\mathbb{B}_{N-d}^q}\right\| \leq \underbrace{\#_{\mathbb{Z}^q}\left(\overline{\mathbb{B}_N^q}\right)}_{=\|\mathbb{P}_N^q\|} \leq \left\|\overline{\mathbb{B}_{N+d}^q}\right\|, \tag{7.50}$$

which is equivalent to

$$\frac{\pi^{\frac{q}{2}}}{\Gamma(\frac{q}{2}+1)}(N-d)^q \leq \underbrace{\#_{\mathbb{Z}^q}\left(\overline{\mathbb{B}_N^q}\right)}_{=\|\mathbb{P}_N^q\|} \leq \frac{\pi^{\frac{q}{2}}}{\Gamma(\frac{q}{2}+1)}(N+d)^q. \tag{7.51}$$

Following the Gaussian approach, we therefore obtain the following fundamental result in the theory of lattice points inside spheres.

Theorem 7.10. *For $N \to \infty$,*

$$\#_{\mathbb{Z}^q}\left(\overline{\mathbb{B}_N^q}\right) = \frac{\pi^{\frac{q}{2}}}{\Gamma(\frac{q}{2}+1)} N^q + O\left(N^{q-1}\right). \tag{7.52}$$

This result can be easily generalized to arbitrary lattice Λ in $\mathbb{R}^q$ without changing the order $O(N^{q-1})$ of the discrepancy, i.e.,

$$\#_\Lambda(\overline{\mathbb{B}_N^q}) - \frac{\pi^{\frac{q}{2}}}{\Gamma(\frac{q}{2}+1)} N^q = O(N^{q-1}),\ N \to \infty. \tag{7.53}$$

In analogy to the 2D case, the problem of determining

$$\alpha_q = \inf\left\{\gamma : \#_{\mathbb{Z}^q}\left(\overline{\mathbb{B}_{\sqrt{N}}^q}\right) = \left\|\overline{\mathbb{B}_{\sqrt{N}}^q}\right\| + O(N^\gamma)\right\} \tag{7.54}$$

is known as the *lattice point problem for the sphere* (for $q = 2$, as we know, *lattice point problem for the circle*). It is clear that

$$\alpha_q \leq \frac{q-1}{2}. \tag{7.55}$$

Moreover, it is not hard to conclude that

$$\alpha_q \geq \frac{q-2}{2}. \tag{7.56}$$

In more detail (see, e.g., F. Fricker [1982]), we have

$$\#_{\mathbb{Z}^q}\left(\overline{\mathbb{B}_{\sqrt{N}}^q}\right) = \left\|\overline{\mathbb{B}_{\sqrt{N}}^q}\right\| + \Omega\left(N^{\frac{q-2}{2}}\right),\ N \to \infty, \tag{7.57}$$

where, as usual, Ω means that there is a constant C and arbitrarily large values of N so that

$$\left|\#_{\mathbb{Z}^q}\left(\overline{\mathbb{B}_{\sqrt{N}}^q}\right) - \left\|\overline{\mathbb{B}_{\sqrt{N}}^q}\right\|\right| \geq CN^{\frac{q-2}{2}} \tag{7.58}$$

is satisfied (see also W. Freeden [2011]).

Three-Dimensional Lattice Point Problem. For the *lattice point problem for the 3D sphere* we know that

$$\#_{\mathbb{Z}^3}\left(\overline{\mathbb{B}_{\sqrt{N}}^3}\right) - \left\|\overline{\mathbb{B}_{\sqrt{N}}^3}\right\| = O(N). \tag{7.59}$$

In other words, for the lattice point problem

$$\alpha_3 = \inf\left\{\gamma : \#_{\mathbb{Z}^3}\left(\overline{\mathbb{B}_{\sqrt{N}}^3}\right) - \left\|\overline{\mathbb{B}_{\sqrt{N}}^3}\right\| = O(N^\gamma)\right\} \tag{7.60}$$

it is known that $\alpha_3 \leq 1$. In order to improve this result, great efforts must be made (similar to the 2D case). E. Landau [1927] proved $\alpha_3 \leq \frac{3}{4}$,

TABLE 7.2: Improvements in the Upper Limit of the 3D Problem for the Sphere.

$\frac{3}{4}$	E. Landau (1927)
$\frac{11}{16}$	I. M. Winogradow (1955)
$\frac{701}{1020}$	O.M. Fomenko (1961)
$\frac{29}{44}$	F. Chamizo, H. Iwaniec (1995)
$\frac{701}{1020}$	D. R. Heath-Brown (1999).

I.M. Winogradow [1955] and O.M. Fomenko [1961] gave essential improvements, more recent progress is due to F. Chamizo, H. Iwaniec [1995], D.R. Heath-Brown [1999]. Table 7.2 summarizes incremental improvements in the upper limit of the 3D problem for the sphere:

Four-Dimensional Lattice Point Problem. For the *lattice point problem for the 4D sphere* a classical result from Lagrange (that each positive integer can be expressed as the sum of at most four integral squares) helps us to verify an estimate (see, e.g., E. Krätzel [2000]) of the form

$$\#_{\mathbb{Z}^4}\left(\overline{\mathbb{B}^4_{\sqrt{N}}}\right) - \left\|\overline{\mathbb{B}^4_{\sqrt{N}}}\right\| = O\Big(N\ln(N)\Big), \tag{7.61}$$

and it is known that the ln-term in (7.61) cannot be omitted. In fact, A. Walfisz [1960] used some extremely deep analysis to show an estimate of the form

$$\#_{\mathbb{Z}^4}\left(\overline{\mathbb{B}^4_{\sqrt{N}}}\right) - \left\|\overline{\mathbb{B}^4_{\sqrt{N}}}\right\| = O\left(N(\ln(N))^{\frac{2}{3}}\right). \tag{7.62}$$

For more details including an Ω-bound the reader is referred to S.D. Adhikari, Y.-F.S. Petermann [1991].

Higher-Dimensional Lattice Point Problems. It is remarkable that the the *lattice point problem for the higher-dimensional sphere* ($q \geq 4$) behaves in much more regular fashion than the dimensions $q = 2, 3$. For higher dimensions, indeed, the asymptotic growth of the number of points is more regular (see, e.g., E. Grosswald [1985] for exact formulas). In fact, the situation changes drastically. The exact asymptotic order of the "lattice discrepancy" $\#_{\mathbb{Z}^5}\left(\overline{\mathbb{B}^5_{\sqrt{N}}}\right) - \left\|\overline{\mathbb{B}^5_{\sqrt{N}}}\right\|$ is known for a long time.

Theorem 7.11. *For $q \geq 5$, the lattice discrepancy satisfies*

$$\begin{aligned} \#_{\mathbb{Z}^q}\left(\overline{\mathbb{B}^q_{\sqrt{N}}}\right) - \left\|\overline{\mathbb{B}^q_{\sqrt{N}}}\right\| &= \Omega\left(N^{\alpha_q}\right), & (7.63) \\ \#_{\mathbb{Z}^q}\left(\overline{\mathbb{B}^q_{\sqrt{N}}}\right) - \left\|\overline{\mathbb{B}^q_{\sqrt{N}}}\right\| &= O\left(N^{\alpha_q}\right), & (7.64) \end{aligned}$$

where

$$\alpha_q = \frac{q-2}{2}. \tag{7.65}$$

These results can be easily verified for general lattices Λ.

Corollary 7.12. *For $q \geq 5$,*

$$\begin{aligned} \#_{\Lambda}\left(\overline{\mathbb{B}^q_{\sqrt{N}}}\right) - \left\|\overline{\mathbb{B}^q_{\sqrt{N}}}\right\| &= \Omega\left(N^{\alpha_q}\right), & (7.66) \\ \#_{\Lambda}\left(\overline{\mathbb{B}^q_{\sqrt{N}}}\right) - \left\|\overline{\mathbb{B}^q_{\sqrt{N}}}\right\| &= O\left(N^{\alpha_q}\right). & (7.67) \end{aligned}$$

All in all, the impression that the lattice point problem for the higher-dimensional spheres becomes more and more problematic is wrong. The essential difficulties arising in the 2- and 3D cases do not occur for dimensions $q \geq 5$. The dimension $q = 4$ is an intermediate case. The reason is that the decomposition of a positive integer as a sum of integral squares for the dimensions $q \geq 4$, is not as "pathologic" as, for instance, for the dimensions $q = 2$ and $q = 3$.

However, the handling of the Λ-*discrepancy*

$$\#_{\Lambda}\left(\overline{\mathbb{B}^q_{\sqrt{N}}}\right) - \left\|\overline{\mathbb{B}^q_{\sqrt{N}}}\right\| \tag{7.68}$$

in terms of an alternating (Fourier) series expansion

$$\#_{\Lambda}\left(\overline{\mathbb{B}^q_{\sqrt{N}}}\right) - \left\|\overline{\mathbb{B}^q_{\sqrt{N}}}\right\| \quad \sim \quad \sum_{\substack{h \in \Lambda^{-1} \\ h \neq 0}} \int_{\substack{|x| \leq \sqrt{N} \\ x \in \mathbb{R}^q}} e^{-2\pi i h \cdot x} \, dV_{(q)}(x) \tag{7.69}$$

involving the qD-Bessel functions

$$\begin{aligned} \int_{\substack{|x| \leq \sqrt{N} \\ x \in \mathbb{R}^q}} e^{-2\pi i h \cdot x} \, dV_{(q)}(x) &= \int_0^{\sqrt{N}} r^{q-1} \int_{\mathbb{S}^{q-1}} e^{-2\pi i r h \cdot \xi} \, dS_{(q-1)}(\xi) \, dr \\ &= \|\mathbb{S}^{q-1}\| \int_0^{\sqrt{N}} r^{q-1} J_0(q; 2\pi r|h|) \, dr \\ &= N^{\frac{q}{2}} \, \|\mathbb{S}^{q-1}\| \, \frac{J_1(q; 2\pi|h|\sqrt{N})}{2\pi|h|\sqrt{N}}, & (7.70) \end{aligned}$$

i.e.,

$$\#_\Lambda\left(\overline{\mathbb{B}^q_{\sqrt{N}}}\right) - \left\|\overline{\mathbb{B}^q_{\sqrt{N}}}\right\| \quad \sim \quad N^{\frac{q}{2}}\, \|\mathbb{S}^{q-1}\| \sum_{\substack{h\in\Lambda^{-1}\\ h\neq 0}} \frac{J_1(q; 2\pi|h|\sqrt{N})}{2\pi|h|\sqrt{N}}, \qquad (7.71)$$

turns out to be completely different.

Roughly speaking, we are confronted with the following difficulties: The convergence discussion of the series expansion of the right side of (7.69) for the bivariate case ($q = 2$), is "not far away" from the univariate context, whereas the series discussion of the higher dimensions $q \geq 3$ becomes increasingly critical, so that, e.g., summability techniques such as Gauss-Weierstrass summability have to come in to play to guarantee convergence obligations.

As a consequence, number theoretic (e.g., Hardy–Landau-type) lattice point identities in the plane and, consequently, the Shannon-type cardinal series expansion, turn out to be attackable (almost) similarly to the 1D case, while the technical procedures become increasingly problematic for increasing dimensions.

Chapter 8

Preparatory Tools of Fourier Analysis

This chapter presents selected results on the Fourier theory: It starts with 1D-Fourier asymptotics, in particular the method of the stationary phase. Then we go over to multi-dimensional orthonormal periodic polynomials and their role in Fourier (orthogonal) expansions. Our main interest concerning the Fourier transform in the Euclidean space $\mathbb{R}^q$ is the relation between functions being not-necessarily absolutely integrable over $\mathbb{R}^q$ as well as periodic with respect to a lattice $\Lambda \subset \mathbb{R}^q$. As a consequence, we are immediately led to the process of "periodization" as the bridging tool to the Poisson summation formula in $\mathbb{R}^q$. For our purposes of lattice point summation, however, it must be pointed out that the convergence criteria to justify the process of periodization in $\mathbb{R}^q$ is different from those developed in the literature, e.g., by E.M. Stein, G. Weiss [1971]. In fact, these authors verify the Poisson summation formula under the strong assumption of the absolute convergence of all occurring sums. However, this assumption is not relevant for our lattice point and resulting sampling identities because of the specific alternating character of the constituting sums. Besides of the obstacles caused by the alternating character of sums, one of the essential calamities in our lattice point oriented approach to sampling is the divergence of the bilinear series for the Euler–Green function on the fundamental cell $\mathcal{F}_\Lambda$ of the lattice Λ in dimensions $q \geq 2$. Moreover, in contrast to the 1D case, a further serious difficulty is the Fourier inversion formula. Again, the standard L^1-theory is not considerably helpful because of the alternating character of the occurring functions involved in our context. In fact, as a powerful remedy, the multivariate inversion formula of Fourier integrals can be understood in the terminology of certain means (such as Gauss–Weierstrass transform). This aspect is discussed in the framework of regular regions as well as the whole Euclidean space. Anyway, the integral transform for discontinuous functions possessing a "potato-like" regular region as a local support remains critical for Fourier inversion. In this respect, it can be shown that the Hankel transform provides a way out which allows it to handle alternating, not absolutely convergent series expansions in terms of Bessel functions.

8.1 Stationary Point Asymptotics

We repeat some well-known 1D-Fourier asymptotics: To this end we start with certain realizations of the *Riemann–Lebesgue theorem*:

Lemma 8.1. *(1D-Riemann–Lebesgue Theorem). Let F be continuous on the interval $(-1,1)$ such that*

$$\int_{-1}^{1} |F(t)| \; dt < \infty. \tag{8.1}$$

Then

$$\lim_{r\to\infty} \int_{-1}^{1} e^{2\pi i r t} F(t) \; dt = 0. \tag{8.2}$$

An immediate consequence is the following asymptotic relation.

Lemma 8.2. *Let F be continuous in $[-1,1]$ and continuously differentiable in $(-1,1)$ such that*

$$\int_{-1}^{1} |F'(t)| \; dt < \infty. \tag{8.3}$$

Then

$$\lim_{r\to\infty} \int_{-1}^{1} e^{2\pi i r t} F'(t) \; dt = 0. \tag{8.4}$$

Of more interest for our purposes is the following limit relation (cf. C. Müller [1952]).

Riemann–Lebesgue Theorem. Let F be continuous in $[-1,1]$ and continuously differentiable in $(-1,1)$ such that

$$\int_{-1}^{1} |F'(t)| \; dt < \infty. \tag{8.5}$$

Then we have for $r \to \infty$

$$\sqrt{\frac{2r}{\pi}} \int_{-1}^{1} e^{2\pi i r t} F(t)(1-t^2)^{-\frac{1}{2}} \; dt = i^{-\frac{1}{2}} e^{2\pi i r} F(1) + i^{\frac{1}{2}} e^{-2\pi i r} F(-1) + o(1). \tag{8.6}$$

Fourier Asymptotics. In 1D Fourier theory, we find a large number of extensions of the limit relation (8.6). For example, from P.I. Natanson [1961], we are able to borrow the following result.

Theorem 8.3. *Let G be of class $\mathrm{C}^{(m)}([\alpha,\beta])$, $\alpha,\beta \in \mathbb{R}$ with $\alpha < \beta$. Suppose that F is given in the form*

$$F(t) = G(t)(t-\alpha)^{\lambda-1}(\beta-t)^{\mu-1}, \quad t \in [\alpha,\beta], \tag{8.7}$$

$0 < \lambda,\ \mu \leq 1$. *Then, for* $r \to \infty$,

$$\int_\alpha^\beta F(t)e^{irt}\,dt = -A_m(r) + B_m(r) + O\left(\frac{1}{r^m}\right), \tag{8.8}$$

where

$$A_m(r) = \sum_{n=0}^{m-1} \frac{\Gamma(n+\lambda)}{n!\; r^{n+\lambda}} e^{ir\alpha + i\frac{\pi}{2}(n+\lambda-2)} \left(\frac{\partial}{\partial u}\right)^n \left((\beta-u)^{\mu-1}G(u)\right)\Big|_{u=\alpha} \tag{8.9}$$

and

$$B_m(r) = \sum_{n=0}^{m-1} \frac{\Gamma(n+\mu)}{n!\; r^{n+\mu}} e^{ir\beta + i\frac{\pi}{2}(n-\mu)} \left(\frac{\partial}{\partial v}\right)^n \left((v-\alpha)^{\lambda-1}G(v)\right)\Big|_{v=\beta}. \tag{8.10}$$

Remark 8.4. *For* $\lambda = \mu = 1$, *the term* $O(r^{-m})$ *can be replaced by* $o(r^{-m})$.

A consequence is the following asymptotic expansion.

Corollary 8.5. *Suppose that* $F \in \mathrm{C}^{(1)}([\alpha,\beta])$, $H \in \mathrm{C}^{(2)}([\alpha,\beta])$ *such that* $H'(t) > 0$ *for* $t \in [\alpha,\beta]$. *Then, for* $r \to \infty$,

$$\int_\alpha^\beta F(t)e^{irH(t)}\,dt = \frac{1}{ir}\left(\frac{F(\beta)}{H'(\beta)}e^{irH(\beta)} - \frac{F(\alpha)}{H'(\alpha)}e^{irH(\alpha)}\right) + o\left(\frac{1}{r}\right). \tag{8.11}$$

Corollary 8.5 is based on the assumption that the function H does not possess a stationary point, i.e., $\tau \in [\alpha,\beta]$ such that $H'(\tau) = 0$. In fact, the asymptotic relation (8.11) depends only on the values of the endpoints of the interval $[\alpha, b]$.

Stationary Point Asymptotics. The following asymptotic relation (8.14) includes influences from stationary points (see, e.g., P.I. Natanson [1961] for the proof).

Theorem 8.6. *Let* G *be of class* $\mathrm{C}^{(m)}([\alpha,\beta])$. *Suppose that* $\tilde{H}$ *is of class* $\mathrm{C}^{(m)}([\alpha,\beta])$ *with* $\tilde{H}(t) > 0$ *for all* $t \in [\alpha,\beta]$. *Furthermore, we let*

$$F(t) = G(t)(t-a)^{\lambda-1}(\beta-t)^{\mu-1}, \quad t \in [\alpha,\beta], \tag{8.12}$$

$0 \leq \lambda,\ \mu \leq 1$, *and* $H \in \mathrm{C}^{(1)}([\alpha,\beta])$ *with*

$$H'(t) = \tilde{H}(t)(t-\alpha)^{\rho-1}(\beta-t)^{\sigma-1}, \quad t \in [\alpha,\beta], \tag{8.13}$$

$\rho, \sigma \geq 1$. *Then, for* $r \to \infty$,

$$\int_\alpha^\beta F(t)\ e^{ikH(t)}\,dt = -A_m(r) + B_m(r) + O\left(\frac{1}{r^{\frac{m}{\rho}}}\right) + O\left(\frac{1}{r^{\frac{m}{\sigma}}}\right), \tag{8.14}$$

where

$$A_m(r) = -e^{irH(\alpha)} \sum_{n=0}^{m-1} \frac{\psi^{(n)}(0)}{\rho} \frac{\Gamma(\frac{n+\lambda}{\rho})}{n!} e^{i\frac{\pi}{2}\frac{n+\lambda}{\rho}} \frac{1}{r^{\frac{n+\lambda}{\rho}}} \tag{8.15}$$

and

$$B_m(r) = -e^{irH(\beta)} \sum_{n=0}^{m-1} \frac{\chi^{(n)}(0)}{\sigma} \frac{\Gamma(\frac{n+\mu}{\sigma})}{n!} e^{-i\frac{\pi}{2}\frac{n+\mu}{\sigma}} \frac{1}{r^{\frac{n+\mu}{\sigma}}} \tag{8.16}$$

with

$$\psi(u) = F(t)u^{1-\lambda}\frac{dt}{du}, \quad u^\rho(t) = H(t) - H(\alpha), \tag{8.17}$$

and

$$\chi(v) = F(t)v^{1-\mu}\frac{dt}{dv}, \quad v^\sigma(t) = H(\beta) - H(t). \tag{8.18}$$

The following consequences in connection with Theorem 8.6 are significant:

(i) If $\lambda = \mu = 1$, then the "O" symbol can be replaced by the "o" symbol.

(ii) If $\rho = \sigma = 1$, then Theorem 8.6 reduces to Theorem 8.3 (by the use of the substitution $H(t) = t$).

Next we come to the key definition of our considerations in this section.

Definition 8.7. *A point $\tau \in [\alpha, \beta]$ is called a stationary point of order n in $[\alpha, b]$, if the following conditions are satisfied:*

(i) $H \in C^{(n+1)}([\alpha, \beta])$,

(ii) $H(\tau) = \ldots = H^{(n)}(\tau) = 0$,

(iii) $H^{(n+1)}(\tau) \neq 0$.
Note that a stationary point of order 1 is simply called a stationary point.

From Theorem 8.6 we are able to formulate some corollaries.

Corollary 8.8. *Suppose that $F \in C^{(1)}([\alpha, \beta])$ and $H \in C^{(3)}([\alpha, \beta])$. If there is a stationary point $\tau \in (\alpha, \beta)$ such that $H'(\tau) = 0$ and $H''(\tau) > 0$, then for $r \to \infty$*

$$\int_\alpha^\beta F(t)e^{irH(t)}\, dt = F(\tau)\sqrt{\frac{2\pi}{H''(\tau)\, r}}\, e^{irH(\tau)+i\frac{\pi}{4}} + o\left(\frac{1}{r^{\frac{1}{2}}}\right). \tag{8.19}$$

Remark 8.9. *If $\tau = \beta$ and/or $\tau = \alpha$ we obtain the same result, however, equipped with the "univariate solid angle" $\frac{1}{2}$.*

Corollary 8.10. *Suppose that $F \in C^{(1)}([\alpha, \beta])$ and $H \in C^{(3)}([\alpha, \beta])$. If there is a stationary point $\tau \in (a, \beta)$ such that $H'(\tau) = 0$ and $H''(\tau) < 0$, then for $r \to \infty$*

$$\int_\alpha^\beta F(t)e^{irH(t)}\, dt = F(\tau)\sqrt{\frac{-2\pi}{H''(\tau)\, r}}\, e^{irH(\tau)-i\frac{\pi}{4}} + o\left(\frac{1}{r^{\frac{1}{2}}}\right). \tag{8.20}$$

8.2 Periodic Polynomials and Fourier Expansions

The space of all $F \in \mathrm{C}^{(m)}(\mathbb{R}^q)$ that are Λ-periodic, i.e., $F(x) = F(x+g)$ for all $g \in \Lambda$ and $x \in \mathbb{R}^q$, is denoted by $\mathrm{C}_\Lambda^{(m)}(\mathbb{R}^q)$, $0 \leq m \leq \infty$. $\mathrm{L}_\Lambda^p(\mathbb{R}^q)$, $1 \leq p < \infty$, is the space of all $F : \mathbb{R}^q \to \mathbb{C}$ that are Λ-periodic and are Lebesgue–measurable on $\mathcal{F}$ with

$$\|F\|_{\mathrm{L}_\Lambda^p(\mathbb{R}^q)} = \left(\int_{\mathcal{F}_\Lambda} |F(x)|^p \, dV(x) \right)^{\frac{1}{p}} < \infty. \tag{8.21}$$

Clearly,

$$\mathrm{C}_\Lambda^{(0)}(\mathbb{R}^q) \subset \mathrm{L}_\Lambda^p(\mathbb{R}^q). \tag{8.22}$$

Furthermore, $\mathrm{L}_\Lambda^2(\mathbb{R}^q)$ is the completion of $\mathrm{C}_\Lambda^{(0)}(\mathbb{R}^q)$ with respect to the norm $\|\cdot\|_{\mathrm{L}_\Lambda^2(\mathbb{R}^q)}$:

$$\mathrm{L}_\Lambda^2(\mathbb{R}^q) = \overline{\mathrm{C}_\Lambda^{(0)}(\mathbb{R}^q)}^{\|\cdot\|_{\mathrm{L}_\Lambda^2(\mathbb{R}^q)}}. \tag{8.23}$$

Next, for the standard Λ-periodic polynomials (orthonormal in $\mathrm{L}_\Lambda^2(\mathbb{R}^q)$-sense) equivalent conditions for the closure and completeness are formulated within the space $\mathrm{L}_\Lambda^2(\mathbb{R}^q)$ of square-integrable Λ-periodic functions in $\mathbb{R}^q$:
Let Λ be a lattice in $\mathbb{R}^q$. The functions Φ_h, $h \in \Lambda^{-1}$, defined by

$$\Phi_h(x) = \frac{1}{\sqrt{\|\mathcal{F}_\Lambda\|}} \, e(h \cdot x) = \frac{1}{\sqrt{\|\mathcal{F}_\Lambda\|}} \, e^{2\pi i h \cdot x}, \; x \in \mathbb{R}^q, \tag{8.24}$$

are Λ-*periodic*. An easy calculation shows that

$$\int_{\mathcal{F}_\Lambda} \Phi_h(x) \overline{\Phi_{h'}(x)} \, dV(x) = \delta_{h,h'} = \begin{cases} 1 & , \quad h = h' \\ 0 & , \quad h \neq h'. \end{cases} \tag{8.25}$$

In other words, the system $\{\Phi_h\}_{h \in \Lambda^{-1}}$ of multi-dimensional "periodic polynomials" is orthonormal with respect to the $\mathrm{L}_\Lambda^2(\mathbb{R}^q)$-inner product:

$$(\Phi_h, \Phi_{h'})_{\mathrm{L}_\Lambda^2(\mathbb{R}^q)} = \int_{\mathcal{F}_\Lambda} \Phi_h(x) \, \overline{\Phi_{h'}(x)} \;\; dV(x) = \delta_{h,h'}. \tag{8.26}$$

An elementary calculation yields $(\Delta_x + \Delta^\wedge(h))\,\Phi_h(x) = 0$, $\Delta^\wedge(h) = 4\pi^2 h^2$, $h \in \Lambda^{-1}$. We shall say that λ is an *eigenvalue* of the lattice Λ with respect to the operator Δ if there is a non-trivial solution U of the differential equation $(\Delta + \lambda) U = 0$ satisfying the boundary condition of periodicity $U(x+g) = U(x)$ for all $g \in \Lambda$. The function U is then called *eigenfunction* of the lattice Λ with respect to the eigenvalue λ and the operator Δ. Application of the standard multi-dimensional Fourier theory (see, for example, E.M. Stein, G. Weiss [1971]) shows that the functions Φ_h are the only eigenfunctions. Furthermore, the family $\{\Delta^\wedge(h)\}_{h \in \Lambda^{-1}}$ with

$$\Delta^\wedge(h) = 4\pi^2 h^2, \; h \in \Lambda^{-1}, \tag{8.27}$$

constitutes the only eigenvalues of Δ with respect to the lattice Λ (note that we simply write $\Delta^\wedge(h)$ instead of $\Delta^\wedge_\Lambda(h)$ if no confusion is likely to arise). The set of all eigenvalues $\Delta^\wedge(h)$ with respect to Δ is the *spectrum* $\mathrm{Spect}_\Delta(\Lambda)$:

$$\mathrm{Spect}_\Delta(\Lambda) = \left\{\Delta^\wedge(h) : \ \Delta^\wedge(h) = 4\pi^2 h^2, h \in \Lambda^{-1}\right\}. \tag{8.28}$$

Clearly, the system $\{\Phi_h\}_{h\in\Lambda^{-1}}$ is closed and complete in the pre–Hilbert space $(\mathrm{C}^{(0)}_\Lambda(\mathbb{R}^q); \|\cdot\|_{\mathrm{L}^2_\Lambda(\mathbb{R}^q)})$ as well as in the Hilbert space $(\mathrm{L}^2_\Lambda(\mathbb{R}^q); \|\cdot\|_{\mathrm{L}^2_\Lambda(\mathbb{R}^q)})$.
A fundamental result in Fourier analysis is that each $F \in \mathrm{L}^2_\Lambda(\mathbb{R}^q)$ can be represented by its Fourier series in the sense

$$\lim_{N\to\infty} \left\| F - \sum_{\substack{|h|\le N\\ h\in\Lambda^{-1}}} F^\wedge_{\mathcal{F}_\Lambda}(h)\ \Phi_h \right\|_{\mathrm{L}^2_\Lambda(\mathbb{R}^q)} = 0, \tag{8.29}$$

where the Fourier coefficients $F^\wedge_{\mathcal{F}_\Lambda}(h)$ of F are given by

$$F^\wedge_{\mathcal{F}_\Lambda}(h) = \int_{\mathcal{F}_\Lambda} F(x)\ \overline{\Phi_h(x)}\ dV(x), \quad h \in \Lambda^{-1}. \tag{8.30}$$

The *Parseval identity* tells us that, for each $F \in \mathrm{L}^2_\Lambda(\mathbb{R}^q)$,

$$\int_{\mathcal{F}_\Lambda} |F(x)|^2\ dV(x) = \sum_{h\in\Lambda^{-1}} \left|F^\wedge_{\mathcal{F}_\Lambda}(h)\right|^2. \tag{8.31}$$

A useful corollary of (8.29) is that any function $F \in \mathrm{L}^1_\Lambda(\mathbb{R}^q)$ fufilling

$$\sum_{h\in\Lambda^{-1}} \left|F^\wedge_{\mathcal{F}_\Lambda}(h)\right| < \infty \tag{8.32}$$

can be modified on a set of measure zero so that it is in $\mathrm{C}^{(0)}_\Lambda(\mathbb{R}^q)$ and equals the Fourier series $\Sigma_{h\in\Lambda^{-1}} F^\wedge_{\mathcal{F}_\Lambda}(h)\Phi_h$ for all $x \in \mathcal{F}_\Lambda$ (see, e.g., E.M. Stein, G. Weiss [1971]).
Finally, we deal with functions F of class $\mathrm{C}^{(k)}_\Lambda(\mathbb{R}^q)$ satisfying the differentiabilty property $k > \frac{q}{2}$. Then it is easy to see that

$$\int_{\mathcal{F}_\Lambda} (\nabla^\alpha F)(x)\ \overline{\Phi_h(x)}\ dV(x) = (-2\pi i h)^\alpha F^\wedge_{\mathcal{F}_\Lambda}(h), \tag{8.33}$$

whenever $F \in \mathrm{C}^{(k)}_\Lambda(\mathbb{R}^q)$ and the multi-indices $\alpha = (\alpha_1, \ldots, \alpha_q)^T$ are chosen, such that $[\alpha] = \alpha_1 + \ldots + \alpha_q \le k$. Since $\nabla^\alpha F$ is continuous in $\mathcal{F}$ it must belong to $\mathrm{L}^2_\Lambda(\mathbb{R}^q)$. In other words, for all $n \le k$,

$$\sum_{[\alpha]=n} \left(\sum_{h\in\Lambda^{-1}} \left|F^\wedge_{\mathcal{F}_\Lambda}(h)\right|^2 ((2\pi h)^\alpha)^2 \right) < \infty. \tag{8.34}$$

Moreover, there exits a constant C (dependent on k, q), such that

$$C|h|^{2k} \leq \sum_{[\alpha]=k} ((2\pi h)^{\alpha})^2. \tag{8.35}$$

From the Cauchy–Schwarz inequality, we therefore obtain for all $N > 0$

$$\begin{aligned} \sum_{\substack{0<|h|\leq N \\ h\in\Lambda^{-1}}} |F^{\wedge}_{\mathcal{F}_\Lambda}(h)| &\leq \sum_{\substack{0<|h|\leq N \\ h\in\Lambda^{-1}}} |F^{\wedge}_{\mathcal{F}_\Lambda}(h)| \left(\sum_{[\alpha]=k} ((2\pi h)^{\alpha})^2 \right)^{1/2} C^{-1/2} |h|^{-k} \\ &\leq \left(\sum_{\substack{0<|h|\leq N \\ h\in\Lambda^{-1}}} |F^{\wedge}_{\mathcal{F}_\Lambda}(h)|^2 \sum_{[\alpha]=k} ((2\pi h)^{\alpha})^2 \right)^{1/2} \\ &\quad \times \left(\sum_{\substack{0<|h|\leq N \\ h\in\Lambda^{-1}}} |h|^{-2k} \right)^{1/2} C^{-1/2}. \end{aligned} \tag{8.36}$$

Now, if $k > \frac{q}{2}$, the sum $\sum_{0<|h|\leq N} |h|^{-2k}$ is finite for $N \to \infty$, hence, the last expression must be finite.

Theorem 8.11. *If F is of class $\mathrm{C}^{(k)}_{\Lambda}(\mathbb{R}^q)$, where k is an integer with $k > \frac{q}{2}$, then*

$$\sum_{h\in\Lambda^{-1}} |F^{\wedge}_{\mathcal{F}_\Lambda}(h)| < \infty, \tag{8.37}$$

where $F^{\wedge}_{\mathcal{F}_\Lambda}(h)$ given by (8.30) form the Fourier coefficients of F.

8.3 Fourier Transform over Euclidean Spaces

Next we present some basic results on the multi-dimensional Fourier transform on Euclidean spaces, however, restricted to the standard reference space $\mathrm{L}^1(\mathbb{R}^q)$ of (Lebesgue) absolutely integrable functions in the Euclidean space $\mathbb{R}^q$ (as proposed, e.g., by E.M. Stein, G. Weiss [1971]).

Fourier Tranform. $\mathrm{L}^p(\mathbb{R}^q)$, $1 \leq p < \infty$, are the spaces of all measurable functions $F : \mathbb{R}^q \to \mathbb{C}$, such that

$$\|F\|_{\mathrm{L}^p(\mathbb{R}^q)} = \left(\int_{\mathbb{R}^q} |F(x)|^p \, dV(x) \right)^{1/p} < \infty. \tag{8.38}$$

The expression $\|F\|_{\mathrm{L}^p(\mathbb{R}^q)}$ defined by (8.38) is called the L^p*-norm of* F.

Definition 8.12. *The Fourier transform of $F : \mathbb{R}^q \to \mathbb{C}$ at $x \in \mathbb{R}^q$ is defined by*

$$F^{\wedge}_{\mathbb{R}^q}(x) = \int_{\mathbb{R}^q} F(y)\, e^{-2\pi i x\cdot y}\, dV(y). \tag{8.39}$$

If $F \in \mathrm{L}^1(\mathbb{R}^q)$, then it is well-known that the function $F^{\wedge}_{\mathbb{R}^q}$ is uniformly continuous. Moreover, the Riemann–Lebesgue theorem of Fourier analysis shows that $F^{\wedge}_{\mathbb{R}^q}(x) \to 0$ as $|x| \to \infty$.

$\mathrm{L}^1(\mathbb{R}^q)$ is endowed with a *convolution* defined in the following way: If F, G belong to $\mathrm{L}^1(\mathbb{R}^q)$, their convolution $H = F * G$ is the function whose value at $x \in \mathbb{R}^q$ is given by

$$H(x) = \int_{\mathbb{R}^q} F(x-y)\, G(y)\, dV(y). \tag{8.40}$$

Clearly, $H \in \mathrm{L}^1(\mathbb{R}^q)$ and $\|H\|_{\mathrm{L}^1(\mathbb{R}^q)} \le \|F\|_{\mathrm{L}^1(\mathbb{R}^q)}\|G\|_{\mathrm{L}^1(\mathbb{R}^q)}$. The operation is commutative and associative. Even more, if $F \in \mathrm{L}^p(\mathbb{R}^q)$, $1 \le p < \infty$, and $G \in \mathrm{L}^1(\mathbb{R}^q)$, then $H = F * G$ belongs to $\mathrm{L}^p(\mathbb{R}^q)$, and we have

$$\|H\|_{\mathrm{L}^p(\mathbb{R}^q)} \le \|F\|_{\mathrm{L}^p(\mathbb{R}^q)}\, \|G\|_{\mathrm{L}^1(\mathbb{R}^q)}. \tag{8.41}$$

An essential feature is that the Fourier transform of the convolution of two functions is the (pointwise) product of their Fourier transforms. More concretely, if F and G belong to $\mathrm{L}^1(\mathbb{R}^q)$, then

$$(F * G)^{\wedge}_{\mathbb{R}^q} = F^{\wedge}_{\mathbb{R}^q} G^{\wedge}_{\mathbb{R}^q}. \tag{8.42}$$

Many other important operations have simple relations with the Fourier transform:

- If τ_d denotes the *translation* by $d \in \mathbb{R}^q$ (by this we mean the operator mapping the function $F \in \mathrm{L}^1(\mathbb{R}^q)$ into the function $\tau_d F = F(\cdot - d)$), then

$$(\tau_d F)^{\wedge}_{\mathbb{R}^q}(x) = e^{-2\pi i d\cdot x} F^{\wedge}_{\mathbb{R}^q}(x), \quad x \in \mathbb{R}^q \tag{8.43}$$

  and

$$(e^{2\pi i d\cdot x} F)^{\wedge}_{\mathbb{R}^q}(x) = (\tau_d F^{\wedge}_{\mathbb{R}^q})\,(x), \quad x \in \mathbb{R}^q. \tag{8.44}$$

- If $\delta_\tau, \tau > 0$, denotes the *dilation* (by this we mean the operator mapping $F \in \mathrm{L}^1(\mathbb{R}^q)$ into the function $\delta_\tau F = F(\tau\cdot)$), then

$$\tau^q(\delta_\tau F)^{\wedge}_{\mathbb{R}^q}(x) = F^{\wedge}_{\mathbb{R}^q}(\tau^{-1}x), \quad x \in \mathbb{R}^q. \tag{8.45}$$

- If $x \mapsto P(x)$, $x \in \mathbb{R}^q$, is a polynomial in q variables $x = (x_1, \ldots, x_q)^T$ and $P(\nabla)$ is the associated differential operator (i.e., replace $x_i = x \cdot \varepsilon^i$ by $\frac{\partial}{\partial x_i} = \nabla \cdot \varepsilon^i$, $i \in \{1, \ldots, q\}$), then

$$P(\nabla) F^{\wedge}_{\mathbb{R}^q}(x) = (P(-2\pi i\cdot)F)^{\wedge}_{\mathbb{R}^q}(x), \quad x \in \mathbb{R}^q, \tag{8.46}$$

  and

$$(P(\nabla)F)^{\wedge}_{\mathbb{R}^q}(x) = P(-2\pi i x) F^{\wedge}_{\mathbb{R}^q}(x), \quad x \in \mathbb{R}^q. \tag{8.47}$$

Applying Fubini's theorem we immediately have the identity

$$\int_{\mathbb{R}^q} F^{\wedge}_{\mathbb{R}^q}(x)\ G(x)\ dV(x) = \int_{\mathbb{R}^q} F(x)\ G^{\wedge}_{\mathbb{R}^q}(x)\ dV(x) \tag{8.48}$$

provided that F, G belong to $\mathrm{L}^1(\mathbb{R}^q)$.

If $F \in \mathrm{L}^1(\mathbb{R}^q)$, $G \in \mathrm{L}^p(\mathbb{R}^q)$, $1 \leq p \leq 2$, then $H = F * G \in \mathrm{L}^p(\mathbb{R}^q)$ satisfies $H^{\wedge}_{\mathbb{R}^q}(x) = F^{\wedge}_{\mathbb{R}^q}(x)\ G^{\wedge}_{\mathbb{R}^q}(x)$ (for almost every x). Finally, we mention the *(Fourier) inversion formula*

$$F(x) = \int_{\mathbb{R}^q} F^{\wedge}_{\mathbb{R}^q}(y)\ e^{2\pi i x\cdot y}\ dV(y) \tag{8.49}$$

(for almost every x) in its standard framework that $F, F^{\wedge}_{\mathbb{R}^q}$ belong to $\mathrm{L}^1(\mathbb{R}^q)$. The inversion formula briefly reads as follows

$$(F^{\wedge}_{\mathbb{R}^q})^{\wedge}_{\mathbb{R}^q}(-x) = \underbrace{\int_{\mathbb{R}^q} e^{2\pi i x\cdot w} \underbrace{\int_{\mathbb{R}^q} F(y)\ e^{-2\pi i w\cdot y}\ dV(y)}_{=F^{\wedge}_{\mathbb{R}^q}(w)} dV(w)}_{=(F^{\wedge}_{\mathbb{R}^q})^{\wedge}_{\mathbb{R}^q}(-x)} = F(x). \tag{8.50}$$

Once more it should be noted that the conditions $F, F^{\wedge}_{\mathbb{R}^q}$ belonging to $\mathrm{L}^1(\mathbb{R}^q)$ usually are not both satisfied in our lattice point and sampling applications. In fact, typical counterexamples are multivariate "ansatz functions" leading to so-called Hardy–Landau identities (cf. (7.69)).

8.4 Periodization and Classical Poisson Summation Formula

Suppose that a (appropriate) function F is given in Euclidean space $\mathbb{R}^q$. The question is what is its periodic analog? Our purpose is to show that there are at least two ways of obtaining a periodization of a function under consideration:

The *first periodization* is straightforward:

$$x \mapsto S(x) = \sum_{g\in\Lambda} F(x+g). \tag{8.51}$$

Since this (formal) sum S is extended over all points of the lattice Λ it is obviously periodic, i.e.,

$$S(x+g') = \sum_{g\in\Lambda} F(x+g+g') = \sum_{g\in\Lambda} F(x+g) = S(x) \tag{8.52}$$

for all $g' \in \Lambda$.

The *second periodization* is based on the Fourier expansion

$$x \mapsto \sum_{h\in\Lambda^{-1}} S^{\wedge}_{\mathcal{F}_\Lambda}(h)\ \Phi_h(x), \tag{8.53}$$

where

$$S^{\wedge}_{\mathcal{F}_\Lambda}(h) = \int_{\mathcal{F}_\Lambda} S(x)\ \overline{\Phi_h(x)}\ dV(x),\ x \in \mathbb{R}^q. \tag{8.54}$$

The Fourier coefficients $S^{\wedge}_{\mathcal{F}_\Lambda}(h)$, $h \in \Lambda^{-1}$, of the series (8.53) are expressible in the form

$$S^{\wedge}_{\mathcal{F}_\Lambda}(h) = \int_{\mathcal{F}_\Lambda} \sum_{g\in\Lambda} F(x+g)\ \overline{\Phi_h(x)}\ dV(x) = \sum_{g\in\Lambda} \int_{\mathcal{F}_\Lambda+\{g\}} F(x)\ \overline{\Phi_h(x)}\ dV(x), \tag{8.55}$$

where, as usual, $\mathcal{F}_\Lambda + \{g\}$ denotes the translate of $\mathcal{F}_\Lambda$ containing the lattice point $g \in \Lambda$. Since the fundamental cell $\mathcal{F}_\Lambda$ of the lattice Λ defines translates $\mathcal{F}_\Lambda + \{g\}$, $g \in \Lambda$, that are mutually disjoint, their union is the whole Euclidean space $\mathbb{R}^q$. This observation implies the identity

$$\int_{\mathcal{F}_\Lambda} \sum_{g\in\Lambda} F(x+g)\ \overline{\Phi_h(x)}\ dV(x) = \int_{\mathbb{R}^q} F(x)\ \overline{\Phi_h(x)}\ dV(x) = \frac{1}{\sqrt{\|\mathcal{F}_\Lambda\|}} F^{\wedge}_{\mathbb{R}^q}(h). \tag{8.56}$$

The formal conclusion is that the first and second periodization are identical. Hence, we are led to the *Poisson summation formula for the Euclidean space* $\mathbb{R}^q$

$$\sum_{g\in\Lambda} F(x+g) = \sum_{h\in\Lambda^{-1}} \int_{\mathbb{R}^q} F(y)\ \overline{\Phi_h(y)}\ dV(y)\ \Phi_h(x). \tag{8.57}$$

As a matter of fact, the identity (8.57) may be rigorously understood in many ways. The simplest and standard realization (see E.M. Stein, G. Weiss [1971]) is in the following version:

Theorem 8.13. *Suppose that F and $F^{\wedge}_{\mathbb{R}^q}$ satisfy the conditions*

$$F(x) = O\left(\frac{1}{|x|^{q+\varepsilon}}\right), \quad \varepsilon > 0, \quad |x| \to \infty, \tag{8.58}$$

$$F^{\wedge}_{\mathbb{R}^q}(x) = O\left(\frac{1}{|x|^{q+\varepsilon}}\right), \quad \varepsilon > 0, \quad |x| \to \infty \tag{8.59}$$

with

$$F(x) = \int_{\mathbb{R}^q} F^{\wedge}_{\mathbb{R}^q}(y)\ e^{2\pi i x\cdot y}\ dV(y), \quad x \in \mathbb{R}^q. \tag{8.60}$$

Then, F and $F^{\wedge}_{\mathbb{R}^q}$ are continuous in $\mathbb{R}^q$, such that

$$\sum_{g\in\Lambda} F(x+g) = \sum_{h\in\Lambda^{-1}} \int_{\mathbb{R}^q} F(y)\ \overline{\Phi_h(y)}\ dV(y)\ \Phi_h(x). \tag{8.61}$$

In particular, for $x = 0$, we have

$$\sum_{g\in\Lambda} F(g) = \frac{1}{\sqrt{\|\mathcal{F}_\Lambda\|}} \sum_{h\in\Lambda^{-1}} \int_{\mathbb{R}^q} F(y)\overline{\Phi_h(y)}\, dV(y). \tag{8.62}$$

The four series in (8.61) *and* (8.62) *converge absolutely.*

In Chapter 18, we shall provide sufficient conditions for the validity of the Poisson summation formula, such that not all series show the absolute convergence as a necessary presumption, and specific alternating summands occurring in number and sampling theory may be handled appropriately.

8.5 Gauss–Weierstrass Transform over Euclidean Spaces

First we borrow some auxiliary results from C. Müller [1998].

Lemma 8.14. *For $t \geq 0$, $a \in \mathbb{R}$, we have*

$$\|\mathbb{S}^{q-1}\| \int_0^\infty e^{-\pi t r^2} J_0(q; 2\pi a r)\, r^{q-1}\, dr = \left(\frac{1}{t}\right)^{\frac{q}{2}} e^{-\frac{\pi}{t}a^2}. \tag{8.63}$$

In particular with $x, y \in \mathbb{R}^q$ and $a = |x-y|$ we obtain the identity

$$\|\mathbb{S}^{q-1}\| \int_0^\infty e^{-\pi t r^2} J_0(q; 2\pi|x-y|r)\, r^{q-1}\, dr = \left(\frac{1}{t}\right)^{\frac{q}{2}} e^{-\frac{\pi}{t}|x-y|^2}, \tag{8.64}$$

which takes a particular role in the theory of Theta functions (see Section 15.2).

In order to discuss the Gauss–Weierstrass transform in Euclidean spaces $\mathbb{R}^q$, we start from the *Gaussian function* $G : \mathbb{R}^q \to \mathbb{C}$ given by

$$G(x) = e^{2\pi i x\cdot z} e^{-\pi\tau x^2}, \ z \in \mathbb{R}^q, \ \tau > 0. \tag{8.65}$$

An elementary calculation yields

$$\begin{aligned} G^{\wedge}_{\mathbb{R}^q}(x) &= \int_{\mathbb{R}^q} G(y)\, e^{-2\pi i x\cdot y}\, dV_{(q)}(y) \qquad (8.66)\\ &= \int_{\mathbb{R}^q} e^{2\pi i y\cdot z} e^{-\pi\tau y^2} e^{-2\pi i x\cdot y}\, dV_{(q)}(y)\\ &= \int_{\mathbb{R}^q} e^{-\pi\tau y^2} e^{2\pi i (z-x)\cdot y}\, dV(y). \end{aligned}$$

With $y = \rho\eta$ we find that

$$G^{\wedge}_{\mathbb{R}^q}(x) = \int_0^{\infty} \rho^{q-1} e^{-\pi\tau\rho^2} \left(\int_{\mathbb{S}^{q-1}} e^{2\pi i \rho (z-x)\cdot\eta} \, dS_{(q-1)}(\eta) \right) d\rho. \tag{8.67}$$

Remembering the definition of the Bessel function of order 0 we obtain

$$\begin{aligned} \int_{\mathbb{S}^{q-1}} e^{2\pi i \rho((z-x)\cdot\eta)} \, dS_{(q-1)}(\eta) &= \|\mathbb{S}^{q-1}\| \int_{-1}^{1} e^{2\pi i \rho |z-x| s} (1-s^2)^{\frac{q-3}{2}} \, ds \\ &= \|\mathbb{S}^{q-1}\| \, J_0(q; 2\pi|z-x|\rho). \end{aligned} \tag{8.68}$$

For $\tau > 0$, $a \in \mathbb{R}$, we know from Lemma 8.14 that

$$\|\mathbb{S}^{q-1}\| \int_0^{\infty} e^{-\pi\tau r^2} J_0(q; 2\pi a r) \, r^{q-1} \, dr = \tau^{-q/2} e^{-\frac{\pi}{\tau} a^2}. \tag{8.69}$$

In other words, with $x, z \in \mathbb{R}^q$ and $a = |x - z|$, we have

$$\|\mathbb{S}^{q-1}\| \int_0^{\infty} e^{-\pi\tau r^2} J_0(q; 2\pi|x-z|r) \, r^{q-1} \, dr = \tau^{-q/2} e^{-\frac{\pi}{\tau}(x-z)^2}. \tag{8.70}$$

Altogether this shows that

$$G^{\wedge}_{\mathbb{R}^q}(x) = \left(\frac{1}{\tau}\right)^{\frac{q}{2}} e^{-\frac{\pi}{\tau}|x-z|^2}. \tag{8.71}$$

Next we consider the *Weierstrass function* $W : \mathbb{R}^q \to \mathbb{C}$ given by

$$W(x) = \left(\frac{1}{\tau}\right)^{q/2} e^{-\frac{\pi}{\tau}(z \;\; x)^2}, \quad z \in \mathbb{R}^q, \ \tau > 0. \tag{8.72}$$

Now we have

$$\begin{aligned} W^{\wedge}_{\mathbb{R}^q}(-x) &= \left(\frac{1}{\tau}\right)^{q/2} \int_{\mathbb{R}^q} e^{-\frac{\pi}{\tau}(z-y)^2} e^{2\pi i x\cdot y} \, dV_{(q)}(y) \qquad (8.73) \\ &= \left(\frac{1}{\tau}\right)^{q/2} e^{2\pi i x\cdot z} \int_{\mathbb{R}^q} e^{-\frac{\pi}{\tau}(y-z)^2} e^{2\pi i x\cdot(y-z)} \, dV_{(q)}(y). \end{aligned}$$

By use of polar coordinates $y = z + \rho\eta$ the integral can be transformed to

$$\begin{aligned} &\int_0^{\infty} \rho^{q-1} e^{-\frac{\pi}{\tau}\rho^2} \left(\int_{\mathbb{S}^{q-1}} e^{2\pi i \rho(x\cdot\eta)} \, dS_{(q-1)}(\eta) \right) d\rho \qquad (8.74) \\ &= \|\mathbb{S}^{q-1}\| \int_0^{\infty} \rho^{q-1} e^{-\frac{\pi}{\tau}\rho^2} J_0(q; 2\pi|x|\rho) \, d\rho. \end{aligned}$$

The last integral is known from Lemma 8.14. For $x \in \mathbb{R}^q$ we have

$$W^{\wedge}_{\mathbb{R}}(-x) = e^{2\pi i x\cdot z} e^{-\pi\tau x^2} = G(x). \tag{8.75}$$

Summarizing our results, we therefore obtain the following interrelations.

Lemma 8.15. *For all $x \in \mathbb{R}^q$ we have*

$$W^{\wedge}_{\mathbb{R}^q}(-x) = G(x) = \int_{\mathbb{R}^q} \underbrace{G^{\wedge}_{\mathbb{R}^q}(y)}_{=W(y)} e^{2\pi i x\cdot y}\, dV_{(q)}(y), \tag{8.76}$$

$$G^{\wedge}_{\mathbb{R}^q}(x) = W(x) = \int_{\mathbb{R}^q} \underbrace{W^{\wedge}_{\mathbb{R}^q}(y)}_{=G(-y)} e^{2\pi i x\cdot y}\, dV_{(q)}(y). \tag{8.77}$$

We are now in the position to introduce the so-called Gauss transform and Weierstrass transform as convolution integrals involving the kernels G and W, respectively.

Definition 8.16. *For $\tau > 0$, the Gauss transform $\mathcal{G}_\tau(F)$ of $F : \mathbb{R}^q \to \mathbb{C}$ at $x \in \mathbb{R}^q$ is defined by*

$$\mathcal{G}_\tau(F)(x) = \left(\frac{1}{\tau}\right)^{q/2} \int_{\mathbb{R}^q} e^{-\frac{\pi}{\tau}(x-y)^2} F(y)\, dV_{(q)}(y), \tag{8.78}$$

while the Weierstrass transform $\mathcal{W}_\tau(F)$ of $F : \mathbb{R}^q \to \mathbb{C}$ at $x \in \mathbb{R}^q$ is defined by

$$\mathcal{W}_\tau(F)(x) = \int_{\mathbb{R}^q} e^{-\pi\tau y^2} e^{2\pi i x\cdot y} F(y)\, dV_{(q)}(y). \tag{8.79}$$

The relation between the Gauss transform and the Weierstrass transform is characterized by the following identities.

Lemma 8.17. *For $F \in \mathrm{L}^1(\mathbb{R}^q)$, $\tau > 0$, and $z \in \mathbb{R}^q$*

$$\mathcal{W}_\tau(F^{\wedge}_{\mathbb{R}^q})(z) = (\mathcal{W}_\tau(F))^{\wedge}_{\mathbb{R}^q}(z) = \mathcal{G}_\tau(F)(z), \tag{8.80}$$

$$\mathcal{G}_\tau(F^{\wedge}_{\mathbb{R}^q})(z) = (\mathcal{G}_\tau(F))^{\wedge}_{\mathbb{R}^q}(z) = \mathcal{W}_\tau(F)(-z). \tag{8.81}$$

Next our aim is to show that the "integral means" $\mathcal{W}_\tau(F)(x)$ and $\mathcal{G}_\tau(F)(x)$ tend to $F(x)$ as $\tau \to 0$, if suitable conditions are imposed on the position x as well as the function F under consideration. We begin with the Weierstrass transform.

Lemma 8.18. *Suppose that F is continuous and uniformly bounded in $\mathbb{R}^q$ such that, for $z \in \mathbb{R}^q$, the "spherical principal value" of $F^{\wedge}_{\mathbb{R}^q}(z)$, i.e., the limit*

$$\lim_{N\to\infty} \int_{\substack{|x|\leq N\\ x\in\mathbb{R}^q}} F(x) e^{-2\pi i z\cdot x}\, dV_{(q)}(x) \tag{8.82}$$

exists. Then

$$\lim_{\substack{\tau\to 0\\ \tau>0}} \mathcal{W}_\tau(F)(z) = (\mathcal{G}_\tau(F))^{\wedge}_{\mathbb{R}^q}(-z) = \lim_{N\to\infty} \int_{\substack{|x|\leq N\\ x\in\mathbb{R}^q}} F(x) e^{-2\pi i x\cdot z}\, dV_{(q)}(x). \tag{8.83}$$

Next we introduce the notion of the "spherical continuity" at a point $z \in \mathbb{R}^q$.

Definition 8.19. *A function $F : \mathbb{R}^q \to \mathbb{C}$ is called spherically continuous at $z \in \mathbb{R}^q$ if the spherical mean*

$$\Phi(r; z) = \frac{1}{\|\mathbb{S}^{q-1}\|} \int_{\mathbb{S}^{q-1}} F(z + r\xi) \, dS_{q-1}(\xi) \tag{8.84}$$

exists for all values $r > 0$ and is continuous for $r \to 0$ with

$$\Phi(0; z) = F(z). \tag{8.85}$$

This property is particularly significant if F is a "discontinuous function". We come back to this aspect, later on.

Lemma 8.20. *Suppose that $F \in \mathrm{L}^1(\mathbb{R}^q)$ is spherically continuous at $z \in \mathbb{R}^q$. Then*

$$\lim_{\substack{\tau \to 0 \\ \tau > 0}} \mathcal{G}_\tau(F)(z) = \lim_{\substack{\tau \to 0 \\ \tau > 0}} \mathcal{W}_\tau(F^\wedge_{\mathbb{R}^q})(z) = F(z). \tag{8.86}$$

It should be pointed out that Lemma 8.20 allows the interpretation as *pointwise (Gaussian) inversion formula.*

Theorem 8.21. *Suppose that $F \in \mathrm{L}^1(\mathbb{R}^q)$ is spherically continuous at $z \in \mathbb{R}^q$. Furthermore, assume that the "spherical principal value"*

$$\lim_{N \to \infty} \int_{\substack{|y| \leq N \\ y \in \mathbb{R}^q}} e^{2\pi i z \cdot y} F^\wedge_{\mathbb{R}^q}(y) \, dV_{(q)}(y) \tag{8.87}$$

exists. Then we have

$$F(z) = \lim_{N \to \infty} \int_{\substack{|y| \leq N \\ y \in \mathbb{R}^q}} e^{2\pi i z \cdot y} F^\wedge_{\mathbb{R}^q}(y) \, dV_{(q)}(y). \tag{8.88}$$

If the spherical principal value does not exist, the limit relation

$$F(z) = \lim_{\substack{\tau \to 0 \\ \tau > 0}} \int_{\mathbb{R}^q} e^{-\pi \tau y^2} e^{2\pi i z \cdot y} F^\wedge_{\mathbb{R}^q}(y) \, dV_{(q)}(y) \tag{8.89}$$

remains valid as a "mean representation" of F at the point z of spherical continuity.

8.6 Hankel Transform and Discontinuous Integrals

The Fourier transform preserves the spherical symmetry in the sense that orthogonal invariance of the original function in $\mathbb{R}^q$ ensures the same property for the transformed function. In this respect, it is worthy to mention that the orthogonal invariance reduces the Fourier transform to a 1D integral transform. Next, this aspect will be investigated for (spherically) discontinuous functions in balls leading to the so-called Hankel transform. In fact, the Hankel transform turns out to be a helpful tool for evaluating discontinuous integrals occurring in weighted lattice point sums, such as multi-dimensional extensions of the Hardy–Landau identities.

To be more concrete, we start our considerations with a function F_n, $n \in \mathbb{N}_0$, given in the separated form

$$F_n(x) = \Phi_n(r)\ Y_n(q;\xi), \quad x = r\xi, r = |x|, \quad \xi \in \mathbb{S}^{q-1}, \tag{8.90}$$

where $Y_n(q;\cdot)$ is a spherical harmonic of degree n and dimension q and the radial function $r \mapsto \Phi(r)$, $r \geq 0$, is assumed to be continuous, such that

$$\int_0^\infty |\Phi_n(r)|\ r^{q-1}\ dr < \infty. \tag{8.91}$$

Under these assumptions the Fourier transform $(F_n)^\wedge_{\mathbb{R}^q}$ at $y = s\eta$, $s = |y|$, $\eta \in \mathbb{S}^{q-1}$, admits the representation

$$(F_n)^\wedge_{\mathbb{R}^q}(y) = \int_{\mathbb{R}^q} F_n(x) e^{-2\pi i y\cdot x}\ dV_{(q)}(x) = i^n \Psi_n(s)\ Y_n(q;\eta), \tag{8.92}$$

where

$$\Psi_n(s) = \|\mathbb{S}^{q-1}\| \int_0^\infty r^{q-1}\Phi_n(r)\ J_n(q; 2\pi s r)\ dr. \tag{8.93}$$

Note that

$$\int_{\mathbb{S}^{q-1}} e^{-2\pi i r s(\xi\cdot\eta)} Y_n(q;\xi)\ dS_{(q-1)}(\xi) = i^n \|\mathbb{S}^{q-1}\| J_n(q; 2\pi s r) Y_n(q;\eta). \tag{8.94}$$

In other words, the Fourier transform

$$(F_n)^\wedge_{\mathbb{R}^q}(y) = i^n \Psi_n(s) Y_n(q;\eta) \tag{8.95}$$

again shows a splitting into a radial and an angular part (as the function F_n itself given by (8.90)).

In what follows, we replace the continuity of Φ_n on $(0,\infty)$ by the so-called symmetric continuity.

Definition 8.22. *Suppose that Φ_n is symmetrically continuous on $(0,\infty)$, i.e., for all $r > 0$*

$$\lim_{\substack{s\to 0\\ s>0}} \frac{1}{2}\left(\Phi_n(r+s) + \Phi_n(r-s)\right) = \Phi_n(r). \tag{8.96}$$

Then the integral transform Ψ_n as defined by (8.93) is called the Hankel transform of Φ_n of degree n.

Finally we mention the following inversion formula.

Lemma 8.23. *Suppose that Φ is symmetrically continuous on $(0,\infty)$. Moreover, assume that (8.91) holds true. If*

$$\Psi_n(s) = \|\mathbb{S}^{q-1}\| \int_0^\infty r^{q-1}\Phi_n(r) J_n(q; 2\pi sr)\, dr, \tag{8.97}$$

then

$$\Phi_n(r) = \|\mathbb{S}^{q-1}\| \int_0^\infty s^{q-1}\Psi_n(s) J_n(q; 2\pi sr)\, ds. \tag{8.98}$$

The spherical symmetry, in fact, reduces the complexity, as the integral transform is of 1D nature. Moreover, it is handled within the framework of the theory of Bessel functions.

Finally, we investigate an example that is particularly important to our approach concerned with Hardy–Landau identities of analytic theory of numbers: We especially let

$$F_n(x) = \Phi_n(r)\ Y_n(q;\xi), \quad x = r\xi,\ \xi \in \mathbb{S}^{q-1}, \tag{8.99}$$

where $Y_n(q;\cdot)$ is a member of $\mathrm{Harm}_n(\mathbb{S}^{q-1})$ and $\Phi_n : r \mapsto \Phi_n(r)$ is given by

$$\Phi_n(r) = \begin{cases} r^n & , \quad 0 \le r < R \\ \frac{1}{2}r^n & , \quad r = R \\ 0 & , \quad r > R. \end{cases} \tag{8.100}$$

The function F_n clearly is of class $\mathrm{L}^1(\mathbb{R}^q)$, it is discontinuous at $r = R$, but everywhere spherically continuous. As already known, the Fourier inversion formula holds true, i.e.,

$$F(x) = \int_{\mathbb{R}^q} e^{2\pi i x\cdot y}\ F^{\wedge}_{\mathbb{R}^q}(y)\ dV_{(q)}(y), \tag{8.101}$$

provided that the integral exists as a spherical principal value. The Hankel transform is given by

$$\Psi_n(s) = \|\mathbb{S}^{q-1}\| \int_0^R r^{n+q-1} J_n(q; 2\pi sr)\, dr. \tag{8.102}$$

As Φ_n is symmetrically continuous, Lemma 8.23 is applicable, and we find

$$\Phi_n(r) = \|\mathbb{S}^{q-1}\| \int_0^\infty s^{q-1}\Psi_n(s) J_n(q; 2\pi sr)\, ds, \tag{8.103}$$

if the integral exists (which should be discussed in more detail): In order to guarantee the existence of (8.103) we remember (cf. (6.78))

$$\frac{d}{dr}\left(r^{n+q-1} J_n(q;r)\right) = r^{n+q-1} J_{n+1}(q;r) \tag{8.104}$$

such that

$$\int_0^R r^{n+q-1} J_n(q; 2\pi sr)\, dr = \frac{R^{n+q-1}}{2\pi s} J_{n+1}(q; 2\pi sR). \tag{8.105}$$

Consequently, we find

$$\Psi_n(s) = \|\mathbb{S}^{q-1}\| \, \frac{R^{n+q-1}}{2\pi s} J_{n+1}(q; 2\pi sR). \tag{8.106}$$

The existence of the integral (8.103) is a consequence of the asymptotic relation (6.28) of the Bessel functions

$$J_n(q; 2\pi sr) = \frac{2}{\|\mathbb{S}^{q-1}\|} \left(\frac{1}{Rs}\right)^{\frac{q-1}{2}} \cos\left(2\pi s - \left(n\frac{\pi}{2} + (q-1)\frac{\pi}{4}\right)\right) + O\left(s^{-\frac{q-1}{2}}\right). \tag{8.107}$$

In fact, after a simple calculation we obtain

$$\begin{aligned} & s^{q-1}\Psi_n(s) J_n(q; 2\pi sr) \\ &= \frac{R^{n+q-1}}{2\pi} s^{q-2} J_{n+1}(q; 2\pi sR) J_n(q; 2\pi sr) \\ &= \frac{C}{s}\left(\sin\left(2\pi s(r+R) - 2\left(n\frac{\pi}{2} + (q-1)\frac{\pi}{4}\right) + \sin\left(2\pi s(r-R)\right)\right)\right) \\ &\quad + O\left(\frac{1}{s^2}\right) \end{aligned} \tag{8.108}$$

with a certain explicitly calculable constant C (whose value is not important in our context). This shows that the integral (8.103) is convergent.
Finally, in connection with (8.103) and (8.106), we are able to deduce from Lemma 8.23.

Lemma 8.24. *For $R > 0$,*

$$\begin{aligned} & \int_0^\infty s^{q-2} J_{n+1}(q; 2\pi sR)\; J_n(q; 2\pi sr)\, ds \\ &= \frac{2\pi}{\|\mathbb{S}^{q-1}\|} \frac{1}{R^{q-1}} \left(\frac{r}{R}\right)^n \begin{cases} 1, & r < R, \\ \frac{1}{2}, & r = R, \\ 0, & r > R. \end{cases} \end{aligned} \tag{8.109}$$

The formula (8.109) constitutes a (radially symmetric) expression of a so-called *Weber–Schlafheitlin discontinuous integral* (for more details on discontinuous integrals involving the theory of Bessel functions the reader is referred, e.g., to the monograph G.N. Watson [1944]).

Part IV

Multivariate Euler-Type Summation Formulas over Regular Regions

Chapter 9

Euler–Green Function and Euler-Type Summation Formula

Next we are interested in the development of the Euler-type summation formula. Auxiliary tools are the constituting ingredients of the Euler–Green function with respect to the $(L + \lambda)$-operator and its resulting properties. Later on, we use the concept of Euler–Green functions to establish Euler summation formulas with respect to iterated $(L + \lambda)$-operators and regular regions $\mathcal{G} \in \mathbb{R}^q$.

9.1 Euler–Green Function

As already known, for the operator L in $\mathbb{R}^q$ given by $L = \Delta + 4\pi i a \cdot \nabla$, $a \in \mathbb{R}^q$, an elementary calculation shows that $(L + L^\wedge(h))\, e^{2\pi i h \cdot x} = 0$, $x \in \mathbb{R}^q$, holds true, where $L^\wedge(h) = 4\pi^2 h^2 - 4\pi i h \cdot 2\pi i a = 4\pi^2((h+a)^2 - a^2)$. Analogously to the case of the Laplace operator, λ is called an *eigenvalue* of the lattice Λ with respect to the operator L if there is a non-trivial solution U of the differential equation $(L+\lambda)\, U = 0$ satisfying the boundary condition of periodicity $U(x+g) = U(x)$ for all $x \in \mathbb{R}^q$ and $g \in \Lambda$. The function U is called an *eigenfunction* of the lattice Λ with respect to the eigenvalue λ and the operator L. In analogy to the 1D case, we are able to verify that the spectrum is given by

$$\mathrm{Spect}_L(\Lambda) = \{L^\wedge(h) : L^\wedge(h) = 4\pi^2((h+a)^2 - a^2),\ h \in \Lambda^{-1}\}. \qquad (9.1)$$

In accordance with the principles of Hilbert's approach (starting with the 1D background worked out in D. Hilbert [1912]), the *Euler–Green function* $G(L + \lambda; \cdot)$ (sometimes also called Λ-*lattice function*) may be understood as the Green function for the differential operator $L + \lambda$, $\lambda \in \mathbb{R}$, in Euclidean space $\mathbb{R}^q$ corresponding to the boundary condition of periodicity with regard to the lattice Λ.

Obviously, we have

$$\mathrm{Spect}_{L+\lambda}(\Lambda) = \{(L+\lambda)^\wedge(h) : (L+\lambda)^\wedge(h) = L^\wedge(h) - \lambda, h \in \Lambda^{-1}\}. \qquad (9.2)$$

Since the Fourier series of $G(L+\lambda;\cdot)$ formally given by

$$G(L+\lambda;x) \sim \frac{1}{\|\mathcal{F}_\Lambda\|} \sum_{\substack{(L+\lambda)^\wedge(h)\neq 0\\ h\in\Lambda^{-1}}} \frac{1}{-(L+\lambda)^\wedge(h)}\, e^{2\pi i h\cdot x} \tag{9.3}$$

does not show the necessary pointwise convergence property in $\mathbb{R}^q, q \geq 2$, as it was the case for its 1D counterpart, we are required to introduce $G(L+\lambda;\cdot)$ by its defining properties as Green's function in accordance with the Hilbert theory:

Definition 9.1. *$G(L+\lambda;\cdot) : \mathbb{R}^q \setminus \Lambda \to \mathbb{C}$, $\lambda \in \mathbb{R}$ fixed, is called Euler–Green function (or Λ-lattice function) for the operator $L+\lambda, L = \Delta + 4\pi i a\cdot\nabla$, $a \in \mathbb{R}^q$, if it satisfies the following properties:*

(i) For all $x \in \mathbb{R}^q \setminus \Lambda$ and $g \in \Lambda$

$$G(L+\lambda;x+g) = G(L+\lambda;x). \tag{9.4}$$

(ii) For all $x \in \mathbb{R}^q \setminus \Lambda$

$$\overline{G(L+\lambda;x)} = G(L+\lambda;-x). \tag{9.5}$$

(iii) $G(L+\lambda;\cdot)$ is twice continuously differentiable for all $x \in \mathbb{R}^q \setminus \Lambda$. For $\lambda \notin \mathrm{Spect}_L(\Lambda)$,

$$(L+\lambda)G(L+\lambda;x) = 0. \tag{9.6}$$

For $\lambda \in \mathrm{Spect}_L(\Lambda)$,

$$(L+\lambda)G(L+\lambda;x) = -\frac{1}{\|\mathcal{F}_\Lambda\|} \sum_{\substack{(L+\lambda)^\wedge(h)= 0\\ h\in\Lambda^{-1}}} e^{2\pi i h\cdot x}. \tag{9.7}$$

(iv) In a neighborhood of the origin,

$$G(L;x) + \frac{(2\pi)^{-\frac{q}{2}}\left(\sqrt{-\lambda-4\pi^2a^2}\right)^{\frac{q-2}{2}} e^{-2\pi i a\cdot x}}{|x|^{\frac{q-2}{2}}} K_{\frac{q-2}{2}}\left(\sqrt{-\lambda-4\pi^2a^2}\ |x|\right) \tag{9.8}$$

is continuously differentiable (so that $G(L;x)$ takes the singularity of the Kelvin function $K_{\frac{q-2}{2}}$ of order $\frac{q-2}{2}$ in the origin, hence, in all lattice points $h \in \Lambda$ because of condition (ii)).

(v) For all $h \in \Lambda^{-1}$ with $(L+\lambda)^\wedge(h) = 0$

$$\int_{\mathcal{F}_\Lambda} G(L+\lambda;x)\ e^{-2\pi i h\cdot x}\ dV(x) = 0. \tag{9.9}$$

For all $h \in \Lambda^{-1}$ and $\lambda \in \mathbb{R}$ we readily see by virtue of the second Green formula that

$$\int_{\mathcal{F}_\Lambda} G(L+\lambda;x)\, e^{-2\pi i h\cdot x}\, dV(x) = \begin{cases} 0, & (L+\lambda)^\wedge(h) = 0, \\ -\frac{1}{(L+\lambda)^\wedge(h)}, & (L+\lambda)^\wedge(h) \neq 0, \end{cases} \tag{9.10}$$

such that $G(L+\lambda;x)$ possesses the required Fourier coefficients. Consequently, for each number $\lambda \in \mathbb{R}$, there exists one and only one Euler–Green function $G(L+\lambda;\cdot)$ satisfying the constituting conditions *(i)–(v)* listed under Definition 9.1 (the proof follows by the same arguments as for the operator $\Delta + \lambda$ by virtue of the Fredholm theory of weakly singular integral equations (see W. Freeden [2011])).

9.2 Euler-Type Summation Formulas over Regular Regions Based on Euler–Green Functions

Next we are concerned with the *Euler-type summation formula*, i.e., the Euler summation formula with respect to the operator $L+\lambda$ in Euclidean space $\mathbb{R}^q$: Suppose that $\mathcal{G} \subset \mathbb{R}^q$ is a regular region. Let H be a function of class $C^{(2)}(\overline{\mathcal{G}}), \overline{\mathcal{G}} = \mathcal{G} \cup \partial\mathcal{G}$. Then, for every $\lambda \in \mathbb{R}$ and (sufficiently small) $\varepsilon > 0$, Green's theorem in connection with the Gauss theorem in $\mathbb{R}^q$ gives the integral identity

$$\int_{\substack{x\in\overline{\mathcal{G}} \\ x\notin \mathbb{B}^q_\varepsilon+\Lambda}} \left\{H(x)\,(L+\lambda)G(L+\lambda;x) - G(L+\lambda;x)\left(\overline{L}+\lambda\right)H(x)\right\}\, dV(x) \tag{9.11}$$

$$= \int_{\substack{x\in\partial\mathcal{G} \\ x\notin \mathbb{B}^q_\varepsilon+\Lambda}} \left\{H(x)\left(\frac{\partial}{\partial\nu} + 4\pi i a\cdot\nu(x)\right)G(L+\lambda;x) - G(L+\lambda;x)\frac{\partial}{\partial\nu}H(x)\right\}\, dS(x)$$

$$+ \sum_{\substack{g\in\overline{\mathcal{G}} \\ g\in\Lambda}} \int_{\substack{|x-g|=\varepsilon \\ x\in\overline{\mathcal{G}}}} \left\{H(x)\left(\frac{\partial}{\partial\nu} + 4\pi i a\cdot\nu(x)\right)G(L+\lambda;x) - G(L+\lambda;x)\frac{\partial}{\partial\nu}H(x)\right\}\, dS(x),$$

where $\frac{\partial}{\partial\nu}$ denotes the derivative in the direction of the outer normal ν on $\partial\mathcal{G}$. Using the differential equation (i.e., Condition *(iii)* of Definition 9.1) we obtain

$$\int_{\substack{x\in\overline{\mathcal{G}} \\ x\notin \mathbb{B}^q_\varepsilon+\Lambda}} H(x)\,(L+\lambda)G(L+\lambda;x)\, dV(x) \tag{9.12}$$

$$= -\frac{1}{\|\mathcal{F}_\Lambda\|} \int_{\substack{x\in\overline{\mathcal{G}} \\ x\notin \mathbb{B}^q_\varepsilon+\Lambda}} \sum_{\substack{(L+\lambda)^\wedge(h)=0 \\ h\in\Lambda^{-1}}} H(x)\, e^{2\pi i h\cdot x}\, dV(x),$$

where the sum on the right side is to be extended over all $h \in \Lambda^{-1}$ satisfying $(L+\lambda)^\wedge(h) = 0$. Letting ε converge to 0 and observing the characteristic singularity of the Euler–Green function (i.e., Condition *(iv)* of Definition 9.1) we obtain the desired summation formula by virtue of standard potential theoretic results (see Eq. (5.6)) analogously to the operator $\Delta+\lambda$ (as discussed by W. Freeden [2011]).

Theorem 9.2. *(Euler-Type Summation Formula for the Operator $L+\lambda$). Let Λ be an arbitrary lattice in $\mathbb{R}^q$. Let $\mathcal{G} \subset \mathbb{R}^q$ be a regular region. Let H be a twice continuously differentiable function on $\overline{\mathcal{G}}$. Then, for $\lambda \notin \mathrm{Spect}_L(\Lambda)$, we have*

$$\begin{aligned}
\sum_{\substack{g\in\overline{\mathcal{G}}\\ g\in\Lambda}}{}' H(g) & \qquad\qquad (9.13)\\
= & \int_{\mathcal{G}} G(L+\lambda;x)\ (\overline{L}+\lambda)H(x)\ dV(x)\\
+ & \int_{\partial\mathcal{G}} H(x)\left(\frac{\partial}{\partial\nu}+4\pi i a\cdot\nu(x)\right) G(L+\lambda;x)\ dS(x)\\
- & \int_{\partial\mathcal{G}} G(L+\lambda;x)\frac{\partial}{\partial\nu}H(x)\ dS(x),
\end{aligned}$$

while, for $\lambda \in \mathrm{Spect_L}(\Lambda)$, we have

$$\begin{aligned}
\sum_{\substack{g\in\overline{\mathcal{G}}\\ g\in\Lambda}}{}' H(g) = & \frac{1}{\|\mathcal{F}_\Lambda\|}\sum_{\substack{(L+\lambda)^\wedge(h)=0\\ h\in\Lambda^{-1}}}\int_{\mathcal{G}} H(x)\ e^{2\pi i h\cdot x}\ dV(x) \qquad (9.14)\\
& +\int_{\mathcal{G}} G(L+\lambda;x)\ (\overline{L}+\lambda)H(x)\ dV(x)\\
& +\int_{\partial\mathcal{G}}\left(\left(\frac{\partial}{\partial\nu}+4\pi i a\cdot\nu(x)\right) G(L+\lambda;x)\right)\ H(x)\ dS(x)\\
& -\int_{\partial\mathcal{G}} G(L+\lambda;x)\frac{\partial}{\partial\nu}H(x)\ dS(x),
\end{aligned}$$

where $\frac{\partial}{\partial\nu}$ denotes the directional derivative in the direction of the outer normal ν on $\partial\mathcal{G}$ and, as always, the symbol $\sum{}' \ldots$ means that

$$\sum_{\substack{g\in\overline{\mathcal{G}}\\ g\in\Lambda}}{}' H(g) = \sum_{\substack{g\in\overline{\mathcal{G}}\\ g\in\Lambda}} \alpha(g)\ H(g), \qquad (9.15)$$

with $\alpha(g)(=\alpha_{\mathcal{G}}(g))$ being the solid angle subtended at $g\in\overline{\mathcal{G}}$ by the surface $\partial\mathcal{G}$ (note that

$$\alpha_{\mathcal{G}}(g) = \begin{cases} 1, & g\in\mathcal{G}\\ \frac{1}{2}, & g\in\partial\mathcal{G}\\ 0, & g\notin\overline{\mathcal{G}}, \end{cases} \qquad (9.16)$$

provided that $\mathcal{G}\subset\mathbb{R}^q$ is a regular region with a continuously differentiable boundary $\partial\mathcal{G}$ (cf. Figure 5.1)).

From an algorithmic point of view it should be noted that the formulas (9.13) and (9.14) express a sum in terms of integrals over $\mathcal{G}$ and its boundary $\partial\mathcal{G}$ involving the derivatives of the function H up to the second order. The difference between the two cases (9.13) and (9.14) shows that "resonance phenomena" occur in the case of eigenvalues in the form of an additional finite sum of certain Fourier transforms over $\mathcal{G}$ determined by the operator L.

Remark 9.3. *The classical case $a = 0$, $\lambda = 0$, i.e., the Euler summation formula corresponding to the Laplacian Δ, turns out to be a particular resonance case. Indeed, the formula for the Laplacian is an immediate multi-dimensional generalization of the 1D Euler summation formula, where $G(\Delta;\cdot)$ takes the multi-dimensional role of the Bernoulli polynomial of degree 2 (see Introduction).*

Analogously to the one-dimensional case we are able to deduce from Theorem 9.2 the identity

$$\begin{aligned}
&\sum_{\substack{g-y\in\overline{\mathcal{G}}\\ g\in\Lambda}}{}' e^{2\pi i a\cdot g}F(g-y) \qquad (9.17)\\
&= \frac{1}{\|\mathcal{F}_\Lambda\|}\ e^{2\pi i a\cdot y} \sum_{\substack{4\pi^2((h+a)^2-a^2)=\lambda\\ h\in\Lambda^{-1}}} e^{2\pi i h\cdot y}\int_{\mathcal{G}} F(x)e^{2\pi i x\cdot(a+h)}dV(x)\\
&+ \int_{\substack{x-y\in\mathcal{G}\\ x\in\mathbb{R}^q}} G(L+\lambda;x)\ (\overline{L}+\lambda)\ e^{2\pi i a\cdot x}F(x-y)\ dV(x)\\
&- \int_{\substack{x-y\in\partial\mathcal{G}\\ x\in\mathbb{R}^q}} \left(\left(\frac{\partial}{\partial\nu}+4\pi i a\cdot\nu(x)\right)G(L+\lambda;x)\right)\ e^{2\pi i a\cdot x}F(x-y)\ dS(x)\\
&+ \int_{\substack{x-y\in\partial\mathcal{G}\\ x\in\mathbb{R}^q}} G(L+\lambda;x)\frac{\partial}{\partial\nu}e^{2\pi i a\cdot x}F(x-y)\ dS(x),
\end{aligned}$$

provided that F is a twice continuously differentiable function on $\overline{\mathcal{G}}$. By using the elementary identities

$$\Delta_x e^{2\pi i a\cdot x}F(x-y) = e^{2\pi i a\cdot x}\left(\Delta+4\pi i a\cdot\nabla_x-4\pi^2a^2\right)F(x-y), \qquad (9.18)$$

$$a\cdot\nabla_x e^{2\pi i a\cdot x}F(x-y) = e^{2\pi i a\cdot x}\left(a\cdot\nabla_x+2\pi i a^2\right)\ F(x-y), \qquad (9.19)$$

and

$$\left(\Delta_x-4\pi i a\cdot\nabla_x\right)e^{2\pi i a\cdot x}F(x-y) \ = \ e^{2\pi i a\cdot x}\left(\Delta_x+4\pi^2a^2\right)F(x-y), \quad (9.20)$$

we are able to reformulate the summation formula (9.17) in such a way that it becomes useful for our desired purposes of Shannon-type sampling.

Corollary 9.4. *Let Λ be an arbitrary lattice in $\mathbb{R}^q$. Let $\mathcal{G}\subset\mathbb{R}^q$ be a regular region. Let F be twice continuously differentiable on $\overline{\mathcal{G}}$. Then, for $\lambda\in\mathbb{R}, a, y\in$*

$\mathbb{R}^q$, *we have*

$$\sum_{\substack{g-y\in\overline{\mathcal{G}}\\ g\in\Lambda}}{}' \; e^{2\pi i a\cdot g}F(g-y) \tag{9.21}$$

$$= \frac{1}{\|\mathcal{F}_\Lambda\|} \; e^{2\pi i a\cdot y} \sum_{\substack{4\pi^2((h+a)^2-a^2)=\lambda\\ h\in\Lambda^{-1}}} e^{2\pi i h\cdot y}\int_{\mathcal{G}} F(x)e^{2\pi i x\cdot(a+h)}dV(x)$$

$$+ \int_{\substack{x-y\in\mathcal{G}\\ x\in\mathbb{R}^q}} G(L+\lambda;x)\; e^{2\pi i a\cdot x}\left(\Delta_x+\lambda+4\pi^2a^2\right)F(x-y)\; dV(x)$$

$$-\int_{\substack{x-y\in\partial\mathcal{G}\\ x\in\mathbb{R}^q}} e^{2\pi i a\cdot x}G(L+\lambda;x)\frac{\partial}{\partial\nu_x}F(x-y)\; dS(x)$$

$$+\int_{\substack{x-y\in\partial\mathcal{G}\\ x\in\mathbb{R}^q}} e^{2\pi i a\cdot x}F(x-y)\left(\frac{\partial}{\partial\nu_x}+4\pi i a\cdot\nu(x)\right)G(L+\lambda;x)\; dS(x).$$

9.3 Iterated Euler–Green Function

The iterated Λ-Euler–Green functions, i.e., the Λ-Euler–Green functions corresponding to operators $(L+\lambda)^m$, $\lambda\in\mathbb{R}$, $m\in\mathbb{N}$, with $L=\Delta+4\pi i a\cdot\nabla$ can be based on the Λ–Euler–Green functions corresponding to operators $L+\lambda$, $\lambda\in\mathbb{R}$, by forming convolution integrals in the usual way as known from mathematical physics (see, e.g., the classical approach initiated by D. Hilbert [1912], R. Courant, D. Hilbert [1924]).

Definition 9.5. *The function $G((L+\lambda)^m;\cdot)$, $m=1,2,\ldots;\ \lambda\in\mathbb{R}$, defined by*

$$\begin{aligned} G\left((L+\lambda)^1;x\right) &= G((L+\lambda);x), \\ G((L+\lambda)^m;x) &= \int_{\mathcal{F}_\Lambda} G\left((L+\lambda)^{m-1};z\right)G\left((L+\lambda)^1;x-z\right)\; dV_{(q)}(z), \end{aligned} \tag{9.22}$$

$m=2,3,\ldots$, *is called the Λ-Euler–Green function (or simply Λ–lattice function) for the operator $(L+\lambda)^m$, $\lambda\in\mathbb{R}$.*

Obviously, for all $x\notin\Lambda$ and $g\in\Lambda$,

$$G((L+\lambda)^m;x+g)=G((L+\lambda)^m;x) \tag{9.23}$$

is satisfied, i.e., $G((L+\lambda)^m;\cdot)$ is Λ-periodical. From the singularity properties of the Kelvin function and in analogy to the known techniques of potential theory (see, e.g., W. Freeden [2011]) it can be deduced that

$$G((L+\lambda)^m;x)=\begin{cases} O\left(|x|^{2m-q}\ln|x|\right), & q=2m,\\ O\left(|x|^{2m-q}\right), & q>2m.\end{cases} \tag{9.24}$$

The differential equation

$$(L+\lambda)G\left((L+\lambda)^m;x\right) = G\left((L+\lambda)^{m-1};x\right), \quad x \notin \Lambda, \tag{9.25}$$

$m = 2, 3, \ldots$, represents a recursion relation relating the Λ-Euler–Green function for the operator $(L+\lambda)^m$ to the Λ-Euler–Green function for $(L+\lambda)^{m-1}$. The series expansion of $G\left((L+\lambda)^m;\cdot\right)$ in terms of eigenfunctions, which is equivalent to the (formal) Fourier (orthogonal) expansion (see W. Freeden [2011] for analogous considerations for the operator $(\Delta+\lambda)^m, m = 2, 3, \ldots$), is given by

$$G\left((L+\lambda)^m;x\right) \sim \frac{1}{\sqrt{\|\mathcal{F}\|}} \sum_{\substack{(L+\lambda)^\wedge(h)\neq 0\\ h\in\Lambda^{-1}}} \frac{\Phi_h(x)}{-((L+\lambda)^m)^\wedge(h)}, \quad m = 2, 3, \ldots. \tag{9.26}$$

Especially, for $m > \frac{q}{2}$, it follows that there is a constant $C > 0$, such that

$$\begin{aligned} &\left| \sum_{\substack{(L+\lambda)^\wedge(h)\neq 0\\ h\in\Lambda^{-1}}} \frac{\Phi_h(x)}{-((L+\lambda)^m)^\wedge(h)} \right| \qquad (9.27)\\ &= \left| \frac{1}{\sqrt{\|\mathcal{F}_\Lambda\|}} \sum_{\substack{4\pi^2((h+a)^2-a^2)\neq\lambda\\ h\in\Lambda^{-1}}} \frac{e^{2\pi i h\cdot x}}{(\lambda - 4\pi^2((h+a)^2-a^2))^m} \right| \\ &\leq C \sum_{h\in\Lambda^{-1}} \frac{1}{(1+h^2)^m} \\ &< \infty. \end{aligned}$$

Therefore, it is clear that the Fourier series converges absolutely and uniformly in $\mathbb{R}^q$, and $G\left((L+\lambda)^m;\cdot\right)$ is continuous in $\mathbb{R}^q$ provided that the integer m is chosen in such a way that $m > \frac{q}{2}$.

Summarizing our results we obtain the following bilinear series of the Λ-Euler–Green function $G\left((L+\lambda)^m;\cdot\right)$.

Lemma 9.6. *For $m > \frac{q}{2}$, the Λ-Euler–Green function $G\left((L+\lambda)^m;\cdot\right)$ is continuous in $\mathbb{R}^q$, and its bilinear series is given in pointwise way by*

$$G\left((L+\lambda)^m;x-y\right) = \sum_{\substack{(L+\lambda)^\wedge(h)\neq 0\\ h\in\Lambda^{-1}}} \frac{\Phi_h(x)\,\overline{\Phi_h(y)}}{-((L+\lambda)^m)^\wedge(h)}, \tag{9.28}$$

i.e.,

$$G((L+\lambda)^m;x-y) = \frac{1}{\|\mathcal{F}_\Lambda\|} \sum_{\substack{4\pi^2((h+a)^2-a^2)\neq\lambda\\ h\in\Lambda^{-1}}} \frac{e^{2\pi i h\cdot(x-y)}}{(\lambda - 4\pi^2((h+a)^2-a^2))^m}, \tag{9.29}$$

$x, y \in \mathbb{R}^q$.

9.4 Euler-Type Summation Formulas over Regular Regions Based on Iterated Euler–Green Functions

Suppose that $\mathcal{G} \subset \mathbb{R}^q$ is a regular region. As usual, for $\lambda \in \mathbb{R}$ and $a, y \in \mathbb{R}^q$, let $L + \lambda = \Delta + 4\pi i a \cdot \nabla + \lambda$ and $\overline{L} + \lambda = \Delta - 4\pi i a \cdot \nabla + \lambda$, and $\frac{\partial}{\partial \nu}$ denotes the normal derivative on $\partial\mathcal{G}$ directed exterior to $\mathcal{G}$. Let us assume that the function H under consideration is of class $\mathrm{C}^{(2k)}\left(\overline{\mathcal{G}}\right)$, $k \in \mathbb{N}$. Then we are able to deduce from the combined application of the Green and Gauss theorems, thereby making use of the differential equation (9.25), that

$$
\begin{aligned}
&\int_{\substack{x\in\overline{\mathcal{G}}\\ x\notin\mathbb{B}^q_\varepsilon+\Lambda}} G((L+\lambda)^{k+1};x)\ (\overline{L}+\lambda)^{k+1}H(x)\ dV(x) \\
&\quad - \int_{\substack{x\in\overline{\mathcal{G}}\\ x\notin\mathbb{B}^q_\varepsilon+\Lambda}} (L+\lambda)G((L+\lambda)^{k+1};x)\ (\overline{L}+\lambda)^{k}H(x)\ dV(x) \\
&= \int_{\substack{x\in\partial\mathcal{G}\\ x\notin\mathbb{B}^q_\varepsilon+\Lambda}} G((L+\lambda)^{k+1};x)\frac{\partial}{\partial\nu}\ (\overline{L}+\lambda)^{k}H(x)\ dS(x) \\
&\quad - \int_{\substack{x\in\partial\mathcal{G}\\ x\notin\mathbb{B}^q_\varepsilon+\Lambda}} \left(\left(\frac{\partial}{\partial\nu}+4\pi i a\cdot\nu(x)\right)G((L+\lambda)^{k+1};x)\right)(\overline{L}+\lambda)^{k}H(x)\ dS(x) \\
&\quad + \sum_{\substack{g\in\overline{\mathcal{G}}\\ g\in\Lambda}} \int_{\substack{|x-g|=\varepsilon\\ x\in\overline{\mathcal{G}}}} G((L+\lambda)^{k+1};x)\left(\frac{\partial}{\partial\nu}\ (\overline{L}+\lambda)^{k}H(x)\right)\ dS(x) \\
&\quad - \sum_{\substack{g\in\overline{\mathcal{G}}\\ g\in\Lambda}} \int_{\substack{|x-g|=\varepsilon\\ x\in\overline{\mathcal{G}}}} \left(\left(\frac{\partial}{\partial\nu}+4\pi i a\cdot\nu(x)\right)G((L+\lambda)^{k+1};x)\right)\ (\overline{L}+\lambda)^{k}H(x)\ dS(x)
\end{aligned}
\tag{9.30}
$$

for every (sufficiently small) $\varepsilon > 0$. From potential theoretic arguments (see W. Freeden [2011] for analogous arguments concerning iterated Helmholtz operators $(\Delta + \lambda)^k$), we know that the integrals over all hyperspheres around the lattice points tend to 0 as $\varepsilon \to 0$. Thus we are allowed to formulate a recursion formula in the form

$$
\begin{aligned}
&\int_{\mathcal{G}} G((L+\lambda)^{k+1};x)\ (\overline{L}+\lambda)^{k+1}H(x)\ dV(x) \\
&= \int_{\mathcal{G}} G((L+\lambda)^{k};x)(\overline{L}+\lambda)^{k}H(x)\ dV(x) \\
&\quad + \int_{\partial\mathcal{G}} G((L+\lambda)^{k+1};x)\left(\frac{\partial}{\partial\nu}(\overline{L}+\lambda)^{k}H(x)\right)\ dS(x) \\
&\quad - \int_{\partial\mathcal{G}} \left(\left(\frac{\partial}{\partial\nu}+4\pi i a\cdot\nu(x)\right)G((L+\lambda)^{k+1};x)\right)(\overline{L}+\lambda)^{k}H(x)\ dS(x).
\end{aligned}
\tag{9.31}
$$

By continuing the recursion we find for H being a function of class $\mathrm{C}^{(2m)}\left(\overline{\mathcal{G}}\right)$, that

$$
\begin{aligned}
&\int_{\mathcal{G}} G((L+\lambda);x)(\overline{L}+\lambda)H(x)\,dV(x) \qquad (9.32)\\
&= \int_{\mathcal{G}} G((L+\lambda)^m;x)(\overline{L}+\lambda)^m H(x)\,dV(x)\\
&\quad + \sum_{k=1}^{m-1} \int_{\partial\mathcal{G}} \left(\left(\frac{\partial}{\partial\nu} + 4\pi i a\cdot\nu(x)\right) G((L+\lambda)^{k+1};x)\right)(\overline{L}+\lambda)^k H(x)\,dS(x)\\
&\quad - \sum_{k=1}^{m-1} \int_{\partial\mathcal{G}} G((L+\lambda)^{k+1};x)\left(\frac{\partial}{\partial\nu}(L+\lambda)^k H(x)\right)\,dS(x).
\end{aligned}
$$

In connection with the Euler-type summation formula (Theorem 9.2) we therefore obtain the Euler-type summation formula with respect to $(L+\lambda)^m$, $\lambda \in \mathbb{R}$, $m \in \mathbb{N}$, which is the key formula in our forthcoming approach to sampling. In fact, it is our point of departure for non-bandlimited as well as bandlimited Shannon sampling methodologies.

Theorem 9.7. *(Euler-Type Summation Formula for the Operator $(L+\lambda)^m$). Let $\mathcal{G} \subset \mathbb{R}^q$ be a regular region. Suppose that H is of class $\mathrm{C}^{(2m)}\left(\overline{\mathcal{G}}\right)$. Then, for $\lambda \in \mathbb{R}$, and $a, y \in \mathbb{R}^q$, we have*

$$
\begin{aligned}
\sum_{\substack{g\in\overline{\mathcal{G}}\\ g\in\Lambda}}{}' H(g) &= \frac{1}{\sqrt{\|\mathcal{F}\|}} \sum_{\substack{(L+\lambda)^\wedge(h)=0\\ h\in\Lambda^{-1}}} \int_{\mathcal{G}} H(x)\Phi_h(x)\,dV(x) \qquad (9.33)\\
&\quad + \int_{\mathcal{G}} G((L+\lambda)^m;x)(\overline{L}+\lambda)^m H(x)\,dV(x)\\
&\quad + \sum_{k=0}^{m-1} \int_{\partial\mathcal{G}} \left(\left(\frac{\partial}{\partial\nu} + 4\pi i a\cdot\nu(x)\right) G((L+\lambda)^{k+1};x)\right)(\overline{L}+\lambda)^k H(x)\,dS(x)\\
&\quad - \sum_{k=0}^{m-1} \int_{\partial\mathcal{G}} G((L+\lambda)^{k+1};x)\left(\frac{\partial}{\partial\nu}(\overline{L}+\lambda)^k H(x)\right)\,dS(x).
\end{aligned}
$$

Replacing $H(x)$ by $e^{2\pi i a\cdot x}F(x-y), x \in \overline{\mathcal{G}}$, and using the already known elementary identities (9.18), (9.19), and (9.20), we are able to reformulate the Euler–type summation formula (Theorem 9.7) in such a way that it relates to iterations of the Helmholtz operator $\Delta + \lambda + 4\pi^2 a^2$.

Corollary 9.8. *Let Λ be an arbitrary lattice in $\mathbb{R}^q$. Let $\mathcal{G} \subset \mathbb{R}^q$ be a regular region. Let F be $2m$-times continuously differentiable on $\overline{\mathcal{G}}$. Then, for $\lambda \in \mathbb{R}$*

and $a, y \in \mathbb{R}^q$, we have

$$\sum_{\substack{g-y\in\overline{\mathcal{G}}\\ g\in\Lambda}}{}' e^{2\pi i a\cdot g} F(g-y) \tag{9.34}$$

$$= \frac{1}{\|\mathcal{F}_\Lambda\|} e^{2\pi i a\cdot y} \sum_{\substack{4\pi^2\left((h+a)^2-a^2\right)=\lambda\\ h\in\Lambda^{-1}}} e^{2\pi i h\cdot y} \int_{\mathcal{G}} F(x)\ e^{2\pi i x\cdot(a+h)} dV(x)$$

$$+ \int_{\substack{x-y\in\mathcal{G}\\ x\in\mathbb{R}^q}} G((L+\lambda)^m;x)\ e^{2\pi i a\cdot x} \left(\Delta_x+\lambda+4\pi^2a^2\right)^m F(x-y)\ dV(x)$$

$$- \sum_{k=0}^{m-1} \int_{\substack{x-y\in\partial\mathcal{G}\\ x\in\mathbb{R}^q}} e^{2\pi i a\cdot x} G((L+\lambda)^{k+1};x) \frac{\partial}{\partial\nu_x} \left(\Delta_x+\lambda+4\pi^2a^2\right)^k F(x-y)\ dS(x)$$

$$+ \sum_{k=0}^{m-1} \int_{\substack{x-y\in\partial\mathcal{G}\\ x\in\mathbb{R}^q}} e^{2\pi i a\cdot x} \left(\Delta_x+\lambda+4\pi^2a^2\right)^k F(x-y)$$

$$\times \left(\frac{\partial}{\partial\nu_x} + 4\pi i a\cdot\nu(x)\right) G((L+\lambda)^{k+1};x)\ dS(x).$$

The close relation of the identity (9.34) to the well-known Euler summation formulas for the Helmholtz operator $\Delta+\lambda$ (as presented in W. Freeden [2011]) is best seen by successively specializing the parameters $\lambda\in\mathbb{R}$ and $a, y\in\mathbb{R}^q$. For $y=0$ and a arbitrary, we are confronted with the following situation.

Corollary 9.9. *Let $\mathcal{G}\subset\mathbb{R}^q$ be a regular region. Let F be $2m$-times continuously differentiable on $\overline{\mathcal{G}}$. Then, for $\lambda\in\mathbb{R}$ and $a\in\mathbb{R}^q$, we have*

$$\sum_{\substack{g\in\overline{\mathcal{G}}\\ g\in\Lambda}}{}' e^{2\pi i a\cdot g} F(g) = \frac{1}{\|\mathcal{F}_\Lambda\|} \sum_{\substack{4\pi^2\left((h+a)^2-a^2\right)=\lambda\\ h\in\Lambda^{-1}}} \int_{\mathcal{G}} F(x) e^{2\pi i x\cdot(a+h)} dV(x) \tag{9.35}$$

$$+ \int_{\substack{x\in\mathcal{G}\\ x\in\mathbb{R}^q}} G((L+\lambda)^m;x) e^{2\pi i a\cdot x} \left(\Delta_x+\lambda+4\pi^2a^2\right)^m F(x)\ dV(x)$$

$$- \sum_{k=0}^{m-1} \int_{\substack{x\in\partial\mathcal{G}\\ x\in\mathbb{R}^q}} e^{2\pi i a\cdot x} G((L+\lambda)^{k+1};x) \frac{\partial}{\partial\nu_x} \left(\Delta_x+\lambda+4\pi^2a^2\right)^k F(x)\ dS(x)$$

$$+ \sum_{k=0}^{m-1} \int_{\substack{x\in\partial\mathcal{G}\\ x\in\mathbb{R}^q}} e^{2\pi i a\cdot x} \left(\Delta_x+\lambda+4\pi^2a^2\right)^k F(x)$$

$$\times \left(\frac{\partial}{\partial\nu_x} + 4\pi i a\cdot\nu(x)\right) G((L+\lambda)^{k+1};x)\ dS(x).$$

For $a=0$, i.e., $L=\Delta$, and y arbitrary we have the well-known case for the Helmholtz operator $\Delta+\lambda$.

Corollary 9.10. *Let Λ be an arbitrary lattice in $\mathbb{R}^q$. Let $\mathcal{G} \subset \mathbb{R}^q$ be a regular region. Let F be $2m$-times continuously differentiable on $\overline{\mathcal{G}}$. Then, for $\lambda \in \mathbb{R}$, $y \in \mathbb{R}^q$, and $L = \Delta$, we have*

$$
\begin{aligned}
&\sum_{\substack{g-y\in\overline{\mathcal{G}}\\ g\in\Lambda}}{}' \; F(g-y) \qquad (9.36)\\
&= \frac{1}{\|\mathcal{F}_\Lambda\|} \sum_{\substack{4\pi^2h^2=\lambda\\ h\in\Lambda^{-1}}} e^{2\pi i h\cdot y} \int_{\mathcal{G}} F(x) e^{2\pi i x\cdot h} dV(x)\\
&+ \int_{\substack{x-y\in\mathcal{G}\\ x\in\mathbb{R}^q}} G((L+\lambda)^m;x)\,(\Delta_x+\lambda)^m\, F(x-y)\; dV(x)\\
&- \sum_{k=0}^{m-1} \int_{\substack{x-y\in\partial\mathcal{G}\\ x\in\mathbb{R}^q}} G((L+\lambda)^{k+1};x) \frac{\partial}{\partial\nu_x}\,(\Delta_x+\lambda)^k\, F(x-y)\; dS(x)\\
&+ \sum_{k=0}^{m-1} \int_{\substack{x-y\in\partial\mathcal{G}\\ x\in\mathbb{R}^q}} (\Delta_x+\lambda)^k\, F(x-y) \frac{\partial}{\partial\nu_x} G((L+\lambda)^{k+1};x)\; dS(x).
\end{aligned}
$$

For $a = y = 0$ we have the *Euler summation formula for iterated Helmholtz operators,* which leads to multi-dimensional metaharmonic theory (as provided by W. Freeden [2011]).

Corollary 9.11. *Let $\mathcal{G} \subset \mathbb{R}^q$ be a regular region. Suppose that H is of class $\mathrm{C}^{(2m)}\left(\overline{\mathcal{G}}\right)$, $m \in \mathbb{N}$. Then we have*

$$
\begin{aligned}
&\sum_{\substack{g\in\overline{\mathcal{G}}\\ g\in\Lambda}}{}' \; F(g) \qquad (9.37)\\
&= \frac{1}{\sqrt{\|\mathcal{F}\|}} \sum_{\substack{(\Delta+\lambda)^\wedge(h)=0\\ h\in\Lambda^{-1}}} \int_{\mathcal{G}} F(x)\Phi_h(x)\; dV(x)\\
&\quad + \int_{\mathcal{G}} G((\Delta+\lambda)^m;x)(\Delta+\lambda)^m F(x)\; dV(x)\\
&\quad + \sum_{k=0}^{m-1} \int_{\partial\mathcal{G}} \left(\frac{\partial}{\partial\nu}\right) G((\Delta+\lambda)^{k+1};x)(\Delta+\lambda)^k F(x)\; dS(x)\\
&\quad - \sum_{k=0}^{m-1} \int_{\partial\mathcal{G}} G((L+\lambda)^{k+1};x) \left(\frac{\partial}{\partial\nu}(\Delta+\lambda)^k F(x)\right)\; dS(x).
\end{aligned}
$$

In addition, for $\lambda = 0$, the formula (9.33) constitutes the *Euler summation formula for iterated Laplace operators* Δ^m, which compares a volume integral with a lattice point sum over $\mathcal{G}$ under explicit knowledge of the remainder term in integral form. This summation formula plays an important role in multi-dimenensional cubature methods such as Romberg integration (see W. Freeden, C. Gerhards [2017], W. Freeden, M. Gutting [2017] and the references therein).

Corollary 9.12. *Let $\mathcal{G} \subset \mathbb{R}^q$ be a regular region. Suppose that F is of class $\mathrm{C}^{(2m)}\left(\overline{\mathcal{G}}\right)$, $\overline{\mathcal{G}} = \mathcal{G} \cup \partial\mathcal{G}$. Then*

$$
\begin{aligned}
\sum_{\substack{g\in\overline{\mathcal{G}} \\ g\in\Lambda}}{}' F(g) \quad = \quad & \frac{1}{\|\mathcal{F}\|}\int_{\mathcal{G}} F(x)\, dV(x) \qquad (9.38) \\
& + \int_{\mathcal{G}} G\left(\Delta^m; x\right) \Delta^m F(x)\, dV(x) \\
& + \sum_{k=0}^{m-1} \int_{\partial\mathcal{G}} \left(\frac{\partial}{\partial\nu} G\left(\Delta^{k+1}; x\right)\right) \Delta^k F(x)\, dS(x) \\
& - \sum_{k=0}^{m-1} \int_{\partial\mathcal{G}} G\left(\Delta^{k+1}; x\right) \left(\frac{\partial}{\partial\nu} \Delta^k F(x)\right)\, dS(x).
\end{aligned}
$$

Altogether, the generalization of the Euler-type summation formula to an iterated operator $(L+\lambda)^m$, $\lambda \in \mathbb{R}$, opens innovative perspectives to subtle questions of the convergence of a multi-dimensional series and/or the integral, because the operator $(L+\lambda)^m$, can be closely adapted to alternating or oscillating properties of the summand and/or integrand. In this respect it should be remarked that, for the series convergence of alternating summands in one dimension, direct techniques are well-known (for example, Leibniz criterion). In multi-dimensional summation problems the situation is different (cf. W. Freeden [2011]), because the standard one-dimensional concept of the alternating or oscillating series is not applicable. According to our approach, however, the choice of the "wave number" $\lambda \in \mathbb{R}$ within the operator $(L+\lambda)^m$ may be used to adapt the summation formula to oscillating properties of the summand constituting the series. These aspects will be of great advantage in deriving special identities in analytic theory of numbers. The weighted Hardy–Landau identities discussed later in this book are typical examples of this procedure.

Remark 9.13. *Of course, the Euler summation formulas can be formulated to more general, not neccessarily elliptic differential operators (see V.K. Ivanow [1963], W. Freeden [1982]), W. Freeden, J. Fleck [1987]). For example, applicants often use the summation formula in the context of hyperbolic differential operators. Our work, however, is restricted to iterated $(L+\lambda)$-operators and arbitrary lattices, which especially relate to phenomena of oscillation.*

Part V

Bivariate Lattice Point/Ball Summation and Shannon-Type Sampling

Chapter 10

Hardy–Landau-Type Lattice Point Identities (Constant Weight)

As already known, the *1D Hardy–Landau identity* (see (4.18)) can be written (in terms of a constant weight function $F = 1$) as follows

$$\sum_{\substack{|g|\leq R \\ g\in\mathbb{Z}}}{}' \underbrace{F(g)}_{=1} = \sum_{\substack{|g|\leq R \\ g\in\mathbb{Z}}}{}' 1 = 2R \sum_{h\in\mathbb{Z}} \frac{J_1(1; 2\pi hR)}{2\pi hR}, \tag{10.1}$$

where R is an arbitrary, but fixed positive number and (in accordance with the nomenclature of Bessel functions)

$$\frac{J_1(1; 2\pi hR)}{2\pi hR} = \frac{\sin(2\pi hR)}{2\pi hR} = \mathrm{sinc}(2\pi hR). \tag{10.2}$$

In what follows we are interested in the *2D Hardy–Landau identity* (as initiated in several papers by G.H. Hardy [1915], E. Landau [1915], G.H. Hardy, E. Landau [1924], E. Landau [1924]). In fact, our work starts with an extension of the classical 2D Hardy–Landau identity for arbitrary lattices Λ

$$\sum_{\substack{|g|\leq R \\ g\in\Lambda}}{}' 1 = \lim_{N\to\infty} \frac{R}{\|\mathcal{F}_\Lambda\|} \sum_{\substack{|h|\leq N \\ h\in\Lambda^{-1}}} \frac{J_1(2; 2\pi|h|R)}{|h|}, \tag{10.3}$$

however, in a way different from the traditional procedures known from analytic number theory. Instead, we exclusively base our verification on tools of the theory of Bessel functions as proposed by C. Müller [1954c] for the bivariate case.

10.1 Integral Mean Asymptotics for the Euler–Green Function

The following integral estimates (first formulated in the PhD-thesis of W. Freeden [1975] for the the Helmholtz operator $\Delta + \lambda$) play an important part in the discussion of series over alternating functions in $\mathbb{R}^2$.

Theorem 10.1. *For all lattices $\Lambda \subset \mathbb{R}^2$, and for all values $\lambda \in \mathbb{R}$ and vectors $a, y \in \mathbb{R}^2$, the estimates*

$$\int_{\substack{|x-y|=N \\ x\in\mathbb{R}^2}} |G(L+\lambda;x)| \; dS(x) = O(N) \tag{10.4}$$

and

$$\int_{\substack{|x-y|=N \\ x\in\mathbb{R}^2}} \left| \frac{\partial G(L+\lambda;x)}{\partial \nu} \right| \; dS(x) = O(N) \tag{10.5}$$

hold true for $N \to \infty$.

Proof. We start by remembering that there is a positive constant D dependent on the lattice Λ, such that $|g - g'| \geq D$ holds for all points $g, g' \in \Lambda$ with $g \neq g'$. We set

$$\delta = \frac{1}{2}\min(1, \mathrm{D}). \tag{10.6}$$

Then we notice that, for $\lambda \in \mathbb{R}$ and $a, y \in \mathbb{R}^2$ fixed but arbitrary, there exists a constant C such that the estimates

$$|G(L+\lambda; x+y)| \leq C |\ln|x+y-g||, \tag{10.7}$$

and

$$\left| \nabla_x G(L+\lambda; x+y) - \frac{1}{2\pi} \frac{x+y-g}{|x+y-g|^2} \right| \leq C \frac{1}{|x+y-g|} \tag{10.8}$$

hold uniformly with respect to $x+y \in \overline{\mathbb{B}^2_\delta(g)}, g \in \Lambda$. For $x+y \in \mathbb{R}^2$ we denote the distance of $x+y \in \mathbb{R}^q$ to the lattice Λ by

$$D(x+y;\Lambda) = \mathrm{dist}(x+y;\Lambda) = \min_{g\in\Lambda} |x+y-g|. \tag{10.9}$$

It is clear that there is a constant B (depending on $\lambda \in \mathbb{R}$, $a, y \in \mathbb{R}^2$, and δ), such that the estimates

$$|G(L+\lambda;x+y)| \leq B, \tag{10.10}$$

$$|\nabla_x G(L+\lambda;x+y)| \leq B \tag{10.11}$$

are valid for all $x \in \mathbb{R}^2$ with

$$D(x+y;\Lambda) = \mathrm{dist}(x+y;\Lambda) \geq \delta. \tag{10.12}$$

Moreover, for all $w \in \mathbb{R}^2$, we are able to deduce from the well-known Gaussian estimate (cf. Theorem 7.10) that

$$\#_\Lambda \left(\overline{\mathbb{B}^2_{N+\delta}(w)} \setminus \mathbb{B}^2_{N-\delta}(w) \right) = \sum_{\substack{N-\delta \leq |g-w| \leq N+\delta \\ g\in\Lambda}} 1 = O(N) \tag{10.13}$$

for δ fixed, suitably chosen and $N \to \infty$. Thus, for all $w \in \mathbb{R}^2$, it follows that

$$\sum_{\substack{|g-w|=N \\ g\in\Lambda}} 1 = O(N), \quad N \to \infty. \tag{10.14}$$

We use the different results for the cases $D(x+y;\Lambda) > \delta$ and $D(x+y;\Lambda) \leq \delta$ to obtain an estimate of the form

$$\begin{aligned} &\int_{\substack{|x|=N \\ x\in\mathbb{R}^2}} |G(L+\lambda; x+y)| \, dS(x) \\ &= \int_{\substack{|x|=N \\ D(x+y;\Lambda)>\delta \\ x\in\mathbb{R}^2}} |G(L+\lambda; x+y)| \, dS(x) \\ &\quad + \int_{\substack{|x|=N \\ D(x+y;\Lambda)\leq\delta \\ x\in\mathbb{R}^2}} |G(L+\lambda; x+y)| \, dS(x). \end{aligned} \tag{10.15}$$

From (10.10) and (10.11), it follows that

$$\int_{\substack{|x|=N \\ D(x+y;\Lambda)>\delta \\ x\in\mathbb{R}^2}} |G(L+\lambda; x+y)| \, dS(x) \leq B \, \|\mathbb{S}^1\| \, N. \tag{10.16}$$

Furthermore, because of the characteristic singularity of $G(L+\lambda; x+y)$, the estimate

$$\begin{aligned} &\int_{\substack{|x|=N \\ D(x+y;\Lambda)\leq\delta \\ x\in\mathbb{R}^2}} |G(L+\lambda; x+y)| \, dS(x) \\ &= O\left(\#_\Lambda\left(\overline{\mathbb{B}^2_{N+\delta}} \setminus \mathbb{B}^2_{N-\delta}\right)\right) \end{aligned} \tag{10.17}$$

for $N \to \infty$. In connection with (10.13) and Theorem 7.10, we therefore find

$$\int_{\substack{|x|=N \\ D(x+y;\Lambda)\leq\delta \\ x\in\mathbb{R}^2}} |G(L+\lambda; x+y)| \, dS(x) = O(N) \tag{10.18}$$

for $N \to \infty$.
Concerning the second part of Theorem 10.1, we again split the integral such that

$$\begin{aligned} &\int_{\substack{|x|=N \\ x\in\mathbb{R}^2}} \left|\frac{\partial}{\partial\nu} G(L+\lambda; x+y)\right| \, dS(x) \\ &= \int_{\substack{|x|=N \\ D(x+y;\Lambda)>\delta \\ x\in\mathbb{R}^2}} \left|\frac{\partial}{\partial\nu} G(L+\lambda; x+y)\right| \, dS(x) \\ &\quad + \int_{\substack{|x|=N \\ D(x+y;\Lambda)\leq\delta \\ x\in\mathbb{R}^2}} \left|\frac{\partial}{\partial\nu} G(L+\lambda; x+y)\right| \, dS(x). \end{aligned} \tag{10.19}$$

Then, by virtue of estimates known from the concept of solid angles (see Section 6.3 for details), we obtain from the Gaussian estimate, i.e., Theorem 7.10,

$$\begin{aligned} \int_{\substack{|x|=N \\ x\in\mathbb{R}^2}} \left| \frac{\partial}{\partial \nu} G(L+\lambda; x+y) \right| \, dS(x) & \qquad (10.20) \\ = \; O(N) + O\left(\#_\Lambda \left(\overline{\mathbb{B}^2_{N+\delta}} \setminus \mathbb{B}^2_{N-\delta} \right) \right) & \\ = \; O(N) & \end{aligned}$$

for $N \to \infty$. Altogether, this is the desired result of Theorem 10.1. □

10.2 Hardy–Landau-Type Identity

Our point of departure is the infinitely often differentiable function $F_R : \mathbb{R}^2 \to \mathbb{R}$, $R > 0$ fixed, given by

$$F_R(x-y) = \frac{J_1(2\pi|x-y|R)}{|x-y|}, \quad x \in \mathbb{R}^2, \tag{10.21}$$

where $y \in \mathbb{R}^2$ is arbitrary, but fixed and, in accordance with the nomenclature of Bessel functions (note that, in the sense of our convention, $J_1(\cdot) = J_1(2;\cdot)$). Well-known asymptotic estimates known from the 2D-theory of Bessel functions (Lemma 6.28) tell us that

$$\frac{J_1(2\pi|x-y|R)}{|x-y|} = O\left(|x-y|^{-\frac{3}{2}}\right) \tag{10.22}$$

and

$$\left| \nabla_x \frac{J_1(2\pi|x-y|R)}{|x-y|} \right| = O\left(|x-y|^{-\frac{3}{2}}\right) \tag{10.23}$$

are valid for $|x-y| \to \infty$.

Consequently we are able to deduce from Theorem 10.1 that

$$\int_{\substack{|x-y|=N \\ x\in\mathbb{R}^2}} e^{2\pi i a\cdot x} G(L+\lambda; x) \frac{\partial}{\partial \nu_x} F_R(x-y) \, dS(x) = O\left(|x-y|^{-\frac{1}{2}}\right) \tag{10.24}$$

and

$$\begin{aligned} \int_{\substack{|x-y|=N \\ x\in\mathbb{R}^2}} e^{2\pi i a\cdot x} F_R(x-y) \left(\frac{\partial}{\partial \nu_x} + 4\pi i a \cdot \nu(x) \right) G(L+\lambda; x) \, dS(x) & \qquad (10.25) \\ = O\left(|x-y|^{-\frac{1}{2}}\right). & \end{aligned}$$

In connection with the Euler-type summation formula (Corollary 9.8) we therefore obtain

$$\sum_{\substack{|g-y|\le N\\ g\in\Lambda}} e^{2\pi i a\cdot g} F_R(g-y) \tag{10.26}$$

$$= \frac{1}{\|\mathcal{F}_\Lambda\|}\ e^{2\pi i a\cdot y} \sum_{\substack{4\pi^2((h+a)^2-a^2)=\lambda\\ h\in\Lambda^{-1}}} e^{2\pi i h\cdot y} \int_{|x|\le N} F_R(x)\ e^{2\pi i x\cdot(a+h)} dV(x)$$

$$+ \int_{\substack{|x-y|\le N\\ x\in\mathbb{R}^2}} G(L+\lambda;x)\ e^{2\pi i a\cdot x}\ \left(\Delta_x+\lambda+4\pi^2a^2\right) F_R(x-y)\ dV(x)$$

$$+\ o(1), \quad N\to\infty.$$

An elementary calculation in the context of Bessel functions yields

$$\Delta_x \frac{J_1(2\pi|x-y|R)}{|x-y|} + 4\pi^2R^2\frac{J_1(2\pi|x-y|R)}{|x-y|} = 4\pi R\ \frac{J_2(2\pi|x-y|R)}{|x-y|^2}. \tag{10.27}$$

Hence, the differential equation (10.27) relates the Bessel function J_1 to the Bessel function J_2 (of course, both of dimension $q=2$). The differential equation (10.27) indicates to use the "wave number"

$$\lambda = 4\pi^2(R^2-a^2) \tag{10.28}$$

as an adaptive parameter to the oscillating properties of the function F_R defined by (10.21). In fact, from the well-known asymptotic behavior of the 2D-Bessel function J_2, we are able to deduce that, under the special choice $\lambda = 4\pi^2(R^2-a^2)$,

$$(\Delta_x+\lambda+4\pi^2a^2)\frac{J_1(2\pi|x-y|R)}{|x-y|} = O\left(|x-y|^{-\frac{5}{2}}\right) \tag{10.29}$$

i.e.,

$$(\Delta_x+\lambda+4\pi^2a^2)\frac{J_1(2\pi|x-y|R)}{|x-y|} = O\left(|x-y|^{-(2+\varepsilon)}\right),\quad \varepsilon=\frac{1}{2}, \tag{10.30}$$

for $|x-y|\to\infty$. It follows that the integral

$$\int_{\mathbb{R}^2} G(L+\lambda;x)\ e^{2\pi i a\cdot x}\ \left(\Delta_x+\lambda+4\pi^2a^2\right) F_R(x-y)\ dV(x) \tag{10.31}$$

$$= \lim_{N\to\infty}\int_{\substack{|x-y|\le N\\ x\in\mathbb{R}^2}} G(L+\lambda;x)\ e^{2\pi i a\cdot x}\ \left(\Delta_x+\lambda+4\pi^2a^2\right) F_R(x-y)\ dV(x)$$

is absolutely convergent. As an immediate consequence of (10.31) we see by observing (10.28) that

$$\lim_{N\to\infty}\left(\sum_{\substack{|g-y|\le N\\ g\in\Lambda}} e^{2\pi i a\cdot g}F_R(g-y)\right. \tag{10.32}$$

$$\left. -\frac{1}{\|\mathcal{F}_\Lambda\|}\, e^{2\pi i a\cdot y}\sum_{\substack{4\pi^2\left((h+a)^2-a^2\right)\\ =4\pi^2(R^2-a^2)\\ h\in\Lambda^{-1}}} e^{2\pi i h\cdot y}\int_{\substack{|x|\le N\\ x\in\mathbb{R}^2}} F_R(x)\, e^{2\pi i x\cdot(a+h)}dV(x)\right)$$

$$= \int_{\mathbb{R}^2} G(L+4\pi^2(R^2-a^2);x)\, e^{2\pi i a\cdot x}\, \left(\Delta_x+4\pi^2R^2\right) F_R(x-y)\, dV(x).$$

The identity (10.32), therefore, enables us to conclude that the limit

$$\lim_{N\to\infty}\left(\sum_{\substack{|g-y|\le N\\ g\in\Lambda}} e^{2\pi i a\cdot g}\frac{J_1(2\pi|g-y|R)}{|g-y|}\right. \tag{10.33}$$

$$\left. -\frac{1}{\|\mathcal{F}_\Lambda\|}\, e^{2\pi i a\cdot y}\sum_{\substack{|h+a|=R\\ h\in\Lambda^{-1}}} e^{2\pi i h\cdot y}\int_{\substack{|x|\le N\\ x\in\mathbb{R}^2}} \frac{J_1(2\pi|x|R)}{|x|}e^{2\pi i x\cdot(a+h)}dV(x)\right)$$

exists in the indicated spherical sense, and it is equal to

$$\lim_{N\to\infty}\left(\sum_{\substack{|g-y|\le N\\ g\in\Lambda}} e^{2\pi i a\cdot g}\,\frac{J_1(2\pi|g-y|R)}{|g-y|}\right. \tag{10.34}$$

$$\left. -\frac{2\pi}{\|\mathcal{F}_\Lambda\|}\, e^{2\pi i a\cdot y}\sum_{\substack{|h+a|=R\\ h\in\Lambda^{-1}}} e^{2\pi i h\cdot y}\int_0^N \frac{J_1(2\pi rR)}{r}\, J_0(2\pi|h+a|r)\, dr\right).$$

Moreover, in connection with (10.27), we have

$$\lim_{N\to\infty}\left(\sum_{\substack{|g-y|\le N\\ g\in\Lambda}} e^{2\pi i a\cdot g}\frac{J_1(2\pi|g-y|R)}{|g-y|}\right. \tag{10.35}$$

$$\left. -\frac{2\pi}{\|\mathcal{F}_\Lambda\|}\, e^{2\pi i a\cdot y}\sum_{\substack{|h+a|=R\\ h\in\Lambda^{-1}}} e^{2\pi i h\cdot y}\int_0^N \frac{J_1(2\pi rR)}{r}\, J_0(2\pi|h+a|r)\, dr\right)$$

$$= 4\pi R\int_0^\infty \frac{J_2(2\pi rR)}{r^2}\left(\int_{\substack{|x-y|=r\\ x\in\mathbb{R}^2}} G\left(L+4\pi^2(R^2-a^2);x\right)\, e^{2\pi i a\cdot x}\, dS(x)\right)\, dr.$$

The last integral can be written in the form

$$4\pi R \int_0^\infty \frac{J_2(2\pi rR)}{r^2} \int_{\substack{|x-y|=r\\ x\in\mathbb{R}^2}} G\left(L+4\pi^2(R^2-a^2);x\right)\, dS(x)\, dr \tag{10.36}$$

$$= -\frac{2\pi}{\|\mathcal{F}_\Lambda\|} \sum_{\substack{|h+a|\neq R\\ h\in\Lambda^{-1}}} \frac{4\pi R\, e^{2\pi i(h+a)\cdot y}}{-4\pi^2(h-a)^2+4\pi^2\mathbb{R}^2} \int_0^\infty \frac{J_2(2\pi rR)}{r} J_0(2\pi|h+a|R)\, dr.$$

Observing the identity (see, e.g., G.N. Watson [1944])

$$\frac{4\pi R}{4\pi^2(h-a)^2-4\pi^2R^2} \int_0^\infty \frac{J_2(2\pi rR)}{r} J_0(2\pi|h+a|r)\, dr \tag{10.37}$$

$$= \int_0^\infty J_1(2\pi rR)\, J_0(2\pi|h+a|r)\, dr \qquad .$$

we obtain in the context of Bessel functions

$$\lim_{N\to\infty} \left(\sum_{\substack{|g-y|\leq N\\ g\in\Lambda}} e^{2\pi i a\cdot g} \frac{J_1(2\pi|g-y|R)}{|g-y|} \right. \tag{10.38}$$

$$\left. - \frac{2\pi}{\|\mathcal{F}_\Lambda\|} e^{2\pi i a\cdot y} \sum_{\substack{|h+a|=R\\ h\in\Lambda^{-1}}} e^{2\pi i h\cdot y} \int_0^N \frac{J_1(2\pi rR)}{r} J_0(2\pi|h+a|r)\, dr \right)$$

$$= \frac{2\pi}{\|\mathcal{F}_\Lambda\|} e^{2\pi i a\cdot y} \sum_{\substack{|h+a|\neq R\\ h\in\Lambda^{-1}}} e^{2\pi i h\cdot y} \int_0^\infty \frac{J_1(2\pi rR)}{r} J_0(2\pi|h+a|r)\, dr.$$

Now, for arbitrary $R>0$, the Hankel transform of discontinuous integrals (see Eq. (8.109)) guarantees the existence of all occuring integrals such that

$$\lim_{N\to\infty} \sum_{\substack{|g-y|\leq N\\ g\in\Lambda}} e^{2\pi i a\cdot g} \frac{J_1(2\pi|g-y|R)}{|g-y|} \tag{10.39}$$

$$= \frac{2\pi}{\|\mathcal{F}_\Lambda\|} e^{2\pi i a\cdot y} \sum_{\substack{|h+a|=R\\ h\in\Lambda^{-1}}} e^{2\pi i h\cdot y} \int_0^\infty \frac{J_1(2\pi rR)}{r} J_0(2\pi|h+a|r)\, dr$$

$$+ \frac{2\pi}{\|\mathcal{F}_\Lambda\|} e^{2\pi i a\cdot y} \sum_{\substack{|h+a|\neq R\\ h\in\Lambda^{-1}}} e^{2\pi i h\cdot y} \int_0^\infty \frac{J_1(2\pi rR)}{r} J_0(2\pi|h+a|r)\, dr.$$

Thus, we have the following identity

$$\lim_{N\to\infty} \sum_{\substack{|g-y|\leq N\\ g\in\Lambda}} e^{2\pi i a\cdot g} \frac{J_1(2\pi|g-y|R)}{|g-y|} \tag{10.40}$$

$$= \frac{2\pi}{\|\mathcal{F}_\Lambda\|} e^{2\pi i a\cdot y} \sum_{h\in\Lambda^{-1}} e^{2\pi i h\cdot y} \int_0^\infty \frac{J_1(2\pi rR)}{r} J_0(2\pi|h+a|r)\, dr.$$

It follows from the theory of discontinuous integrals (see Lemma 8.24) that

$$2\pi R \int_0^\infty \frac{J_1(2\pi r R)}{r} J_0(2\pi|h+a|r)\, dr = \begin{cases} 1, & |h+a| < R, \\ \frac{1}{2}, & |h+a| = R, \\ 0, & |h+a| > R. \end{cases} \tag{10.41}$$

Consequently, in the notation of (10.21), we obtain

$$\lim_{N\to\infty} \sum_{\substack{|g-y|\le N \\ g\in\Lambda}} e^{2\pi i a\cdot(g-y)}\ F_R(g-y) \tag{10.42}$$

$$= \frac{1}{\|\mathcal{F}_\Lambda\|} \sum_{h\in\Lambda^{-1}} e^{2\pi i h\cdot y} \int_{\mathbb{R}^2} F_R(x)\ e^{2\pi i x\cdot(a+h)}\ dV(x)\ .$$

Altogether, we arrive at the following extension of the Hardy–Landau identity (cf. C. Müller [1954c]), i.e., the *Hardy–Landau-type identity.*

Theorem 10.2. *For $R > 0, a, y \in \mathbb{R}^2$, and arbitrary lattices $\Lambda \subset \mathbb{R}^2$ we have*

$$\lim_{N\to\infty} R \sum_{\substack{|g-y|\le N \\ g\in\Lambda}} e^{2\pi i a\cdot g}\ \frac{J_1(2\pi|g-y|R)}{|g-y|} \tag{10.43}$$

$$= \frac{1}{\|\mathcal{F}_\Lambda\|}\ e^{2\pi i a\cdot y} \sum_{\substack{|a+h|\le R \\ h\in\Lambda^{-1}}}{}' \ e^{2\pi i h\cdot y}. \tag{10.44}$$

Replacing $\Lambda \subset \mathbb{R}^2$ by its inverse lattice $\Lambda^{-1} \subset \mathbb{R}^2$ we finally obtain

Theorem 10.3. *(Hardy–Landau-Type Identity (Constant Weight)) For $a, y \in \mathbb{R}^2, R > 0$, and arbitrary lattices $\Lambda \subset \mathbb{R}^2$*

$$\sum_{\substack{|a+g|\le R \\ g\in\Lambda}}{}' \ e^{2\pi i(a+g)\cdot y} = \frac{R}{\|\mathcal{F}_\Lambda\|} \lim_{N\to\infty} \sum_{\substack{|h-y|\le N \\ h\in\Lambda^{-1}}} e^{2\pi i a\cdot h}\ \frac{J_1(2\pi|h-y|R)}{|h-y|}. \tag{10.45}$$

Remark 10.4. *If $g = 0$ is the only lattice point of Λ contained in the circle $\mathbb{B}_R^2(-a)$ we actually end up with a special sampling case, i.e., the sampling of the function $y \mapsto e^{2\pi i a\cdot y}, y \in \mathbb{R}^2$, which is similar to a 1D result already known from A. Papoulis [1971]*

$$e^{2\pi i a\cdot y} = \lim_{N\to\infty} \sum_{\substack{|h-y|\le N \\ h\in\Lambda}} e^{2\pi i a\cdot h}\ \frac{R}{\|\mathcal{F}_\Lambda\|}\ \frac{J_1(2\pi|h-y|R)}{|h-y|}, \tag{10.46}$$

i.e.,

$$e^{2\pi i a\cdot y} = \lim_{N\to\infty} \sum_{\substack{|h-y|\le N \\ h\in\Lambda}} e^{2\pi i a\cdot h}\ \frac{1}{\|\mathcal{F}_\Lambda\|} \int_{\substack{|w|\le R \\ w\in\mathbb{R}^2}} e^{2\pi i(h-y)\cdot w}\ dV(w). \tag{10.47}$$

10.3 Discrepancy Asymptotics

Next we deal with the *basic version of a Hardy–Landau-type identity,* i.e., $a = y = 0$, in Theorem 10.3. This actually yields the *Hardy–Landau identity* for arbitrary lattices Λ in the form (see G.H. Hardy [1915], E. Landau [1915], and G.H. Hardy, E. Landau [1924]).

$$\sum_{\substack{|g|\leq R\\ g\in\Lambda}}{}' 1 = \lim_{N\to\infty} \frac{R}{\|\mathcal{F}_\Lambda\|} \sum_{\substack{|h|\leq N\\ h\in\Lambda^{-1}}} \frac{J_1(2\pi|h|R)}{|h|} . \tag{10.48}$$

For radii R chosen in such a way that the origin 0 is the only lattice point in $\mathbb{B}^2_R$, this formula also allows the interpretation as a sampling identity of the constant function:

$$1 = \lim_{N\to\infty} \sum_{\substack{|h|\leq N\\ h\in\Lambda^{-1}}} 1 \, \frac{R}{\|\mathcal{F}_\Lambda\|} \frac{J_1(2\pi|h|R)}{|h|}. \tag{10.49}$$

Observing $J_1(r) = \frac{r}{2} + \ldots$ we are able to rewrite the identity (10.48) in its standard form

$$\sum_{\substack{|g|\leq R\\ g\in\Lambda}}{}' 1 = \frac{\pi}{\|\mathcal{F}_\Lambda\|} R^2 + \lim_{N\to\infty} \frac{R}{\|\mathcal{F}_\Lambda\|} \sum_{\substack{0<|h|\leq N\\ h\in\Lambda^{-1}}} \frac{J_1(2\pi|h|R)}{|h|} \tag{10.50}$$

comparing the number of lattice points inside a circle around the origin of radius R with the area of the circle under explicit specification of the remaining *Hardy–Landau discrepancy series expansion*:

$$\lim_{N\to\infty} \frac{R}{\|\mathcal{F}_\Lambda\|} \sum_{\substack{0<|h|\leq N\\ h\in\Lambda^{-1}}} \frac{J_1(2\pi|h|R)}{|h|} = \sum_{\substack{|g|\leq R\\ g\in\Lambda}}{}' 1 - \frac{\pi}{\|\mathcal{F}_\Lambda\|} R^2 . \tag{10.51}$$

Using the well-known asymptotic relation from the theory of Bessel functions (see Lemma 6.29)

$$J_1(2\pi|h|R) = \frac{(|h|R)^{-\frac{1}{2}}}{\pi} \cos\left(2\pi|h|R - \frac{3\pi}{4}\right) + O\left((|h|R)^{-\frac{3}{2}}\right), \tag{10.52}$$

we are able to state that, apart from an additive remainder term, the Hardy–Landau series

$$\frac{R}{\|\mathcal{F}\|} \sum_{\substack{0<|h|\\ h\in\Lambda^{-1}}} \frac{J_1(2\pi|h|R)}{|h|} \tag{10.53}$$

is asymptotically equal to the 2D alternating sum

$$\frac{R^{\frac{1}{2}}}{\pi\|\mathcal{F}\|} \sum_{\substack{0<|h| \\ h\in\Lambda^{-1}}} \frac{\cos\left(2\pi|h|R - \frac{3\pi}{4}\right)}{|h|^{\frac{3}{2}}} . \tag{10.54}$$

In a certain sense, all investigations involving the number theoretical study of the classical 2D Hardy–Landau identity, which have been done during the past century, are based on (partial sums of) the "discrepancy replacement series" (10.54). Nonetheless, the statement that the alternating sum (10.54) is of the order $O(R^{\frac{1}{2}})$ is wrong (cf. E. Landau [1927]), and Hardy's conjecture $O(R^{\frac{1}{2}+\varepsilon})$ for *every* $\varepsilon > 0$ remains a challenge for future work.

Going back to the Hardy–Landau-type identity (10.48) we obtain in analogy to (10.51)

$$\begin{aligned} \sum_{\substack{|a+g|\le R \\ g\in\Lambda}}{}' e^{2\pi i(a+g)\cdot y} &= \frac{\pi}{\|\mathcal{F}_\Lambda\|} R^2 \, e^{2\pi i a\cdot y} \\ &\quad + \frac{R}{\|\mathcal{F}_\Lambda\|} \lim_{N\to\infty} \sum_{\substack{0<|h-y|\le N \\ h\in\Lambda^{-1}}} e^{2\pi i a\cdot h} \, \frac{J_1(2\pi|h-y|R)}{|h-y|}. \end{aligned} \tag{10.55}$$

The series

$$\frac{R}{\|\mathcal{F}_\Lambda\|} \lim_{N\to\infty} \sum_{\substack{0<|h-y|\le N \\ h\in\Lambda^{-1}}} e^{2\pi i a\cdot h} \, \frac{J_1(2\pi|h-y|R)}{|h-y|} \tag{10.56}$$

is asymptotically equal to the 2D alternating sum

$$\frac{R^{\frac{1}{2}}}{\pi\|\mathcal{F}\|} \sum_{\substack{0<|h-y|\le N \\ h\in\Lambda^{-1}}} e^{2\pi i a\cdot h} \, \frac{\cos\left(2\pi|h-y|R - \frac{3\pi}{4}\right)}{|h|^{\frac{3}{2}}} . \tag{10.57}$$

The identity (10.57) can be used for the invesigation of weighted radial and angular lattice point distributions and the phenomenon of asymptotically non-uniform distribution of lattice points in the plane (see C. Müller, A. Dressler [1972], W. Freeden [1978b, 2011], and the references therein) corresponding to arbitrary "center points" $y \in \mathbb{R}^2$.

Chapter 11

Hardy–Landau-Type Lattice Point Identities (General Weights)

In similarity to the 1D theory, we are interested in establishing bivariate weighted lattice point identities, so that Shannon-type sampling formulas follow by an appropriate integration process. However, it turns out that the validity of generally weighted Hardy–Landau identities on regular regions in $\mathbb{R}^q$ is a mathematical challenge showing a diversity of open problems even in the plane.

This chapter gives some insight into possible strategies and viable methodologies for the bivariate case. Before we start with our investigations, some remarks should be made to understand the specific calamities:

The problems to develop general F-weighted 2D Hardy–Landau-type lattice point identities over general regular regions $\mathcal{G} \subset \mathbb{R}^2$

$$\frac{1}{\|\mathcal{F}_\Lambda\|} \lim_{N\to\infty} \sum_{\substack{|h-y|\leq N \\ h\in\Lambda^{-1}}} e^{2\pi i a\cdot h} \underbrace{\int_{\substack{x\in\mathcal{G} \\ x\in\mathbb{R}^2}} F(x)\, e^{-2\pi i x\cdot(h-y)}\, dV(x)}_{=F_{\mathcal{G}}^{\wedge}(h-y)}$$

$$\sim \sum_{\substack{a+g\in\overline{\mathcal{G}} \\ g\in\Lambda}}{}' e^{2\pi i y\cdot(a+g)} F(a+g), \quad a, y \in \mathbb{R}^2, \tag{11.1}$$

are twofold, namely first to guarantee the pointwise convergence of the series expansion

$$\frac{1}{\|\mathcal{F}_\Lambda\|} \lim_{N\to\infty} \sum_{\substack{|h-y|\leq N \\ h\in\Lambda^{-1}}} e^{2\pi i a\cdot h} \underbrace{\int_{\substack{x\in\mathcal{G} \\ x\in\mathbb{R}^2}} F(x)\, e^{-2\pi i x\cdot(h-y)}\, dV(x)}_{=F_{\mathcal{G}}^{\wedge}(h-y)}, \quad a, y \in \mathbb{R}^2 \tag{11.2}$$

and second to identify (11.2) with

$$\sum_{\substack{a+g\in\overline{\mathcal{G}} \\ g\in\Lambda}}{}' e^{2\pi i y\cdot(a+g)} F(a+g), \quad a, y \in \mathbb{R}^2 \tag{11.3}$$

in the pointwise sense.

Following the approach presented in W. Freeden [1975], the essential tool for the solution process is the Euler-type summation formula which enables us to guarantee the pointwise convergence of (11.2), at least under certain restrictions and in spherical summation

$$\sum_{h\in\Lambda^{-1}} \ldots = \lim_{N\to\infty} \sum_{\substack{|h-y|\leq N \\ h\in\Lambda^{-1}}} \ldots . \tag{11.4}$$

C. Müller [1954a,c] only gave results in analytic and geometric theory of numbers in the context of a constant weight function (i.e., $F = 1$), and the restriction to circles $\mathcal{G} = \mathbb{B}^2_R, R > 0$, where the convergence of the series (11.2) can be especially realized by estimates known from the theory of Bessel functions (as we have seen in Chapter 10). However, it seems that the convergence of the series (11.2) for "regular regions" $\mathcal{G} \subset \mathbb{R}^2$ and arbitrary twice continuously differentiable functions F, i.e., "general weights" on $\overline{\mathcal{G}}$, can be guaranteed only under the restrictive assumption that asymptotic procedures of the stationary phase (see Section 8.1) and/or asymptotic relations of the metaharmonic theory for $|y| \to \infty$ (see Section 6.3) are applicable to the curve integral

$$-\frac{1}{4\pi^2 y^2}\int_{\partial\mathcal{G}} F(a)\left(\frac{\partial}{\partial\nu_a}e^{-2\pi i a\cdot y}\right)\, dS(a),\ F \in \mathrm{C}^{(2)}\left(\overline{\mathcal{G}}\right). \tag{11.5}$$

In fact, we already know from W. Freeden [1975] that the integral (11.5) is the crucial expression in all asymptotic investigations of the bandlimited function $F^{\wedge}_{\mathcal{G}}$ given by

$$F^{\wedge}_{\mathcal{G}}(y) = \int_{\substack{x\in\mathcal{G} \\ x\in\mathbb{R}^2}} F(x)e^{-2\pi i x\cdot y}\, dV(x). \tag{11.6}$$

This can be easily seen from the second Green formula

$$\begin{aligned} F^{\wedge}_{\mathcal{G}}(y) &= -\frac{1}{4\pi^2 y^2}\int_{\substack{a\in\mathcal{G} \\ a\in\mathbb{R}^2}} e^{-2\pi i a\cdot y}\left(\Delta_a F(a)\right)\, dV(a) \\ &\quad - \frac{1}{4\pi^2 y^2}\int_{\substack{a\in\partial\mathcal{G} \\ a\in\mathbb{R}^2}} F(a)\left(\frac{\partial}{\partial\nu_a}e^{-2\pi i a\cdot y}\right)\, dS(a) \\ &\quad + \frac{1}{4\pi^2 y^2}\int_{\substack{a\in\partial\mathcal{G} \\ a\in\mathbb{R}^2}} \left(\frac{\partial}{\partial\nu_a}F(a)\right)e^{-2\pi i a\cdot y}\, dS(a) \end{aligned} \tag{11.7}$$

for $|y| \to \infty$. Obviously, under the assumption that the weight function F is twice continuously differentiable on $\overline{\mathcal{G}}$, the integrals

$$\frac{1}{4\pi^2 y^2}\int_{\substack{a\in\mathcal{G} \\ a\in\mathbb{R}^2}} e^{-2\pi i a\cdot y}\left(\Delta_a F(a)\right)\, dV(a) \tag{11.8}$$

and

$$\frac{1}{4\pi^2 y^2}\int_{\substack{a\in\partial\mathcal{G} \\ a\in\mathbb{R}^2}} \left(\frac{\partial}{\partial\nu_a}F(a)\right)e^{-2\pi i a\cdot y}\, dS(a) \tag{11.9}$$

are of the order $O(|y|^{-2})$ for $|y| \to \infty$. However, this is generally not the case for the remaining integral on the right side of (11.7), i.e., the curve integral (11.5). A remedy is the stationary phase and/or asymptotic relations of the metaharmonic theory, which indeed guarantee

$$-\frac{1}{4\pi^2 y^2} \int_{\partial\mathcal{G}} F(a) \left(\frac{\partial}{\partial \nu_a} e^{-2\pi i a \cdot y} \right) dS(a) = O(|y|^{-\frac{3}{2}}) \tag{11.10}$$

instead of $O(|y|^{-1})$. However, these techniques, i.e., the stationary phase and/or asymptotic relations of the metaharmonic theory, limit the generality of the geometric situation. So, in the literature, only extensions of the Hardy–Landau identity for a constant weight function (i.e., $F = 1$) to regular (convex) regions around the origin with smooth boundary curves $\partial\mathcal{G}$ possessing a non-vanishing curvature have been studied, yet (see V.K. Ivanow [1962]).

Nonetheless, the overall intention in a number theoretic context is a twofold generalization of the classical 2D Hardy–Landau identity as far as possible in the following sense:

- the generalization from a "circle geometry" to a "general" geometry $\mathcal{G}$.
- the generalization from a constant to a "general", e.g., twice continuously differentiable weight function F.

Unfortunately, the authors of this work have to confess that their lattice point approach does not succeed in both extreme generalizations, and it may be questioned from the methodological point of today's view, whether the Hardy–Landau identity can be extended at all to general geometries such as regular regions $\mathcal{G}$. Anyway, some advances can be made in the approach of this chapter.

A compromise in lattice point summation was already formulated by W. Freeden [1975], where general weights for the circular geometry in spherical summation of the 2D series (11.2) had been realized. In fact, the concept, due to W. Freeden [1975], amounts to the following progress in Hardy–Landau-type lattice point theory:

- the extension to "general" geometries (i.e., regular regions $\mathcal{G}$), but under homogeneous "boundary weights", i.e., $F \in \mathrm{C}^{(2)}(\overline{\mathcal{G}})$ with $F|\partial\mathcal{G} = 0$,
- the extension to "general weights" $F \in \mathrm{C}^{(2)}(\overline{\mathcal{G}})$, but restricted to convex regions $\overline{\mathcal{G}}$ allowing methods from the theory of the stationary phase and/or asymptotic relations of the metaharmonic theory for the boundary curve in (11.5).

In spite of the bundle of diverse calamities, the aforementioned two approaches to bivariate generalizations of Hardy–Landau identities, i.e., Hardy–Landau-type identities, indeed, become attackable. As a matter of fact, reflecting the internal structure of (11.2), (11.3) we are able to give an almost direct proof analogously to the 1D theory for the lattice point theory.

Finally it should be mentioned that the Shannon-type sampling situation, although it will be implied by the bivariate number theoretic Hardy–Landau context as provided in this chapter, is different: Since sampling is obtained by a "smoothing procedure" caused by a 2D-integration of the Hardy–Landau-type identities, it may be expected that more generality for Shannon-type sampling can be realized than it is verifiable for the Hardy–Landau-type lattice point approach. Indeed, the necessary mathematical assumptions in Shannon-type sampling imposed on geometry and weight turn out to be much weaker than in the auxiliary Hardy–Landau-type framework. The details will be presented later on (in Chapter 16).

11.1 Pointwise Fourier Inversion Formula for Regular Regions

Our considerations on weighted Hardy–Landau-type identities use the pointwise *Fourier inversion formula for regular regions* $\mathcal{G} \subset \mathbb{R}^2$. The pointwise Fourier inversion formula plays a particular role in the context of lattice point identities (note that the basic idea of the Fourier inversion formula as presented here is based on a technique already proposed in W. Freeden [1978a]).

Theorem 11.1. *Let $\mathcal{G}$ be a regular region in $\mathbb{R}^2$. Assume that F is a continuously differentiable function in $\overline{\mathcal{G}}$. Then, for $w \in \mathbb{R}^2$, the integral*

$$\int_{\mathbb{R}^2} e^{2\pi i w\cdot x} \int_{\substack{y\in\mathcal{G}\\ y\in\mathbb{R}^2}} F(y)\, e^{-2\pi i x\cdot y}\, dV(y)\, dV(x) \tag{11.11}$$

exists in the spherical sense

$$\int_{\mathbb{R}^2} \cdots = \lim_{N\to\infty} \int_{\substack{|x|\le N\\ x\in\mathbb{R}^2}} \cdots, \tag{11.12}$$

and we have

$$\begin{aligned} &\int_{\mathbb{R}^2} e^{2\pi i w\cdot x} \int_{\substack{y\in\mathcal{G}\\ y\in\mathbb{R}^2}} F(y)\, e^{-2\pi i x\cdot y}\, dV(y)\, dV(x) \\ &= \lim_{N\to\infty} N \int_{\substack{y\in\mathcal{G}\\ y\in\mathbb{R}^2}} F(y) \frac{J_1(2\pi|w-y|N)}{|w-y|}\, dV(y) \\ &= \alpha(w)\, F(w), \end{aligned} \tag{11.13}$$

where, as always, $\alpha(w)$, $w \in \mathbb{R}^2$, more accurately, $\alpha_{\mathcal{G}}(w)$, is the solid angle subtended by $\partial\mathcal{G}$ at w.

Proof. The theory of Bessel functions allows us to write

$$\int_{\substack{|x|\leq N \\ x\in\mathbb{R}^2}} e^{2\pi i(w-y)\cdot z}\, dV(x) = 2\pi \int_0^N J_0(2\pi|y-w|r)\, dr = N\, \frac{J_1(2\pi|y-w|N)}{|y-w|}. \tag{11.14}$$

Using this result we obtain

$$\begin{aligned}\int_{\substack{|x|\leq N \\ x\in\mathbb{R}^2}} e^{2\pi i w\cdot x} \int_{\substack{y\in\mathcal{G} \\ y\in\mathbb{R}^2}} F(y)\, e^{-2\pi i x\cdot y}\, dV(y)\, dV(x) \\ = N \int_{\substack{y\in\mathcal{G} \\ y\in\mathbb{R}^2}} F(y)\, \frac{J_1(2\pi|w-y|N)}{|w-y|}\, dV(y).\end{aligned} \tag{11.15}$$

Observing the identities

$$\nabla_y \ln|w-y| = -\frac{w-y}{|w-y|^2}, \quad y \neq w, \tag{11.16}$$

$$\nabla_y J_0(2\pi|w-y|N) = 2\pi N\, J_1(2\pi|w-y|N)\frac{w-y}{|w-y|}, \quad y \neq w, \tag{11.17}$$

we obtain by simple manipulations for $y \neq w$ and $F \in \mathrm{C}^{(1}(\overline{\mathcal{G}})$

$$\begin{aligned} &N\, F(y)\, \frac{J_1(2\pi|w-y|N)}{|w-y|} \\ &= -\frac{1}{2\pi}\left(\nabla_y \ln|w-y|\right)\cdot\nabla_y\left(F(y)J_0(2\pi|w-y|N)\right) \\ &\quad -\frac{1}{2\pi}J_0(2\pi|w-y|N)\left(\nabla_y F(y)\cdot\frac{w-y}{|w-y|^2}\right).\end{aligned} \tag{11.18}$$

This leads to the identity

$$\begin{aligned} &N\int_{\substack{y\in\mathcal{G} \\ y\in\mathbb{R}^2}} F(y)\frac{J_1(2\pi|w-y|N)}{|w-y|}\, dV(y) \\ &= -\frac{1}{2\pi}\int_{\substack{y\in\mathcal{G} \\ y\in\mathbb{R}^2}} \left(\nabla_y \ln|w-y|\right)\cdot\nabla_y\left(F(y)J_0(2\pi|w-y|N)\right)\, dV(y) \\ &\quad -\frac{1}{2\pi}\int_{\substack{y\in\mathcal{G} \\ y\in\mathbb{R}^2}} J_0(2\pi|w-y|N)\left(\nabla_y F(y)\cdot\frac{w-y}{|w-y|^2}\right)\, dV(y).\end{aligned} \tag{11.19}$$

With the help of the well-known estimate $|J_0(r)| \leq Cr^{-\frac{1}{2}}$, the second integral on the right-hand side of (11.19) can be estimated in the form

$$-\frac{1}{2\pi}\int_{\substack{y\in\mathcal{G} \\ y\in\mathbb{R}^2}} J_0(2\pi|w-y|N)\left(\nabla_y F(y)\right)\cdot\frac{w-y}{|w-y|^2}\, dV(y) = O\left(N^{-\frac{1}{2}}\right) \tag{11.20}$$

for $N \to \infty$. The first integral on the right-hand side of (11.19) can be handled by well-known potential theoretic methods (for example, O.D. Kellogg [1929]). From the first Green theorem, we obtain for $w \notin \overline{\mathcal{G}}$

$$\begin{aligned} & -\frac{1}{2\pi}\int_{y\in\mathcal{G}} (\nabla_y \ln|w-y|)\cdot\nabla_y\left(F(y)J_0(2\pi|w-y|N)\right)\, dV(y) \\ = & -\frac{1}{2\pi}\int_{y\in\partial\mathcal{G}} \left(\frac{\partial}{\partial\nu_y}\ln|w-y|\right) F(y)J_0(2\pi|w-y|N)\, dS(y), \end{aligned} \tag{11.21}$$

whereas for $w \in \overline{\mathcal{G}}$ and sufficiently small $\rho > 0$ we are able to formulate

$$\begin{aligned} & -\frac{1}{2\pi}\int_{\substack{y\in\mathcal{G}\\ |w-y|\geq\rho}} (\nabla_y \ln|w-y|)\cdot\nabla_y\left(F(y)J_0(2\pi|w-y|N)\right)\, dV(y) \\ = & -\frac{1}{2\pi}\int_{\substack{y\in\partial\mathcal{G}\\ |w-y|\geq\rho}} \left(\frac{\partial}{\partial\nu_y}\ln|w-y|\right) F(y)J_0(2\pi|w-y|N)\, dS(y) \\ & -\frac{1}{2\pi}\int_{\substack{|w-y|=\rho\\ y\in\mathcal{G}}} \left(\frac{\partial}{\partial\nu_y}\ln|w-y|\right) F(y)J_0(2\pi|w-y|N)\, dS(y). \end{aligned} \tag{11.22}$$

Letting $\rho \to 0$ we therefore find by use of the standard estimate for Bessel functions that

$$\begin{aligned} -\frac{1}{2\pi}\int_{\substack{y\in\mathcal{G}\\ y\in\mathbb{R}^2}} & (\nabla_y \ln|w-y|)\cdot\nabla_y\left(F(y)J_0(2\pi|w-y|N)\right)\, dV(y) \\ & = \alpha(w)\, F(w) + O\left(N^{-\frac{1}{2}}\right) \end{aligned} \tag{11.23}$$

for $N \to \infty$. Combining (11.20) and (11.23) we are able to conclude from (11.19) that

$$N\int_{\substack{y\in\mathcal{G}\\ y\in\mathbb{R}^2}} F(y)\, \frac{J_1(2\pi|w-y|N)}{|w-y|}\, dV(y) = \alpha(w)F(w) + O\left(N^{-\frac{1}{2}}\right). \tag{11.24}$$

This asymptotic relation shows the desired result. □

11.2 General Geometry and Homogeneous Boundary Weight

Next we are concerned with a first version of a 2D-Hardy–Landau-type identity:

Let $\mathcal{G} \subset \mathbb{R}^2$ be an arbitrarily chosen regular region. We consider the $\mathcal{G}$-*bandlimited function* $F_{\mathcal{G}}^{\wedge}$ defined by

$$F_{\mathcal{G}}^{\wedge}(y) = \int_{\substack{x \in \mathcal{G} \\ x \in \mathbb{R}^2}} F(x)\, e^{-2\pi i y \cdot x}\, dV(x), \quad y \in \mathbb{R}^2, \tag{11.25}$$

where F is assumed to be of class $\mathrm{C}^{(2)}\left(\overline{\mathcal{G}}\right)$. Obviously, $F_{\mathcal{G}}^{\wedge}$ is real analytic, hence, infinitely differentiable in $\mathbb{R}^2$. Therefore, the 2D Euler summation formula with respect to the differential operator $L + \lambda$ (i.e., Corollary 9.8) is applicable to $F_{\mathcal{G}}^{\wedge}$. More concretely, for all positive values N and arbitrary lattices $\Lambda \subset \mathbb{R}^2$, and all parameter vectors $a, y \in \mathbb{R}^2$, the 2D-Euler-type summation formula under the special choice $\lambda = 0$ yields the identity

$$\begin{aligned}
&\sum_{\substack{|g-y| \leq N \\ g \in \Lambda}} e^{2\pi i a \cdot g} F_{\mathcal{G}}^{\wedge}(g-y) \qquad\qquad (11.26)\\
&= \frac{1}{\|\mathcal{F}_\Lambda\|} e^{2\pi i a \cdot y} \int_{\substack{|x| \leq N \\ x \in \mathbb{R}^2}} F_{\mathcal{G}}^{\wedge}(x)\, e^{2\pi i x \cdot a}\, dV(x) \\
&+ \int_{\substack{|x-y| \leq N \\ x \in \mathbb{R}^2}} G(L;x)\, e^{2\pi i a \cdot x}\, (\Delta_x + 4\pi^2 a^2) F_{\mathcal{G}}^{\wedge}(x-y)\, dV(x) \\
&+ \frac{1}{2} \sum_{\substack{|g-y| = N \\ g \in \Lambda}} e^{2\pi i a \cdot g} F_{\mathcal{G}}^{\wedge}(g-y) \\
&- \int_{\substack{|x-y| = N \\ x \in \mathbb{R}^2}} e^{2\pi i a \cdot x} G(L;x) \frac{\partial}{\partial \nu_x} F_{\mathcal{G}}^{\wedge}(x-y)\, dS(x) \\
&+ \int_{\substack{|x-y| = N \\ x \in \mathbb{R}^2}} F_{\mathcal{G}}^{\wedge}(x-y) \left(\frac{\partial}{\partial \nu_x} + 4\pi i a \cdot \nu(x)\right) G(L;x)\, dS(x).
\end{aligned}$$

In the, sequel, we use the identity (11.26) to derive a weighted Hardy–Landau-type identity on (general) $\mathcal{G} \in \mathbb{R}^2$, however, under the restrictive assumption of a homogeneous boundary weight, i.e., $F|\partial\mathcal{G} = 0$. Indeed, under the restriction that $F \in \mathrm{C}^{(2)}(\overline{\mathcal{G}})$ with $F|\partial\mathcal{G} = 0$, the critical curve integral (11.5) vanishes such that the continuity of F and its derivatives $\nabla F, \Delta F$ on $\overline{\mathcal{G}}$ via (11.7) imply the asymptotic result

$$F_{\mathcal{G}}^{\wedge}(y) = O\left(\frac{1}{|y|^2}\right), \quad |y| \to \infty. \tag{11.27}$$

In multi-index nomenclature we find the following asymptotic relations.

Lemma 11.2. *Suppose that $F \in \mathrm{C}^{(2)}\left(\overline{\mathcal{G}}\right)$ with $F|\partial\mathcal{G} = 0$. Then $F_{\mathcal{G}}^{\wedge}$ is infinitely often differentiable in $\mathbb{R}^2$. For all multi-indices $\alpha = (\alpha_1, \alpha_2)^T$ with $\alpha_i \in \mathbb{N}_0,\ i = 1, 2$, we have*

$$(\nabla_y)^\alpha F_{\mathcal{G}}^{\wedge}(y) = O\left(\frac{1}{|y|^2}\right), \quad |y| \to \infty. \tag{11.28}$$

In accordance with the arguments of C.F. Gauss [1826] we obtain from Lemma 11.2 the asymptotic relation

$$\frac{1}{2} \sum_{\substack{|g|=N \\ g\in\Lambda}} F_{\mathcal{G}}^{\wedge}(g) = O\left(N^{-2} \sum_{\substack{|g|=N \\ g\in\Lambda}} 1 \right) = O\left(N^{-1}\right), \quad N \to \infty. \tag{11.29}$$

The already-known 2D-integral estimates (cf. Theorem 10.1) tell us that

$$\int_{\substack{|x-y|=N \\ x\in\mathbb{R}^2}} |G(L;x)| \; dS(x) \;=\; O(N) \tag{11.30}$$

and

$$\int_{\substack{|x-y|=N \\ x\in\mathbb{R}^2}} \left| \frac{\partial}{\partial \nu} G(L;x) \right| dS(x) \;=\; O(N) \tag{11.31}$$

for $N \to \infty$. Hence, in connection with Lemma 11.2 and (11.29), the relations (11.30) and (11.31) show that

$$\int_{\substack{|x-y|=N \\ x\in\mathbb{R}^2}} e^{2\pi i a\cdot x} G(L;x) \frac{\partial}{\partial \nu_x} F_{\mathcal{G}}^{\wedge}(x-y) \; dS(x) = O(N^{-1}), \quad N \to \infty, \tag{11.32}$$

and

$$\int_{\substack{|x-y|=N \\ x\in\mathbb{R}^2}} F_{\mathcal{G}}^{\wedge}(x-y) \left(\frac{\partial}{\partial \nu_x} + 4\pi i a \cdot \nu(x) \right) G(L;x) \; dS(x) = O(N^{-1}), \; N \to \infty. \tag{11.33}$$

From the summation formula (11.26), in connection with (11.29), (11.32), and (11.33), we therefore see that

$$\begin{aligned} & \sum_{\substack{|g-y|\le N \\ g\in\Lambda}} e^{2\pi i a\cdot g} F_{\mathcal{G}}^{\wedge}(g-y) \qquad\qquad (11.34) \\ = & \; \frac{1}{\|\mathcal{F}_\Lambda\|} \, e^{2\pi i a\cdot y} \int_{\substack{|x|\le N \\ x\in\mathbb{R}^2}} F_{\mathcal{G}}^{\wedge}(x) \, e^{2\pi i x\cdot a} \; dV(x) \\ + & \int_{\substack{|x-y|\le N \\ x\in\mathbb{R}^2}} G(L;x) \, e^{2\pi i a\cdot x} (\Delta_x + 4\pi^2 a^2) F_{\mathcal{G}}^{\wedge}(x-y) \; dV(x) \\ + & \; O(N^{-1}), \quad N \to \infty. \end{aligned}$$

It is not difficult to conclude from (11.28) by application of the second Green formula that

$$\begin{aligned} &\int_{\substack{|x-y|\le N\\ x\in\mathbb{R}^2}} G(L;x)\, e^{2\pi i a\cdot x}(\Delta_x + 4\pi^2 a^2) F_{\mathcal{G}}^{\wedge}(x-y)\, dV(y) \\ &= \int_{\substack{|x-y|\le N\\ y\in\mathbb{R}^2}} G(L^2;x)\, e^{2\pi i a\cdot x}(\Delta_x + 4\pi^2 a^2)^2 F_{\mathcal{G}}^{\wedge}(x-y)\, dV(y) \\ &+ \ o(1), \qquad N\to\infty, \end{aligned} \tag{11.35}$$

where

$$G(L^2;x) = \frac{1}{\|\mathcal{F}_\Lambda\|} \sum_{\substack{h\neq 0\\ h\in\Lambda^{-1}}} \frac{e^{2\pi i h\cdot x}}{(4\pi^2(a^2-(a+h)^2))^2} \tag{11.36}$$

and the series on the right hand side of (11.36) associated to the iterated operator $L^2 = L\ L$ is absolutely and uniformly convergent in $\mathbb{R}^2$, so that $G(L^2;\cdot)$ is continuous in $\mathbb{R}^2$. By virtue of Lebesgue's theorem of dominated convergence we obtain

$$\begin{aligned} &\int_{\substack{|x-y|\le N\\ y\in\mathbb{R}^2}} G(L^2;x)\, e^{2\pi i a\cdot x}\ (\Delta_x + 4\pi^2 a^2) F_{\mathcal{G}}^{\wedge}(x-y)\, dV(y) \\ &= \frac{1}{\|\mathcal{F}_\Lambda\|} \sum_{\substack{h\neq 0\\ h\in\Lambda^{-1}}} \frac{1}{(4\pi^2(h+a)^2-a^2)^2} \\ &\qquad\qquad \times \int_{\substack{|x-y|\le N\\ y\in\mathbb{R}^2}} e^{2\pi i x\cdot(a+h)}(\Delta_x + 4\pi^2 a^2)^2 F_{\mathcal{G}}^{\wedge}(x-y)\, dV(y). \end{aligned} \tag{11.37}$$

Now we are able to apply the Fourier inversion formula (cf. Theorem 11.1). More specifically, we have for all $h\in\Lambda^{-1}$ with $h\neq 0$

$$\begin{aligned} &\int_{\substack{|x-y|\le N\\ y\in\mathbb{R}^2}} e^{2\pi i x\cdot(a+h)}(\Delta_x + 4\pi^2 a^2)^2 F_{\mathcal{G}}^{\wedge}(x-y)\, dV(y) \\ &= \int_{\substack{|x-y|\le N\\ y\in\mathbb{R}^2}} e^{2\pi i x\cdot(a+h)} \int_{\mathcal{G}} (4\pi^2 a^2 - 4\pi^2 z^2)^2 F(z) e^{-2\pi i z\cdot(x-y)}\, dV(z)\, dV(x) \\ &= e^{2\pi i y\cdot(a+h)} \int\limits_{\substack{|x|\le N\\ x\in\mathbb{R}^2}} e^{2\pi i x\cdot(a+h)} \int\limits_{\mathcal{G}} \tilde{F}(z) e^{-2\pi i z\cdot x} dV(z)\, dV(x), \end{aligned} \tag{11.38}$$

where, for brevity, we use the auxiliary function $\tilde{F}:\overline{\mathcal{G}}\to\mathbb{R}$ given by

$$\tilde{F}(z) = (4\pi a^2 - 4\pi^2 z^2)^2 F(z),\ z\in\overline{\mathcal{G}}. \tag{11.39}$$

We are allowed to deduce from the Fourier inversion formula (Theorem 11.1) that, for every $\varepsilon > 0$, there exists an integer $M(=M(\varepsilon))$ such that, for all $N\geq M$,

$$\left| \int_{\substack{|x|\le N\\ x\in\mathbb{R}^2}} e^{2\pi i\ x\cdot(a+h)} \left(\int_{\substack{z\in\mathcal{G}\\ z\in\mathbb{R}^2}} \tilde{F}(y)\ e^{-2\pi i x\cdot z}\ dV(z) \right) dV(x) \right| \le \varepsilon \tag{11.40}$$

holds uniformly for all $h \in \Lambda^{-1}$ with $|h| \geq R_0$, where R_0 is chosen in such a way that $R_0 > \sup_{x\in\overline{\mathcal{G}}} |x|$. Therefore, for all $N \geq M$, we have

$$\begin{aligned} &\left| \sum_{\substack{|h|\geq R_0\\ h\in\Lambda^{-1}}} \frac{1}{4\pi^2(a^2-(a+h)^2)} \int_{\substack{|x|\leq N\\ x\in\mathbb{R}^2}} e^{2\pi i x\cdot(a+h)} (\Delta_x + 4\pi^2 a^2)^2 F_{\mathcal{G}}^{\wedge}(x)\, dV(x) \right| \\ &\qquad \leq \varepsilon\, C \sum_{\substack{|h|\geq R_0\\ h\in\Lambda^{-1}}} \frac{1}{4\pi^2(a^2-(a+h)^2)} \qquad (11.41)\\ &\qquad \leq \varepsilon\, C'. \end{aligned}$$

As a consequence, for $N \to \infty$, we are able to conclude that

$$\begin{aligned} &\sum_{\substack{|g-y|\leq N\\ g\in\Lambda}} e^{2\pi i a\cdot g} \int_{\mathcal{G}} F(x)\, e^{-2\pi i x\cdot(g-y)}\, dV(x) \qquad (11.42)\\ &= \frac{1}{\|\mathcal{F}_\Lambda\|} e^{2\pi i a\cdot y} \int_{\substack{|x|\leq N\\ x\in\mathbb{R}^2}} e^{2\pi i x\cdot a} \int_{\mathcal{G}} F(z)\, e^{-2\pi i z\cdot x}\, dV(z)\, dV(x)\\ &\quad + \frac{1}{\|\mathcal{F}_\Lambda\|} \sum_{\substack{0<|h|\leq R_0\\ h\in\Lambda^{-1}}} \frac{e^{2\pi i y\cdot(a+h)}}{(4\pi^2(a^2-(a+h)^2))^2}\\ &\quad \times \int_{\substack{|x|\leq N\\ x\in\mathbb{R}^2}} e^{2\pi i x\cdot(a+h)} \int_{\mathcal{G}} (4\pi^2 a^2 - 4\pi^2 z^2)^2 F(z) e^{-2\pi i z\cdot x}\, dV(z)\, dV(x)\\ &\quad + o(1), \qquad N\to\infty. \end{aligned}$$

Finally, the application of the inversion formula (i.e., Theorem 11.1) to the remaining finite set of integrals, i.e., the execution of the limit $N \to \infty$, guarantees the following lattice point identity.

Theorem 11.3. *Let Λ be an arbitrary lattice in $\mathbb{R}^2$. Let $\mathcal{G} \subset \mathbb{R}^2$ be an arbitrary regular region. Suppose that $F \in \mathrm{C}^{(2)}\left(\overline{\mathcal{G}}\right)$ satisfies the homogeneous boundary condition $F|\partial\mathcal{G} = 0$. Then, for all $a, y \in \mathbb{R}^2$, the series*

$$\sum_{g\in\Lambda} e^{2\pi i a\cdot g} \int_{\mathcal{G}} F(x)\, e^{-2\pi i x\cdot(g-y)}\, dV(x) \qquad (11.43)$$

exists in spherical sense

$$\sum_{g\in\Lambda} \cdots = \lim_{N\to\infty} \sum_{\substack{|g-y|\leq N\\ g\in\Lambda}} \cdots . \qquad (11.44)$$

Summarizing our considerations we are led to the following conclusion:

Corollary 11.4. *Let Λ be a lattice in $\mathbb{R}^2$. Suppose that $F \in C^{(2)}\left(\overline{\mathcal{G}}\right)$ satisfies the homogeneous boundary condition $F|\partial\mathcal{G} = 0$. Then we have*

$$\lim_{N\to\infty} \sum_{\substack{|g-y|\leq N\\ g\in\Lambda}} e^{2\pi i a\cdot(g-y)}\, F^{\wedge}_{\mathcal{G}}(g-y) \tag{11.45}$$

$$= \frac{1}{\|\mathcal{F}_\Lambda\|} \sum_{h\in\Lambda^{-1}} e^{2\pi i h\cdot y} \int_{\mathbb{R}^2} F^{\wedge}_{\mathcal{G}}(x)\, e^{2\pi i x\cdot(a+h)}\, dV(x),$$

i.e.,

$$\frac{1}{\|\mathcal{F}_\Lambda\|} \sum_{\substack{a+h\in\overline{\mathcal{G}}\\ h\in\Lambda^{-1}}}{}' e^{2\pi i y\cdot(a+h)} F(a+h) \tag{11.46}$$

$$= \sum_{g\in\Lambda} e^{2\pi i a\cdot g} \int_{\mathcal{G}} F(x) e^{-2\pi i x\cdot(g-y)} dV(x).$$

Replacing $\Lambda \subset \mathbb{R}^2$ by its inverse lattice $\Lambda^{-1} \subset \mathbb{R}^2$ we obtain the following extended version of the 2D Poisson summation formula on a regular region $\mathcal{G}$.

Corollary 11.5. *(Hardy–Landau-type identity on a regular region $\mathcal{G}$ with vanishing weight on the boundary $\partial\mathcal{G}$). Let Λ be a lattice in $\mathbb{R}^2$. Let $\mathcal{G} \subset \mathbb{R}^2$ be an arbitrary regular region. Suppose that F is twice continuously differentiable on $\overline{\mathcal{G}}$ with $F|\partial\mathcal{G} = 0$. Then, for $a, y \in \mathbb{R}^2$, we have*

$$\sum_{\substack{a+g\in\overline{\mathcal{G}}\\ g\in\Lambda}}{}' e^{2\pi i y\cdot(a+g)} F(a+g) \tag{11.47}$$

$$= \lim_{N\to\infty} \frac{1}{\|\mathcal{F}_\Lambda\|} \sum_{\substack{|h-y|\leq N\\ h\in\Lambda^{-1}}} e^{2\pi i a\cdot h} \int_{\mathcal{G}} F(x) e^{-2\pi i x\cdot(h-y)}\, dV(y).$$

11.3 Circles and General Weights

Next we discuss weighted lattice point sums in circles, where the weights are generated by an arbitrary twice continuously differentiable function $F : \overline{\mathbb{B}^2_R} \to \mathbb{R}$, $R > 0$. In order to verify a weighted version of the Hardy–Landau-type identity on circles $\mathbb{B}^2_R$ (cf. W. Freeden [1975]), we again base our activities on the 2D Euler-type summation formula for general lattices $\Lambda \subset \mathbb{R}^2$, but now on the bandlimited function $F^{\wedge}_{\mathbb{B}^2_R}$ given by

$$F^{\wedge}_{\mathbb{B}^2_R}(y) = \int_{\substack{|a|\leq R\\ a\in\mathbb{R}^2}} F(a)\, e^{-2\pi i a\cdot y}\, dV(a), \quad y \in \mathbb{R}^2, \tag{11.48}$$

where F is assumed to be of class $C^{(2)}(\overline{\mathbb{B}^2_R})$, i.e., we now especially choose $\overline{\mathcal{G}} = \overline{\mathbb{B}^2_R}$.

The 2D Euler-type summation formula (cf. Corollary 9.8) yields for every $N > 0$, $\lambda \in \mathbb{R}$, and all $a, y \in \mathbb{R}^2$

$$\sum_{\substack{|g-y|\leq N \\ g\in\Lambda}} e^{2\pi i a\cdot g}\, F^{\wedge}_{\mathbb{B}^2_R}(x-y) \tag{11.49}$$

$$= \frac{1}{\|\mathcal{F}_\Lambda\|}\; e^{2\pi i a\cdot y} \sum_{\substack{4\pi^2((h+a)^2-a^2)=\lambda \\ h\in\Lambda^{-1}}} e^{2\pi i\, h\cdot y} \int_{\substack{|x|\leq N \\ x\in\mathbb{R}^2}} F^{\wedge}_{\mathbb{B}^2_R}(x) e^{2\pi i\, x\cdot(a+h)} dV(x)$$

$$+ \int_{\substack{|x-y|\leq N \\ x\in\mathbb{R}^q}} G(L+\lambda;x)\, e^{2\pi i a\cdot x} \left(\Delta_x + \lambda + 4\pi^2 a^2\right) F^{\wedge}_{\mathbb{B}^2_R}(x-y)\, dV(x)$$

$$+ \frac{1}{2} \sum_{\substack{|g-y|=N \\ g\in\Lambda}} e^{2\pi i a\cdot g}\, F^{\wedge}_{\mathbb{B}^2_R}(g-y)$$

$$- \int_{\substack{|x-y|=N \\ x\in\mathbb{R}^q}} e^{2\pi i a\cdot x} G(L+\lambda;x) \frac{\partial}{\partial\nu_x} F^{\wedge}_{\mathbb{B}^2_R}(x-y)\, dS(x)$$

$$+ \int_{\substack{|x-y|=N \\ x\in\mathbb{R}^q}} e^{2\pi i a\cdot x} F^{\wedge}_{\mathbb{B}^2_R}(x-y) \left(\frac{\partial}{\partial\nu_x} + 4\pi i a\cdot\nu(x)\right) G(L+\lambda;x)\, dS(x).$$

From the second Green theorem (cf. (11.7)) we obtain for all $y \in \mathbb{R}^2\backslash\{0\}$ by observation of the continuity of the first- and second-order derivative of F on $\overline{\mathbb{B}^2_R}$

$$\begin{aligned} F^{\wedge}_{\mathbb{B}^2_R}(y) &= \frac{i}{2\pi y^2} \int_{\substack{|a|=R \\ a\in\mathbb{R}^2}} F(a)\, (\nu(a)\cdot y)\, e^{-2\pi i y\cdot a}\, dS(a) \\ &\quad + O\left(\frac{1}{|y|^2}\right). \end{aligned} \tag{11.50}$$

The critical integral on the right side of (11.50) can be formulated using polar coordinates $a = R\eta$, $y = |y|\xi$ as an integral over the unit circle $\mathbb{S}^1$ in $\mathbb{R}^2$

$$\begin{aligned} &\int_{\substack{|a|=R \\ a\in\mathbb{R}^2}} F(a)\, (\nu(a)\cdot y)\, e^{-2\pi i y\cdot a}\, dS(a) \\ &= R|y| \int_{\eta\in\mathbb{S}^1} F(R\eta)\, (\eta\cdot\xi)\, e^{-2\pi i R|y|(\xi\cdot\eta)}\, dS(\eta). \end{aligned} \tag{11.51}$$

The right side defines an entire solution of the Helmholtz equation (in the variable $y \in \mathbb{R}^2$). Hence, the metaharmonic theory (i.e., Corollary 6.40) and/or the method of stationary phase (see Corollary 8.10) provides an asymptotic estimate of the same order, namely

$$\int_{\eta\in\mathbb{S}^1} F(R\eta)\, (\eta\cdot\xi)\, e^{-2\pi i R|y|(\xi\cdot\eta)}\, dS(\eta) \;=\; O\left(|y|^{-\frac{1}{2}}\right),\; |y|\to\infty. \tag{11.52}$$

In connection with (11.50) and (11.51), we therefore obtain from (11.52)

$$F^{\wedge}_{\mathbb{B}^2_R}(y) = O\left(|y|^{-\frac{3}{2}}\right),\; |y|\to\infty. \tag{11.53}$$

Lemma 11.6. *For all multi-indices $\alpha = (\alpha_1, \alpha_2)^T$ with $\alpha_i \in \mathbb{N}_0$, $i = 1, 2$, we have*

$$(\nabla_y)^\alpha F^\wedge_{\mathbb{B}^2_R}(y) \; = \; O\left(|y|^{-\frac{3}{2}}\right), \; |y| \to \infty. \tag{11.54}$$

In connection with the already-mentioned result of C.F. Gauss [1826], it thus follows that

$$\frac{1}{2} \sum_{\substack{|g|=N \\ g\in\Lambda}} F^\wedge_{\mathbb{B}^2_R}(g) = O\left(N^{-\frac{3}{2}} \sum_{\substack{|g|=N \\ g\in\Lambda}} 1 \right) = O\left(N^{-\frac{1}{2}}\right), \quad N \to \infty. \tag{11.55}$$

The relations (11.30) and (11.31), in connection with Lemma 11.6, show that

$$\int_{\substack{|x-y|=N \\ x\in\mathbb{R}^2}} e^{2\pi i a\cdot x} G(L;x) \frac{\partial}{\partial \nu_x} F^\wedge_{\mathbb{B}^2_R}(x-y)\, dS(x) = O(N^{-\frac{1}{2}}), \tag{11.56}$$

and

$$\int_{\substack{|x-y|=N \\ x\in\mathbb{R}^2}} e^{2\pi i a\cdot x} F^\wedge_{\mathbb{B}^2_R}(x-y) \left(\frac{\partial}{\partial \nu_x} + 4\pi i a \cdot \nu(x) \right) G(L;x)\, dS(x) = O(N^{-\frac{1}{2}}), \tag{11.57}$$

for $N \to \infty$.

Altogether, for every $\lambda \in \mathbb{R}$, and all $a, y \in \mathbb{R}^2$, it follows from (11.49) in connection with (11.55), (11.56), and (11.57) that

$$\begin{aligned} &\sum_{\substack{|g-y|\leq N \\ g\in\Lambda}} e^{2\pi i a\cdot g}\, F^\wedge_{\mathbb{B}^2_R}(x-y) \qquad (11.58) \\ &= \frac{1}{\|\mathcal{F}_\Lambda\|}\, e^{2\pi i a\cdot y} \sum_{\substack{4\pi^2((h+a)^2-a^2)=\lambda \\ h\in\Lambda^{-1}}} e^{2\pi i h\cdot y} \int_{\substack{|x|\leq N \\ x\in\mathbb{R}^2}} F^\wedge_{\mathbb{B}^2_R}(x)\; e^{2\pi i x\cdot(a+h)} dV(x) \\ &+ \int_{\substack{|x-y|\leq N \\ x\in\mathbb{R}^q}} G(L+\lambda;x)\, e^{2\pi i a\cdot x} \left(\Delta_x + \lambda + 4\pi^2 a^2\right) F^\wedge_{\mathbb{B}^2_R}(x-y)\, dV(x) \\ &+ O(N^{-\frac{1}{2}}), \quad N \to \infty. \end{aligned}$$

Choosing especially $\lambda = 4\pi^2 \left(R^2 - a^2\right)$ we are able to conclude by virtue of the Green theorem that

$$\left(\Delta_x + 4\pi^2 R^2\right)\, F^\wedge_{\mathbb{B}^2_R}(x-y) \; = \; O\left(|x-y|^{-\frac{5}{2}}\right). \tag{11.59}$$

Therefore, the absolute convergence of the integral

$$\int_{\mathbb{R}^q} G(L + 4\pi^2 \left(R^2 - a^2\right); x)\, e^{2\pi i a\cdot x} \left(\Delta_x + 4\pi^2 R^2\right) F^\wedge_{\mathbb{B}^2_R}(x-y)\, dV(x) \tag{11.60}$$

can be guaranteed, and we have (cf. (11.35))

$$\int_{\substack{|x-y|\le N\\ x\in\mathbb{R}^q}} G\left(L+4\pi^2\left(R^2-a^2\right);x\right)\ e^{2\pi i a\cdot x}\left(\Delta_x+4\pi^2R^2\right)F^{\wedge}_{\mathbb{B}^2_R}(x-y)\,dV(x)$$
$$= \frac{1}{\|\mathcal{F}_\Lambda\|}\sum_{\substack{|h+a|\neq R\\ h\in\Lambda^{-1}}}\frac{1}{(4\pi^2(R^2-(h+a)^2))^2} \tag{11.61}$$
$$\times\int_{\substack{|x-y|\le N\\ y\in\mathbb{R}^2}} e^{2\pi i x\cdot(a+h)}(\Delta_x+4\pi^2R^2)^2F^{\wedge}_{\mathbb{B}^2_R}(x-y)\,dV(y).$$

It can be readily seen that, for all $h\in\Lambda^{-1}$ with $|h+a|\neq R$,

$$\int_{\substack{|x-y|\le N\\ y\in\mathbb{R}^2}} e^{2\pi i x\cdot(a+h)}(\Delta_x+4\pi^2R^2)^2F^{\wedge}_{\mathbb{B}^2_R}(x-y)\,dV(y) \tag{11.62}$$
$$= \int_{\substack{|x-y|\le N\\ y\in\mathbb{R}^2}} e^{2\pi i x\cdot(a+h)}\int_{\substack{|x|\le R\\ \xi\in\mathbb{R}^2}}(4\pi^2R^2-4\pi^2z^2)^2F(z)e^{-2\pi i z\cdot(x-y)}\,dV(z)\,dV(x)$$
$$= e^{2\pi i y\cdot(a+h)}\int_{\substack{|x|\le N\\ x\in\mathbb{R}^2}} e^{2\pi i x\cdot(a+h)}\int_{\substack{|x|\le R\\ \xi\in\mathbb{R}^2}}(4\pi^2R^2-4\pi^2z^2)^2F(z)e^{-2\pi i z\cdot x}dV(z)\,dV(x).$$

We introduce an auxiliary function, namely $\tilde{F}:\overline{\mathbb{B}^2_R}\to\mathbb{R}$ given by

$$\tilde{F}(z)=(4\pi R^2-4\pi^2z^2)^2F(z),\quad z\in\overline{\mathbb{B}^2_R}. \tag{11.63}$$

The Fourier inversion formula (Theorem 11.1) tells us that, for every $\varepsilon>0$, there exists an integer $M(=M(\varepsilon))$ such that, for all $N\ge M$,

$$\left|\int_{\substack{|x|\le N\\ x\in\mathbb{R}^2}} e^{2\pi i\ x\cdot(a+h)}\left(\int_{\substack{|z|\le R\\ z\in\mathbb{R}^2}}\tilde{F}(y)\ e^{-2\pi i x\cdot z}\ dV(z)\right)\ dV(x)\right|\le\varepsilon \tag{11.64}$$

holds uniformly for all $h\in\Lambda^{-1}$ with $|a+h|\ge R_0>R$. Therefore, for all $N\ge M$, we have

$$\left|\sum_{\substack{|h+a|\ge R_0\\ h\in\Lambda^{-1}}}\frac{1}{(4\pi^2(R^2-(h+a)^2))^2}e^{2\pi i y\cdot(a+h)}\right. \tag{11.65}$$
$$\left.\times\int_{\substack{|x-y|\le N\\ y\in\mathbb{R}^2}} e^{2\pi i\ x\cdot(a+h)}\left(\int_{\substack{|z|\le R\\ z\in\mathbb{R}^2}}\tilde{F}(y)\ e^{-2\pi i x\cdot z}\ dV(z)\right)\ dV(x)\right|$$
$$\le\ \varepsilon\,C\sum_{\substack{|h+a|\ge R_0\\ h\in\Lambda^{-1}}}\frac{1}{4\pi^2(a^2-(a+h)^2)}$$
$$\le\ \varepsilon\,C'.$$

In other words, for $N \to \infty$, we are able to show that

$$
\begin{aligned}
&\sum_{\substack{|g-y|\le N\\ g\in\Lambda}} e^{2\pi i a\cdot g}\ F^{\wedge}_{\mathbb{B}^2_R}(x-y) \qquad (11.66)\\
&= \frac{1}{\|\mathcal{F}_\Lambda\|} \sum_{\substack{|h+a|=R\\ h\in\Lambda^{-1}}} e^{2\pi i(a+h)\cdot y} \int_{\substack{|x|\le N\\ x\in\mathbb{R}^2}} F^{\wedge}_{\mathbb{B}^2_R}(x) e^{2\pi i x\cdot(a+h)} dV(x)\\
&\quad + \frac{1}{\|\mathcal{F}_\Lambda\|} \sum_{\substack{|h+a|\neq R\\ |h+a|\le R_0, h\in\Lambda^{-1}}} \frac{e^{2\pi i y\cdot(a+h)}}{(4\pi^2(R^2-(a+h)^2))^2}\\
&\quad \times \int_{\substack{|x|\le N\\ x\in\mathbb{R}^2}} e^{2\pi i x\cdot(a+h)} \int_{\mathcal{G}} (4\pi^2R^2 - 4\pi^2 z^2)^2 F(z) e^{-2\pi i z\cdot x}\, dV(z)\, dV(x)\\
&\quad + o(1), \quad N\to\infty.
\end{aligned}
$$

Integrating by parts in the already-known way we are led to the following statement.

Theorem 11.7. *(Hardy–Landau-type identity on circles $\mathbb{B}^2_R$, $R>0$, and general weight functions). Let Λ be an arbitrary lattice in $\mathbb{R}^2$. Then, for every function $F \in \mathrm{C}^{(2)}(\overline{\mathbb{B}^2_R})$, $R>0$, and all $a,y\in\mathbb{R}^2$, the series*

$$
\sum_{g\in\Lambda} e^{2\pi i a\cdot g} \int_{\substack{|x|\le R\\ x\in\mathbb{R}^2}} F(x) e^{-2\pi i x\cdot(g-y)}\, dV(x) \qquad (11.67)
$$

converges in the spherical sense.

Remark 11.8. *The function F is <u>not</u> assumed to be of class $\mathrm{C}^{(4)}(\overline{\mathbb{B}^2_R})$ (as in W. Freeden [1975]), instead by use of the Fourier inversion formula within the Euler summation process we are able to prove Theorem 11.7 under the weaker assumption that F is twice continuously differentiable in $\overline{\mathbb{B}^2_R}$.*

Summarizing our considerations we come to the following conclusion:

Corollary 11.9. *Let Λ be an arbitrary lattice in $\mathbb{R}^2$. Suppose that F is twice continuously differentiable in $\overline{\mathbb{B}^2_R}$, $R>0$. Then we have*

$$
\begin{aligned}
&\lim_{N\to\infty} \sum_{\substack{|g-y|\le N\\ g\in\Lambda}} e^{2\pi i a\cdot(g-y)}\ F^{\wedge}_{\mathbb{B}^2_R}(g-y) \qquad (11.68)\\
&= \frac{1}{\|\mathcal{F}_\Lambda\|} \sum_{h\in\Lambda^{-1}} e^{2\pi i h\cdot y} \int_{\mathbb{R}^2} F^{\wedge}_{\mathbb{B}^2_R}(x)\ e^{2\pi i x\cdot(a+h)}\ dV(x).
\end{aligned}
$$

By applying the Fourier inversion formula (i.e., Theorem 11.1) we obtain

$$\frac{1}{\|\mathcal{F}_\Lambda\|} \sum_{\substack{|a+h|\leq R \\ g\in\Lambda}}{}' \; e^{2\pi i y\cdot(a+h)} F(a+h) \tag{11.69}$$

$$= \lim_{N\to\infty} \sum_{\substack{|g-y|\leq N \\ h\in\Lambda^{-1}}} e^{2\pi i a\cdot g} \int_{\substack{|x|\leq R \\ x\in\mathbb{R}^2}} F(x)\, e^{-2\pi i x\cdot(g-y)}\, dV(y).$$

Replacing $\Lambda \subset \mathbb{R}^2$ by $\Lambda^{-1} \subset \mathbb{R}^2$, we finally arrive at obtain the following extension of the Hardy–Landau identity.

Corollary 11.10. *Let Λ be an arbitrary lattice in $\mathbb{R}^2$. Suppose that F is twice continuously differentiable in $\overline{\mathbb{B}^2_R}$, $R > 0$. Then, for all $a, y \in \mathbb{R}^2$,*

$$\sum_{\substack{|a+g|\leq R \\ g\in\Lambda}}{}' \; e^{2\pi i y\cdot(a+g)} F(a+g) \tag{11.70}$$

$$= \frac{1}{\|\mathcal{F}_\Lambda\|} \lim_{N\to\infty} \sum_{\substack{|h-y|\leq N \\ h\in\Lambda^{-1}}} e^{2\pi i a\cdot h} \int_{\substack{|x|\leq R \\ x\in\mathbb{R}^2}} F(x)\, e^{-2\pi i x\cdot(h-y)}\, dV(y).$$

Choosing $F = 1$ we are confronted with the generalization of the famous *Hardy–Landau identity on circular disks* attributed to C. Müller [1954c]

$$e^{2\pi i a\cdot y} \sum_{\substack{|g+a|\leq R \\ g\in\Lambda}}{}' \; e^{2\pi i g\cdot y} = \lim_{N\to\infty} \frac{\pi R^2}{\|\mathcal{F}_\Lambda\|} \sum_{\substack{|h-y|\leq N \\ h\in\Lambda^{-1}}} e^{2\pi i a\cdot h} \frac{J_1(2\pi|h-y|R)}{\pi|h-y|R}. \tag{11.71}$$

Once again, for $a = y = 0$ we get the classical Hardy–Landau identity (see G.H. Hardy [1915], E. Landau [1915], G.H. Hardy, E. Landau [1924]) for arbitrary lattices Λ

$$\sum_{\substack{|g|\leq R \\ g\in\Lambda}}{}' \; 1 = \lim_{N\to\infty} \frac{\pi R^2}{\|\mathcal{F}_\Lambda\|} \sum_{\substack{|h|\leq N \\ h\in\Lambda^{-1}}} \frac{J_1(2\pi|h|R)}{\pi|h|R}, \tag{11.72}$$

that expresses the number of lattice points inside a circle around the origin with radius R in terms of a 2D alternating series involving the Bessel function J_1.

11.4 Smooth Convex Regions and General Weights

We deal with weighted lattice point sums in convex regions $\mathcal{G}$ including the origin and possessing a smooth boundary curve $\partial\mathcal{G}$ with non-vanishing curvature (see V.K. Ivanow [1962] for the case of constant weight). As we saw, the validity of a weighted Hardy–Landau-type identity essentially depends on the asymptotic expansion

$$F_{\mathcal{G}}^{\wedge}(y) = \frac{i}{2\pi y^2} \int_{\partial\mathcal{G}} F(a)(\nu(a)\cdot y)\ e^{-2\pi i a\cdot y}\ dS(a) + O\left(\frac{1}{|y|^2}\right), \tag{11.73}$$

for $|y| \to \infty$. In terms of polar coordinates $y = r\xi$, $r = |y|$, $\xi \in \mathbb{S}^1$, the boundary integral on the right of (11.73) can be rewritten in the form

$$\begin{aligned} &\int_{\partial\mathcal{G}} F(a)(\nu(a)\cdot y)\ e^{-2\pi i a\cdot y}\ dS(a) \qquad (11.74)\\ &\quad = |y| \int_{\partial\mathcal{G}} F(a)\cos(\sphericalangle(\nu(a),\xi))\ e^{-2\pi i|y|\ |a|\cos\left(\sphericalangle\ \left(\xi,\frac{a}{|a|}\right)\right)}\ dS(a), \end{aligned}$$

where $\vartheta = \sphericalangle(\nu(a),\xi), a = (a_1,a_2)^T$, is the angle between the vectors $\nu(a)$ and ξ. We fix the unit vector $\xi = (\cos\varphi, \sin\varphi)^T$, $0 \le \varphi < 2\pi$, in the (a_1,a_2)–plane. Then we introduce a new (w_1,w_2)-coordinate system by rotating the (a_1,a_2)–plane in the counterclockwise sense with the angle φ. In doing so we are able to describe the curve $\partial\mathcal{G}$ in a parametric way by choosing as parameter the angle $\vartheta = \sphericalangle(\nu(a),\xi)$, i.e., the angle between the normal ν and the w_1-axis at the point y under consideration. Under these circumstances we have $dS = \cos(\vartheta)\ C(\vartheta)\ d\vartheta$, where $C(\vartheta)$ is the curvature radius of $\partial\mathcal{G}$ and $y\cdot a = |y|w_1(\vartheta)$. Consequently, the integral (11.74) can be rewritten in the form

$$\int_0^{2\pi} \tilde{F}(\vartheta)\ e^{-2\pi i|y|w_1(\vartheta)}\cos^2(\vartheta)\ C(\vartheta)\ d\vartheta, \tag{11.75}$$

where

$$\tilde{F}(\vartheta) = F\left(w_1(\vartheta), w_2(\vartheta)\right),\ \vartheta \in [0,2\pi). \tag{11.76}$$

As before we introduce auxiliary functions $\tilde{F}, \tilde{G}$ by

$$\tilde{F}(\vartheta) = \tilde{F}(\vartheta)\cos^2(\vartheta)C(\vartheta), \quad \tilde{G}(\vartheta) = w_1(\vartheta). \tag{11.77}$$

The derivatives of $\tilde{G}$ are important for the application of the theory of the stationary phase. We readily find for the derivative $\tilde{G}'$

$$\tilde{G}'(\vartheta) = \frac{dw_1}{dS}\frac{dS}{d\vartheta} = -\ C(\vartheta)\sin(\vartheta), \tag{11.78}$$

hence, $\tilde{G}'(\vartheta) = 0$ for $\vartheta = 0, \pi, 2\pi$. The second derivative is given by

$$\tilde{G}''(\vartheta) = -C'(\vartheta)\sin\vartheta - C(\vartheta)\cos(\vartheta), \tag{11.79}$$

so that

$$\tilde{G}''(0) = \tilde{G}''(2\pi) = -C(0) < 0, \quad \tilde{G}''(\pi) = C(\pi) > 0. \tag{11.80}$$

Consequently, the method of the stationary phase (cf. Section 8.1) is applicable. It shows that

$$\int_0^{2\pi} \tilde{F}(\vartheta)\, e^{-2\pi i|y|\tilde{G}(\vartheta)}\, d\vartheta = O\left(|y|^{-\frac{1}{2}}\right), \quad |y| \to \infty. \tag{11.81}$$

Following almost literally our earlier arguments we therefore obtain the following result.

Theorem 11.11. *(Hardy–Landau-type identity on certain convex regions and general weight functions). Let Λ be an arbitrary lattice in $\mathbb{R}^2$. Let $\mathcal{G} \subset \mathbb{R}^2$ be a convex region containing the origin and possessing a boundary curve ∂G, such that its normal field ν is continuously differentiable and its curvature is non-vanishing. Suppose that F is of class $\mathrm{C}^{(2)}(\overline{\mathcal{G}})$. Then, for all $a, y \in \mathbb{R}^2$, the series*

$$\sum_{g\in\Lambda} e^{2\pi i a\cdot g} \int_{\mathcal{G}} F(x)\, e^{-2\pi i x\cdot(g-y)}\, dV(x) \tag{11.82}$$

converges in the indicated spherical sense.

Summarizing our considerations we are led to the following conclusion: Let $\mathcal{G} \subset \mathbb{R}^2$ be a convex region containing the origin and possessing a boundary curve ∂G such that its normal field ν is continuously differentiable and its curvature is non-vanishing. Then the function $F_{\mathcal{G}}^{\wedge}$ obeys the identity

$$\begin{aligned} &\lim_{N\to\infty} \sum_{\substack{|g-y|\le N\\ g\in\Lambda}} e^{2\pi i a\cdot(g-y)}\, F_{\mathcal{G}}^{\wedge}(g-y) \\ &= \frac{1}{\|\mathcal{F}_\Lambda\|} \sum_{h\in\Lambda^{-1}} e^{2\pi i h\cdot y} \int_{\mathbb{R}^2} F_{\mathcal{G}}^{\wedge}(x)\, e^{2\pi i x\cdot(a+h)}\, dV(x). \end{aligned} \tag{11.83}$$

Equivalently, by observing the Fourier inversion formula (i.e., Theorem 11.1), we obtain

$$\begin{aligned} &\frac{1}{\|\mathcal{F}_\Lambda\|} \sum_{\substack{a+h\in\overline{\mathcal{G}}\\ h\in\Lambda^{-1}}}{}' e^{2\pi i y\cdot(a+h)}\, F(a+h) \\ &= \lim_{N\to\infty} \sum_{\substack{|g-y|\le N\\ g\in\Lambda}} e^{2\pi i a\cdot g} \int_{\mathcal{G}} F(x) e^{-2\pi i x\cdot(g-y)} dV(x). \end{aligned} \tag{11.84}$$

Replacing the lattice Λ by its inverse lattice Λ^{-1} we finally have:

Corollary 11.12. *Under the assumptions of Theorem 11.11, we have*

$$\begin{aligned} &\sum_{\substack{a+g\in\overline{\mathcal{G}}\\ g\in\Lambda}}{}' e^{-2\pi i y\cdot(a+g)}\, F(a+g) \\ &= \frac{1}{\|\mathcal{F}_\Lambda\|} \lim_{N\to\infty} \sum_{\substack{|h-y|\le N\\ h\in\Lambda^{-1}}} e^{2\pi i a\cdot h} \int_{\mathcal{G}} F(x) e^{-2\pi i x\cdot(h-y)}\, dV(x). \end{aligned} \tag{11.85}$$

Chapter 12

Bandlimited Shannon-Type Sampling (Preparatory Results)

In what follows, we make a first excursion into bandlimited Shannon-type sampling based on the bivariate results of the preceding sections about Hardy–Landau-type lattice point identites. Over- and undersampling are studied in more detail. In particular, the accelerating character of the convergence of the cardinal series in the case of oversampling is demonstrated by a constructive example.

12.1 From Hardy–Landau-Type Identities to Shannon-Type Sampling

Let $\mathcal{G}$ be a regular region in $\mathbb{R}^2$ as, e.g., specified in Theorem 11.11. Consequently, $\mathcal{G} \subset \mathbb{R}^2$ is assumed to be a convex region containing the origin and possessing a boundary curve $\partial\mathcal{G}$, such that its normal field ν is continuously differentiable and its curvature is non-vanishing. Suppose that Λ is a lattice in $\mathbb{R}^2$ generated by the basis vectors $g_1, g_2 \in \mathbb{R}^2$. Assume that F is of class $\mathrm{C}^{(2)}(\overline{\mathcal{G}})$ (note that the assumptions on the geometry as well as the weight function will be drastically weakened for sampling purposes in the next chapters).

Our bivariate sampling purposes based on the number theoretic presumptions start from the fact that

$$\int_{\mathcal{F}_\Lambda} \sum_{\substack{a+g\in\overline{\mathcal{G}} \\ g\in\Lambda}}{}' F(a+g)e^{-2\pi i y\cdot(a+g)}\, dV(a) = \int_{\mathcal{G}} F(x)\, e^{-2\pi i x\cdot y} dV(x) = F_{\mathcal{G}}^{\wedge}(y). \tag{12.1}$$

For given regular regions $\mathcal{G}, \mathcal{H} \subset \mathbb{R}^2$ with $\mathcal{G}$ of the type specified in Theoerem 11.11 and $\overline{\mathcal{G}}, \overline{\mathcal{H}} \subset \overline{\mathcal{F}_\Lambda}$ we obtain from Corollary 11.12

$$\int_{\mathcal{H}} \sum_{\substack{a+g\in\overline{\mathcal{G}} \\ g\in\Lambda^{-1}}}{}' e^{-2\pi i y\cdot(a+g)} F(a+g)\, dV(a) \tag{12.2}$$

$$= \sum_{h\in\Lambda^{-1}} \underbrace{\int_{\mathcal{G}} F(x)\, e^{-2\pi i x\cdot h}\, dV(x)}_{=F_{\mathcal{G}}^{\wedge}(h)} \frac{1}{\|\mathcal{F}_\Lambda\|} \underbrace{\int_{\mathcal{H}} e^{2\pi i a\cdot(h-y)}\, dV(a)}_{=K_{\mathcal{H}}(h-y)},$$

where the series on the right side is absolutely and uniformly convergent on $\mathbb{R}^2$. The integral on the left side of (12.2) can be calculated in more detail

$$\int_{\mathcal{H}} \sum_{\substack{a+g\in\overline{\mathcal{G}} \\ g\in\Lambda^{-1}}}{}' e^{-2\pi i y\cdot(a+g)} F(a+g)\, dV(a) = \int_{\mathcal{G}\cap\mathcal{H}} F(a) e^{-2\pi i y\cdot a} dV(a) = F_{\mathcal{G}\cap\mathcal{H}}^{\wedge}(y). \tag{12.3}$$

12.2 Over- and Undersampling

In accordance with the formula (12.3) we are led to distinguish two particular situations:

(Oversampling) $\overline{\mathcal{G}}, \overline{\mathcal{H}} \subset \overline{\mathcal{F}_\Lambda}$ such that $\mathcal{G}$ is a subset of $\mathcal{H}$. Suppose that the bandlimited signal with respect to $\mathcal{G}$ is given by (11.6). Then we are able to conclude from (12.3) that

$$F_{\mathcal{G}}^{\wedge}(y) = \sum_{h\in\Lambda^{-1}} F_{\mathcal{G}}^{\wedge}(h) \frac{1}{\|\mathcal{F}_\Lambda\|} K_{\mathcal{H}}(h-y), \tag{12.4}$$

i.e., $F_{\mathcal{G}}^{\wedge}$ can be reconstructed from its lattice point values in Λ^{-1}. In fact, since $\mathcal{H}$ is not necessarily equal to $\mathcal{G}$, the bandwidth for sampling may be larger than required.

Undersampling: $\mathcal{G} \cap \mathcal{H}$ is a strict subset of $\mathcal{G}$. Then we are led to the identity

$$F_{\mathcal{G}\cap\mathcal{H}}^{\wedge}(y) = \sum_{h\in\Lambda^{-1}} F_{\mathcal{G}}^{\wedge}(h) \frac{1}{\|\mathcal{F}_\Lambda\|} K_{\mathcal{H}}(h-y), \tag{12.5}$$

i.e., the values $F_{\mathcal{G}}^{\wedge}(h)$, $h \in \Lambda^{-1}$, sum up to $F_{\mathcal{G}\cap\mathcal{H}}^{\wedge}$, so that $F_{\mathcal{G}}^{\wedge}$ cannot be reconstructed under this assumption. Therefore, the sampling of $F_{\mathcal{G}}^{\wedge}(y)$ by the right side of (12.5) is erroneous. This phenomenon is known from the univariate theory as *aliasing*.

Example 12.1. *Let $\mathcal{G}$ be a regular region as specified by Theorem 11.11 with $\overline{\mathcal{G}} \subset \overline{\mathcal{F}_\Lambda}$. Assume that $\mathcal{H} = \mathbb{B}^2_R \subset \mathcal{F}_\Lambda$. Then we obtain the following latttice point identity*

$$F^{\wedge}_{\mathcal{G}\cap\mathbb{B}^2_R}(y) = \frac{R^2}{\|\mathcal{F}_\Lambda\|} \sum_{h\in\Lambda^{-1}} F^{\wedge}_{\mathcal{G}}(h) \frac{1}{\|\mathcal{F}_\Lambda\|} \frac{J_1(2\pi|h-y|R)}{|h-y|R}, \quad y \in \mathbb{R}^2. \tag{12.6}$$

So, over- and undersampling, respectively, occur dependent on the geometric configuration.

It should be pointed out that 1D oversampling is known to accelerate the convergence of the sampling series (see, e.g., R.J. Marks II [1991]). This phenomenon can be also detected in our 2D framework: In more detail, let $\mathcal{G}$ be a regular region in $\mathbb{R}^2$ satisfying the aforementioned conditions. Moreover, assume that $\mathcal{H}$ is an arbitrary regular region with $\overline{\mathcal{G}} \subsetneq \overline{\mathcal{H}} \subset \overline{\mathcal{F}_\Lambda}$. From Corollary 11.11, we obtain in the usual way

$$\begin{aligned} F^{\wedge}_{\mathcal{G}}(y) &= \int_{\mathcal{H}} F_{\overline{\mathcal{G}}}(a) J_{\overline{\mathcal{G}}}(a) e^{-2\pi i y\cdot a}\, dV(a) \\ &= \sum_{h\in\Lambda^{-1}} F^{\wedge}_{\mathcal{G}}(h) \frac{1}{\|\mathcal{F}_\Lambda\|} \int_{\mathcal{H}} J_{\overline{\mathcal{G}}}(a) e^{2\pi i a\cdot(h-y)}\, dV(a), \end{aligned} \tag{12.7}$$

where $J_{\overline{\mathcal{G}}}$ is assumed to be of class $\mathrm{C}^{(2)}(\overline{\mathcal{H}})$ with $J_{\overline{\mathcal{G}}}(a) = 1$ for all $a \in \overline{\mathcal{G}}$ and $J_{\overline{\mathcal{G}}}(a) = 0$ for all $a \in \partial\mathcal{H}$ and $F_{\overline{\mathcal{G}}}$ is defined by $F_{\overline{\mathcal{G}}}(a) = F(a)$, $a \in \overline{\mathcal{G}}$ and $F_{\overline{\mathcal{G}}}(a) = 0$, $a \notin \overline{\mathcal{G}}$. The second Green theorem yields

$$\begin{aligned} \int_{\mathcal{H}} J_{\overline{\mathcal{G}}}(a)\, e^{2\pi i a\cdot(h-y)}\, dV(a) &= -\frac{1}{4\pi^2|h-y|^2} \int_{\mathcal{H}\setminus\mathcal{G}} \Delta_a J_{\overline{\mathcal{G}}}(a)\, e^{2\pi i a\cdot(h-y)}\, dV(a) \\ &\quad + \frac{1}{4\pi^2|h-y|^2} \int_{\partial\mathcal{H}} \frac{\partial}{\partial\nu} J_{\overline{\mathcal{G}}}(a)\, e^{2\pi i a\cdot(h-y)}\, dS(a). \end{aligned} \tag{12.8}$$

In other words, oversampling involving the series in (12.7) takes advantage of the homogeneous boundary condition $J_{\overline{\mathcal{G}}}|_{\partial\mathcal{H}} = 0$ and the fact $\Delta_a J_{\overline{\mathcal{G}}}(a) = 0$, $a \in \overline{\mathcal{G}}$.

Example 12.2. *We choose $\mathcal{G} = \mathbb{B}^2_R$ and $\mathcal{H} = \mathbb{B}^2_{R+\varepsilon}$, $\varepsilon > 0$ fixed, thereby assuming that $R + \varepsilon < \inf_{x\in\partial\mathcal{F}_\Lambda} |x|$ and $J_{\overline{\mathbb{B}^2_R}}$ is given by*

$$J_{\mathbb{B}^2_R}(a) = \begin{cases} 1, & |a| \leq R, \\ \frac{(R+\varepsilon)^2 - a^2}{(R+\varepsilon)^2 - R^2}, & R < |a| \leq R + \varepsilon. \end{cases} \tag{12.9}$$

Then, for $|h - y| > 0$, we obtain from (12.8) and the asymptotic relations known for Bessel functions (see, e.g., W. Magnus et al. [1966], G.N. Watson

[1944])

$$\int_{\mathbb{B}^2_{R+\varepsilon}} J_{\overline{\mathbb{B}^2_R}}(a) e^{2\pi i a\cdot(h-y)}\, dV(a) = O\left(\frac{1}{|h-y|^{\frac{5}{2}}}\right), \ |h-y| \to \infty \tag{12.10}$$

instead of

$$\int_{\mathbb{B}^2_R} e^{2\pi i a\cdot(h-y)}\, dV(a) = O\left(\frac{1}{|h-y|^{\frac{3}{2}}}\right), \ |h-y| \to \infty, \tag{12.11}$$

which shows the accelerating character of the convergence by the kernel used in (12.7).

Chapter 13

Lattice Ball Euler Summation Formulas and Shannon-Type Sampling

In the research monograph about metaharmonic lattice point theory, W. Freeden [2011] proposed two kinds of summation formulas in $\mathbb{R}^q, q \geq 2$, namely lattice point–generated Euler summation formulas and lattice ball–generated Euler summation formulas. The first kind of formulas are based on the constituting properties of the Euler–Green functions with respect to (iterated) $(\Delta + \lambda)$-operators, while the second kind uses (ball-)averaged integral variants of the Euler–Green functions with respect to $(\Delta + \lambda)$-operators.

It is not surprising that an integral average such as a ball mean implies a smoothing effect. As a consequence, lattice ball summation offers a better convergence behavior in series expansions over infinite lattices than lattice point summation. In particular, associated discrepancy expansions for the circle/sphere problem (cf. (7.40)) are by far easier to handle asymptotically.

Next we extend the lattice ball concept to the operator L in order to derive lattice ball Shannon-type sampling variants in $\mathbb{R}^2$. Our considerations are based on the so-called *Λ-Euler–Green τ-ball mean function with respect to the operator $L+\lambda$*, which is an average of the Λ-Euler–Green function with respect to the operator $L + \lambda$ over balls around lattice points.

13.1 Lattice Ball Euler–Green Function

For Λ being an arbitrary lattice in $\mathbb{R}^2$ we begin with the definition of the Λ-Euler–Green τ-ball mean function with respect to the operator $L + \lambda$.

Definition 13.1. *For sufficiently small $\tau > 0$ (i.e., $\tau \in \mathbb{R}$ with $0 < \tau < \inf_{x \in \partial \mathcal{F}_\Lambda} |x|$), let $G_\tau(L+\lambda; \cdot)$ be the function defined by*

$$G_\tau(L+\lambda; x) = \frac{1}{\|\mathbb{B}^2_\tau\|} \int_{\substack{|y| \leq \tau \\ y \in \mathbb{R}^2}} G(L+\lambda; x-y)\, dV(y), \tag{13.1}$$

where, as usual, $\mathbb{B}^2_\tau$ is the open circle in $\mathbb{R}^2$ around the origin with radius τ so that

$$\|\mathbb{B}^2_\tau\| = \int_{\mathbb{B}^2_\tau} dV(x) = \int_{\substack{|y|\leq\tau \\ y\in\mathbb{R}^2}} dV(y) = \pi\tau^2. \tag{13.2}$$

Then, $G_\tau(L+\lambda;\cdot)$ is called the Λ-Euler–Green τ-ball mean function with respect to the operator $L+\lambda$ in $\mathbb{R}^2$.

From the definition of the Λ-Euler–Green function (Definition 13.1) it is clear that the Λ-*Euler–Green τ-ball mean* is Λ-periodic, i.e.,

$$G_\tau(L+\lambda;x) = G_\tau(L+\lambda;x+g), \quad x\in\mathbb{R}^2,\ g\in\Lambda. \tag{13.3}$$

The theory of Bessel functions tells us that

$$\int_{\mathbb{B}^2_\tau} e^{2\pi ih\cdot x}\, dV(x) = \int_{\substack{|x|\leq\tau \\ x\in\mathbb{R}^2}} e^{2\pi ih\cdot x}\, dV(x) = \tau^2\,\frac{J_1(2\pi|h|\tau)}{|h|\tau} = \tau\,\frac{J_1(2\pi|h|\tau)}{|h|}. \tag{13.4}$$

holds true for all $h\in\Lambda^{-1}$. Moreover, in analogy to the $(\Delta+\lambda)$-Helmholtz theory known from W. Freeden [2011] we are able to calculate the $(L+\lambda)$-derivative of the Λ–Euler (Green) τ–mean.

Theorem 13.2. *Let $G_\tau(L+\lambda;\cdot)$ be the Λ-Euler–Green τ-ball mean function. Then, for all $x\in\mathcal{F}_\Lambda\subset\mathbb{R}^2$ with $|x|\leq\tau$, we have*

$$(L+\lambda)G_\tau(L+\lambda;x) = \frac{1}{\|\mathbb{B}^2_\tau\|} - \frac{\tau^2}{\|\mathcal{F}_\Lambda\|}\sum_{\substack{(L+\lambda)^\wedge(h)=0 \\ h\in\Lambda^{-1}}}\frac{J_1(2\pi|h|\tau)}{|h|\tau}\,e^{2\pi ih\cdot x}, \tag{13.5}$$

whereas, for all $x\in\mathcal{F}_\Lambda\subset\mathbb{R}^2$ with $|x|>\tau$, we have

$$(L+\lambda)G_\tau(L+\lambda;x) = \frac{0}{\|\mathbb{B}^2_\tau\|} - \frac{\tau^2}{\|\mathcal{F}_\Lambda\|}\sum_{\substack{(L+\lambda)^\wedge(h)=0 \\ h\in\Lambda^{-1}}}\frac{J_1(2\pi|h|\tau)}{|h|\tau}\,e^{2\pi ih\cdot x}. \tag{13.6}$$

Theorem 13.2 can be rewritten by use of the *characteristic lattice ball function* $\mathcal{X}_{\overline{\mathbb{B}^2_\tau}+\Lambda}:\mathbb{R}^2\to\mathbb{R}$ given by

$$\mathcal{X}_{\overline{\mathbb{B}^2_\tau}+\Lambda}(x) = \begin{cases} 1, & x\in\bigcup_{g\in\Lambda}\overline{\mathbb{B}^2_\tau+\{g\}}, \\ 0, & \text{otherwise}, \end{cases} \tag{13.7}$$

such that

$$\begin{aligned} &(L+\lambda)\ G_\tau(L+\lambda;x) \\ &= \frac{1}{\|\mathbb{B}^2_\tau\|}\mathcal{X}_{\overline{\mathbb{B}^2_\tau}+\Lambda}(x) - \frac{\tau^2}{\|\mathcal{F}_\Lambda\|}\sum_{\substack{(L+\lambda)^\wedge(h)=0 \\ h\in\Lambda^{-1}}}\frac{J_1(2\pi|h|\tau)}{|h|\tau}\,e^{2\pi ih\cdot x}. \end{aligned} \tag{13.8}$$

13.2 Lattice Ball Euler Summation Formula

Until now, the (pointwise) Euler-type summation formula corresponding to the operator $L+\lambda$, $L=\Delta+4\pi ia\cdot\nabla$, $a\in\mathbb{R}^2$, $\lambda\in\mathbb{R}$, was used to compare values of a given function on a set of lattice points inside and on the boundary surface of a regular region $\mathcal{G}\subset\mathbb{R}^2$ with a certain finite class of bandlimited functions, i.e., Fourier transform integrals over $\mathcal{G}$ in the case of resonance. Next, the Euler-type summation formula corresponding to the operator L will be formulated for a sum of "mean values over (small) balls" with centers located at the lattice points of Λ. In fact, the representation (13.8) may be used to construct a *lattice ball variant of the Euler-type summation formula* based on the Λ-Euler–Green τ-ball mean.

Let $\mathcal{G}$ be a regular region in $\mathbb{R}^2$. Furthermore, let H be a function of class $\mathrm{C}^{(2)}\left(\overline{\mathcal{G}}\right)$. Then the second Green theorem yields for every (sufficiently small) $\varepsilon>0$ and every $a\in\mathbb{R}^2$,

$$\begin{aligned}
&\int_{\substack{x\in\overline{\mathcal{G}}\\ x\notin\mathbb{B}^2_\varepsilon+\Lambda}} \left\{H(x)(L+\lambda)G_\tau(L+\lambda;x)) - G_\tau(L+\lambda;x)(\overline{L}+\lambda)H(x)\right\}\,dV(x) \qquad (13.9)\\
&= \int_{\substack{x\in\partial\mathcal{G}\\ x\notin\mathbb{B}^2_\varepsilon+\Lambda}} H(x)\left(\frac{\partial}{\partial\nu}+4\pi ia\cdot\nu(x)\right)G_\tau(L+\lambda;x)\,dS(x)\\
&\quad - \int_{\substack{x\in\partial\mathcal{G}\\ x\notin\mathbb{B}^2_\varepsilon+\Lambda}} G_\tau(L+\lambda;x)\frac{\partial}{\partial\nu}H(x)\,dS(x)\\
&\quad + \sum_{\substack{g\in\overline{\mathcal{G}}\\ g\in\Lambda}}\int_{\substack{|x-g|=\varepsilon\\ x\in\overline{\mathcal{G}}}} H(x)\left(\frac{\partial}{\partial\nu}+4\pi ia\cdot\nu(x)\right)G_\tau(L+\lambda;x)dS(x),\\
&\quad - \sum_{\substack{g\in\overline{\mathcal{G}}\\ g\in\Lambda}}\int_{\substack{|x-g|=\varepsilon\\ x\in\overline{\mathcal{G}}}} G_\tau(L+\lambda;x)\frac{\partial}{\partial\nu}H(x)dS(x),
\end{aligned}$$

where ν is the outer (unit) normal field. By observation of the differential equation (13.8), we obtain for $\varepsilon\to 0$

$$\begin{aligned}
&\int_{\mathcal{G}} H(x)\left(\frac{1}{\|\mathbb{B}^2_\tau\|}\mathcal{X}_{\overline{\mathbb{B}^2_\tau}+\Lambda}(x) - \frac{\tau^2}{\|\mathcal{F}_\Lambda\|}\sum_{\substack{(L+\lambda)^\wedge(h)=0\\ h\in\Lambda^{-1}}}\frac{J_1(2\pi|h|\tau)}{|h|\tau}\;e^{2\pi ih\cdot x}\right)dV(x)\\
&\quad - \int_{\mathcal{G}} G_\tau\left(L+\lambda;x\right)\,(\overline{L}_x+\lambda)H(x)\,dV(x)\\
&= \int_{\partial\mathcal{G}} H(x)\left(\frac{\partial}{\partial\nu}+4\pi ia\cdot\nu(x)\right)G_\tau\left(L+\lambda;x\right)\,dS(x)\\
&\quad - \int_{\partial\mathcal{G}} G_\tau\left(L+\lambda;x\right)\frac{\partial}{\partial\nu}H(x)\,dS(x)\,. \qquad (13.10)
\end{aligned}$$

In terms of the characteristic lattice ball function (see (13.7)) this leads to

Theorem 13.3. *(Λ-Lattice Ball–Generated Euler-Type Summation Formula for the Operator $(L+\lambda)$). Let Λ be an arbitrary lattice in $\mathbb{R}^2$. Let H be of class $\mathrm{C}^{(2)}(\overline{\mathcal{G}})$. Then, for sufficiently small $\tau > 0$, we have*

$$
\begin{aligned}
&\frac{1}{\|\mathbb{B}^2_\tau\|} \int_{\mathcal{G}} \mathcal{X}_{\overline{\mathbb{B}^2_\tau}+\Lambda}(x) H(x)\, dV(x) && (13.11)\\
&\quad = \frac{\tau^2}{\|\mathcal{F}_\Lambda\|} \sum_{\substack{(L+\lambda)^\wedge(h)=0\\ h\in\Lambda^{-1}}} \frac{J_1(2\pi|h|\tau)}{|h|\tau} \int_{\mathcal{G}} H(x)\ e^{2\pi i h\cdot x}\, dV(x)\\
&\quad + \int_{\mathcal{G}} G_\tau\,(L+\lambda;x)\ (\overline{L}_x+\lambda) H(x)\, dV(x)\\
&\quad - \int_{\partial\mathcal{G}} G_\tau\,(L+\lambda;x)\, \frac{\partial}{\partial\nu} H(x)\, dS(x)\\
&\quad + \int_{\partial\mathcal{G}} H(x) \left(\frac{\partial}{\partial\nu} + 4\pi i a\cdot\nu(x)\right) G_\tau\,(L+\lambda;x)\ dS(x).
\end{aligned}
$$

Theorem 13.3 is the lattice ball counterpart of the Euler-type summation formula related to the Λ-*Euler–Green τ-ball mean* with respect to the operator $L+\lambda$.

From the Fourier expansion of $G\,(L+\lambda;\cdot)$, we are able to determine the series representation of $G_\tau\,(L+\lambda;\cdot)$, namely

$$
G_\tau\,(L+\lambda;x) \sim \frac{1}{\pi}\frac{1}{\|\mathcal{F}_\Lambda\|} \sum_{\substack{(L+\lambda)^\wedge(h)\neq 0\\ h\in\Lambda^{-1}}} \frac{J_1(2\pi|h|\tau)}{|h|\tau}\ \frac{e^{2\pi i h\cdot x}}{(L+\lambda)^\wedge(h)}, \tag{13.12}
$$

where $(L+\lambda)^\wedge(h) = \lambda - L^\wedge(h),\ L^\wedge(h) = 4\pi^2((h+a)^2 - a^2)$. The series of $G_\tau(L+\lambda;\cdot)$ is absolutely and uniformly convergent in $\mathbb{R}^2$. Indeed we have in the pointwise sense

$$
G_\tau\,(L+\lambda;x) = \frac{1}{\|\mathcal{F}_\Lambda\|} \sum_{\substack{(L+\lambda)^\wedge(h)\neq 0\\ h\in\Lambda^{-1}}} \frac{J_1(2\pi|h|\tau)}{\pi|h|\tau}\ \frac{e^{2\pi i h\cdot x}}{(L+\lambda)^\wedge(h)}. \tag{13.13}
$$

Consequently, $G_\tau(L+\lambda;\cdot)$ is a continuous function in $\mathbb{R}^2$, while its counterpart $G(L+\lambda;\cdot)$ in $\mathbb{R}^2$ has logarithmic singularities in the lattice points of Λ.

Observing the elementary manipulations (9.18), (9.19), and (9.20) we immediately obtain the following variant from Theorem 13.3.

Corollary 13.4. *Let Λ be an arbitrary lattice in $\mathbb{R}^2$. Suppose that $\mathcal{G} \subset \mathbb{R}^2$ is a regular region. Let H be of class $\mathrm{C}^{(2)}(\overline{\mathcal{G}})$. Then, for sufficiently small $\tau > 0$,* *for all* $\lambda \in \mathbb{R}$, *and* $a, y \in \mathbb{R}^2$, *we have*

$$\frac{1}{\|\mathbb{B}^2_\tau\|} \int_{x-y\in\mathcal{G}} \mathcal{X}_{\overline{\mathbb{B}^2_\tau}+\Lambda}(x)\ e^{2\pi i a\cdot x}\ H(x-y)\ dV(x) \tag{13.14}$$
$$= \frac{\tau^2}{\|\mathcal{F}_\Lambda\|}\ e^{2\pi i a\cdot y} \sum_{\substack{4\pi^2\left((h+a)^2-a^2\right)=\lambda \\ h\in\Lambda^{-1}}} \frac{J_1(2\pi|h|\tau)}{|h|\tau}\ e^{2\pi i h\cdot y} \int_{\mathcal{G}} H(x)\ e^{2\pi i x\cdot(a+h)}\ dV(x)$$
$$+ \int_{\substack{x-y\in\mathcal{G} \\ x\in\mathbb{R}^2}} G_\tau(L+\lambda;x)\ e^{2\pi i a\cdot x}\ \left(\Delta_x + \lambda + 4\pi^2 a^2\right) H(x-y)\ dV(x)$$
$$- \int_{\substack{x-y\in\partial\mathcal{G} \\ x\in\mathbb{R}^2}} e^{2\pi i a\cdot x}\ G_\tau\left(L+\lambda;x\right) \frac{\partial}{\partial\nu} H(x)\ dS(x)$$
$$+ \int_{\substack{x-y\in\partial\mathcal{G} \\ x\in\mathbb{R}^2}} e^{2\pi i a\cdot x}\ H(x) \left(\frac{\partial}{\partial\nu} + 4\pi i a\cdot\nu(x)\right) G_\tau\left(L+\lambda;x\right)\ dS(x).$$

13.3 Lattice Ball Mean Shannon-Type Sampling

In comparison to the weighted lattice point sums, the weighted lattice ball variants admit the desired two-fold generalizations without any restriction, namely

- the generalization from a constant $F = 1$ to a "general" twice continuously differentiable weight function F,
- the generalization from a "convex geometry" to a "general" regular geometry $\mathcal{G}$.

As a matter of fact, the lattice ball concept together with metaharmonic tools known from W. Freeden [2011] enable us to develop two kinds of lattice-ball-type Poisson-type summation formulas for regular regions $\mathcal{G}$, without any restriction with respect to the boundary curve of the regular region $\mathcal{G}$ and the weight function F on the boundary. Neither the convexity nor the non-zero curvature have to be prescribed. The price to be paid is an additional Bessel "jinc-term" to enforce the convergence of the weighted Hardy–Landau lattice ball series.

Variant 1: From Corollary 13.4 we are immediately able to deduce the lattice ball identity

$$\frac{1}{\pi\tau^2} \int_{\substack{x+a\in\mathcal{G} \\ \mathcal{G}\subset\mathbb{R}^2}} \mathcal{X}_{\overline{\mathbb{B}^2_\tau}+\Lambda}(x)\ e^{2\pi i y\cdot(x+a)}\ F(x+a)\ dV(x) \tag{13.15}$$
$$= \frac{1}{\|\mathcal{F}_\Lambda\|} \lim_{N\to\infty} \sum_{\substack{|h-y|\le N \\ h\in\Lambda^{-1}}} \frac{J_1(2\pi|h|\tau)}{\pi|h|\tau}\ e^{2\pi i h\cdot a} \int_{\mathcal{G}} F(x)\ e^{-2\pi i x\cdot(h-y)}\ dV(x),$$

provided that F is a twice continuously differentiable function on $\overline{\mathcal{G}}$, so that for all $\mathcal{G}, \mathcal{H} \subset \mathbb{R}^2$,

$$\begin{aligned} &\int_{\mathcal{H}} \sum_{a+g\in\overline{\mathcal{G}}}{}' e^{-2\pi i y\cdot(a+g)} F(a+g) \\ &= \lim_{\tau\to 0} \sum_{h\in\Lambda^{-1}} \frac{J_1(2\pi|h-y|\tau)}{\pi|h-y|\tau} F_{\mathcal{G}}^{\wedge}(h) \frac{1}{\|\mathcal{F}_\Lambda\|} K_{\mathcal{H}}(h-y). \end{aligned} \tag{13.16}$$

In other words, we have for all $\mathcal{G}, \mathcal{H} \subset \mathcal{F}_\Lambda$,

$$\begin{aligned} F_{\mathcal{G}\cap\mathcal{H}}^{\wedge}(y) &= \int_{\mathcal{G}\cap\mathcal{H}} e^{-2\pi i y\cdot a} F(a)\, dV(a) \\ &= \lim_{\tau\to 0} \sum_{h\in\Lambda^{-1}} \frac{J_1(2\pi|h-y|\tau)}{\pi|h-y|\tau} F_{\mathcal{G}}^{\wedge}(h) \frac{1}{\|\mathcal{F}_\Lambda\|} K_{\mathcal{H}}(h-y). \end{aligned} \tag{13.17}$$

For sufficiently small $\tau > 0$, we therefore have for all $\mathcal{G}, \mathcal{H} \subset \mathcal{F}_\Lambda$ with $\mathcal{G} \subset \mathcal{H}$

$$F_{\mathcal{G}}^{\wedge}(y) \approx \sum_{h\in\Lambda^{-1}} \frac{J_1(2\pi|h-y|\tau)}{\pi|h-y|\tau} F_{\mathcal{G}}^{\wedge}(h) \frac{1}{\|\mathcal{F}_\Lambda\|} K_{\mathcal{G}}(h-y). \tag{13.18}$$

For sufficiently small $\tau > 0$ and an arbitrary regular region $\mathcal{G}$ with $\overline{\mathcal{G}} \subset \overline{\mathcal{F}_\Lambda}$, we consequently obtain

$$F_{\mathcal{G}}^{\wedge}(y) \approx \sum_{h\in\Lambda^{-1}} \frac{J_1(2\pi|h-y|\tau)}{\pi|h-y|\tau} F_{\mathcal{G}}^{\wedge}(h) \frac{1}{\|\mathcal{F}_\Lambda\|} K_{\mathcal{F}_\Lambda}(h-y). \tag{13.19}$$

The summand of the series on the right side of (13.19) has a convergence order $O(|h-y|)^{-\frac{5}{2}}$ caused by the jinc-term

$$\frac{J_1(2\pi|h-y|\tau)}{\pi|h-y|\tau} = O(|h-y|)^{-\frac{3}{2}} \tag{13.20}$$

and (cf. Lemma 11.2)

$$F_{\mathcal{G}}^{\wedge}(x-y) = O\left(\frac{1}{|x-y|}\right), \quad |x-y| \to \infty. \tag{13.21}$$

Variant 2: For a function $F \in \mathrm{C}^{(2)}\left(\overline{\mathcal{G}}\right)$ with $\mathcal{G}$ being an arbitrary regular region in $\mathbb{R}^2$, an arbitrary lattice Λ in $\mathbb{R}^2$, sufficiently small $\tau > 0$, and sufficiently large N, and all $a, y \in \mathbb{R}^2$, we obtain from Corollary 13.4

$$\frac{1}{\pi\tau^2}\int_{|x-y|\le N} \mathcal{X}_{\overline{\mathbb{B}^2_\tau+\Lambda}}(x)\ e^{2\pi i a\cdot x}\ F^\wedge_{\mathcal{G}}(x-y)\ dV(x) \tag{13.22}$$
$$= \frac{1}{\|\mathcal{F}_\Lambda\|}\ e^{2\pi i a\cdot y}\sum_{\substack{|h|>0\\ h\in\Lambda^{-1}}}\frac{J_1(2\pi|h+a|\tau)}{\pi|h+a|\tau}\ e^{2\pi i h\cdot y}\int_{|x|\le N} F^\wedge_{\mathcal{G}}(x)\ e^{2\pi i x\cdot(a+h)}\ dV(x)$$
$$+ \int_{\substack{|x-y|\le N\\ x\in\mathbb{R}^2}} G_\tau(L;x)\ e^{2\pi i a\cdot x}\ \left(\Delta_x+4\pi^2a^2\right) F^\wedge_{\mathcal{G}}(x-y)\ dV(x)$$
$$- \int_{\substack{|x-y|=N\\ x\in\mathbb{R}^2}} e^{2\pi i a\cdot x}\ G_\tau(L;x)\ \frac{\partial}{\partial\nu}F^\wedge_{\mathcal{G}}(x-y)\ dS(x)$$
$$+ \int_{\substack{|x-y|=N\\ x\in\mathbb{R}^2}} e^{2\pi i a\cdot x}\ F^\wedge_{\mathcal{G}}(x-y)\left(\frac{\partial}{\partial\nu}+4\pi i a\cdot\nu(x)\right) G_\tau(L;x)\ dS(x),$$

with $F^\wedge_{\mathcal{G}}$ given in the usual form (11.6), i.e.,

$$F^\wedge_{\mathcal{G}}(x) = \int_{\substack{y\in\mathcal{G}\\ y\in\mathbb{R}^2}} F(y)\ e^{-2\pi i x\cdot y}\ dV(y), \quad x\in\mathbb{R}^2. \tag{13.23}$$

The function $G_\tau(L;\cdot)$, $0<\tau<\inf_{x\in\partial\mathcal{F}_\Lambda}|x|$, is continuous in $\mathbb{R}^2$. The absolutely and uniformly convergent series expansion of $G_\tau(L;\cdot)$ in $\mathbb{R}^2$

$$G_\tau(L;x) = \frac{1}{\|\mathcal{F}_\Lambda\|}\sum_{\substack{|h|>0\\ h\in\Lambda^{-1}}}\frac{J_1(2\pi|a+h|\tau)}{\pi|a+h|\tau}\ \frac{e^{2\pi i h\cdot x}}{-L^\wedge(h)}. \tag{13.24}$$

enables us to write

$$\int_{|x-y|=N} e^{2\pi i a\cdot x}\ G_\tau(L;x)\ \frac{\partial}{\partial\nu}F^\wedge_{\mathcal{G}}(x-y)\ dS(x) \tag{13.25}$$
$$= \frac{1}{\pi\tau}\frac{1}{\|\mathcal{F}_\Lambda\|}\sum_{\substack{|h|>0\\ h\in\Lambda^{-1}}}\frac{J_1(2\pi|h|\tau)}{-L^\wedge(h)|h|}\int_{\substack{|x-y|=N\\ x\in\mathbb{R}^2}} e^{2\pi i(a+h)\cdot x}\ \frac{\partial}{\partial\nu}F^\wedge_{\mathcal{G}}(x-y)\ dS(x)$$

and

$$\int_{|x-y|=N} e^{2\pi i\ a\cdot x}\ F^\wedge_{\mathcal{G}}(x-y)\left(\frac{\partial}{\partial\nu}+4\pi i a\cdot\nabla\right) G_\tau(L;x)\ dS(x) \tag{13.26}$$
$$= \frac{1}{\|\mathcal{F}_\Lambda\|}\sum_{\substack{|h|>0\\ h\in\Lambda^{-1}}}\frac{1}{-L^\wedge(h)}\ \frac{J_1(2\pi|h|\tau)}{\pi|h|\tau}$$
$$\times\int_{\substack{|x-y|=N\\ x\in\mathbb{R}^2}} e^{2\pi i\ a\cdot x}F^\wedge_{\mathcal{G}}(x-y)\left(\frac{\partial}{\partial\nu_x}+4\pi i\ a\cdot\nabla_x\right)e^{2\pi i h\cdot x}\ dS(x).$$

From Lemma 11.2 we know the asymptotic estimate

$$(\nabla_x)^\alpha F^\wedge_{\mathcal{G}}(x-y)\ =\ O\left(\frac{1}{|x-y|}\right), \quad |x-y|\to\infty, \tag{13.27}$$

for all multi-indices $\alpha = (\alpha_1, \alpha_2)^T$ with $\alpha_i \in \mathbb{N}_0$, $i \in \{1, 2\}$, provided that F is assumed to be of class $\mathrm{C}^{(2)}(\overline{\mathcal{G}})$. Hence, analogously to well-known estimates of the metaharmonic theory (as already used in W. Freeden [2011]), it readily follows that

$$\int_{|x-y|=N} e^{2\pi i\ a\cdot x}\ G_\tau\left(L;x\right) \frac{\partial}{\partial\nu} F_{\mathcal{G}}^{\wedge}(x-y)\ dS(x) = O\left(N^{-\frac{1}{2}}\right) \tag{13.28}$$

and

$$\int_{|x-y|=N} e^{2\pi i\ a\cdot x}\ F_{\mathcal{G}}^{\wedge}(x-y) \left(\frac{\partial}{\partial\nu} + 4\pi i\ a\cdot\nabla\right) G_\tau\left(L;x\right)\ dS(x) = O\left(N^{-\frac{1}{2}}\right) \tag{13.29}$$

hold true for *any kind of* a regular region $\mathcal{G} \subset \mathbb{R}^2$ as $N \to \infty$.

Because of the estimate $|J_1(2\pi|h|\tau)| = O\left(|h|^{-\frac{1}{2}}\right)$ we are allowed to interchange integration and summation so that

$$\begin{aligned} &\int_{\substack{|x-y|\leq N \\ x\in\mathbb{R}^2}} G_\tau(L;x)\ e^{2\pi i\ a\cdot x}\ \left(\Delta_x + 4\pi^2 a^2\right) F_{\mathcal{G}}^{\wedge}(x-y)\ dV(x) \\ &= \frac{1}{\pi\tau}\frac{1}{\|\mathcal{F}_\Lambda\|} \sum_{\substack{|h|>0 \\ h\in\Lambda^{-1}}} \frac{J_1(2\pi|h|\tau)}{-L^\wedge(h)|h|} \int_{\substack{|x-y|\leq N \\ x\in\mathbb{R}^2}} e^{2\pi i(a+h)\cdot x} \left(\Delta_x + 4\pi^2 a^2\right) F_{\mathcal{G}}^{\wedge}(x-y)\ dV(x). \end{aligned} \tag{13.30}$$

We know that

$$\begin{aligned} &\int_{\substack{|x-y|\leq N \\ x\in\mathbb{R}^2}} e^{2\pi i(a+h)\cdot x} \left(\Delta_x + 4\pi^2 a^2\right) F_{\mathcal{G}}^{\wedge}(x-y)\ dV(x) \\ &= \int_{\substack{|x-y|\leq N \\ x\in\mathbb{R}^2}} e^{2\pi i(a+h)\cdot x} \int_{\substack{z\in\mathcal{G} \\ z\in\mathbb{R}^2}} \underbrace{F(z)(4\pi^2 a^2 - 4\pi^2 z^2)}_{=\tilde{F}(z)} e^{-2\pi i(x-y)\cdot z}\ dV(z)\ dV(x). \end{aligned} \tag{13.31}$$

Therefore, from the already-known process involving the Fourier inversion formula, we finally obtain

Theorem 13.5. *Let Λ be an arbitrary lattice in $\mathbb{R}^2$. Let $\mathcal{G} \subset \mathbb{R}^2$ be a regular region. Suppose that F is of class $\mathrm{C}^{(2)}(\overline{\mathcal{G}})$. Then, for all sufficiently small (fixed) values $\tau > 0$ and for all $a, y \in \mathbb{R}^2$, the limit*

$$\lim_{N\to\infty} \frac{1}{\pi\tau^2} \int_{\substack{|x-y|\leq N \\ x\in\mathbb{R}^2}} \mathcal{X}_{\overline{\mathbb{B}^2_\tau + \Lambda}}(x) e^{2\pi i\ a\cdot x} \underbrace{\int_{\mathcal{G}} F(z)\ e^{-2\pi i(x-y)\cdot z}\ dV(z)}_{=F_{\mathcal{G}}^{\wedge}(x-y)}\ dV(x) \tag{13.32}$$

exists, and we have

$$\begin{aligned} &\frac{1}{\|\mathcal{F}_\Lambda\|} \sum_{\substack{a+h\in\overline{\mathcal{G}} \\ h\in\Lambda^{-1}}}{}' \frac{J_1(2\pi|a+h|\tau)}{\pi|a+h|\tau} e^{2\pi i(a+h)\cdot y} F(a+h) \\ &= \lim_{N\to\infty} \frac{1}{\pi\tau^2} \int_{\substack{|x-y|\leq N \\ x\in\mathbb{R}^2}} \mathcal{X}_{\overline{\mathbb{B}^2_\tau + \Lambda}}(x) e^{2\pi i a\cdot x} \int_{\mathcal{G}} F(z)\ e^{-2\pi i(x-y)\cdot z}\ dV(z)\ dV(x). \end{aligned} \tag{13.33}$$

Replacing $\Lambda \subset \mathbb{R}^2$ by its inverse lattice $\Lambda^{-1} \subset \mathbb{R}^2$, we finally arrive at

Corollary 13.6. *Under the assumptions of Theorem 13.5*

$$\sum_{\substack{a+g\in\overline{\mathcal{G}}\\ g\in\Lambda^{-1}}}{}' \frac{J_1(2\pi|g+a|\tau)}{\pi\tau|g+a|}\, e^{-2\pi i(a+g)\cdot y} F(a+g) \tag{13.34}$$

$$= \frac{1}{\|\mathcal{F}_\Lambda\|}\frac{1}{\pi\tau^2}\int_{\mathbb{R}^2} \mathcal{X}_{\overline{\mathbb{B}^2_\tau}+\Lambda^{-1}}(x-y) F^{\wedge}_{\mathcal{G}}(x)\; e^{2\pi i a\cdot(x-y)}\; dV(x).$$

As an immediate result it follows for $\mathcal{G}\subset\mathcal{F}_\Lambda$ that

$$\underbrace{\int_{\mathcal{G}} \frac{J_1(2\pi|a|\tau)}{\pi\tau|a|} F(a) e^{-2\pi i y\cdot a} dV(a)}_{=(F^\tau)^{\wedge}_{\mathcal{G}}(y)} \tag{13.35}$$

$$= \frac{1}{\|\mathcal{F}_\Lambda\|}\frac{1}{\pi\tau^2}\int_{\mathbb{R}^2} \mathcal{X}_{\overline{\mathbb{B}^2_\tau}+\Lambda^{-1}}(x-y)\; F^{\wedge}_{\mathcal{G}}(x) \int_{\mathcal{F}_\Lambda} e^{2\pi a\cdot(x-y)}\; dV(a)\; dV(x).$$

For small $\tau > 0$, $(F^\tau)^{\wedge}_{\mathcal{G}}(y)$ is approximately equal to $F^{\wedge}_{\mathcal{G}}(y)$, so that sampling over lattice balls approximately reduces to ordinary sampling over lattice points.

Finally, it should be mentioned that, in addition to ball-generated means, Gauss–Weierstrass and Abel–Poisson means of the Euler–Green functions can be used to formulate associated Euler-type summation formulas (see W. Freeden, M. Gutting [2017] and the references therein) and to derive summability variants of Shannon-type sampling formulas. More details on Gauss–Weierstrass means will be presented in Chapter 15.

Part VI

Multivariate Poisson-Type Summation Formulas over Regular Regions

Chapter 14

Gauss–Weierstrass Mean Euler-Type Summation Formulas and Shannon-Type Sampling

In what follows we come to a weighted variant of the Fourier transform, namely the Gauss–Weierstrass integral transform in its particular application to regular regions.

In the usual $L^1(\mathbb{R}^q)$–nomenclature, the theory of the integral transforms is well-understood (see Chapter 7). In our approach, however, we are essentially inspired by C. Müller [1998] to transfer the theory of Fourier transforms (defined as spherical principal values) to so-called spherically continuous functions. In doing so, we are led to structures and settings that are adequate for our pointwise efforts in lattice point theory involved with discontinuous functions relative to boundaries of regular regions.

14.1 Gauss–Weierstrass Transform over Regular Regions

We remember the *Gaussian function* $G : \mathbb{R}^q \to \mathbb{C}$ (cf. (8.65)) given by

$$G(x) = e^{2\pi i x\cdot z} e^{-\pi\tau x^2}, \quad z \in \mathbb{R}^q,\ \tau > 0 \tag{14.1}$$

From Lemma 8.15 we know that, for all $x \in \mathbb{R}^q$, we have

$$W^{\wedge}_{\mathbb{R}^q}(-x) = G(x) = \int_{\mathbb{R}^q} \underbrace{G^{\wedge}_{\mathbb{R}^q}(y)}_{=W(y)} e^{2\pi i x\cdot y}\, dV_{(q)}(y), \tag{14.2}$$

$$G^{\wedge}_{\mathbb{R}^q}(x) = W(x) = \int_{\mathbb{R}^q} \underbrace{W^{\wedge}_{\mathbb{R}^q}(y)}_{=G(-y)} e^{2\pi i x\cdot y}\, dV_{(q)}(y), \tag{14.3}$$

where $W : \mathbb{R}^q \to \mathbb{C}$ given by

$$W(x) = \left(\frac{1}{\tau}\right)^{q/2} e^{-\frac{\pi}{\tau}(z-x)^2}, \quad z \in \mathbb{R}^q,\ \tau > 0. \tag{14.4}$$

is the *Weierstrass function*.

Now, we introduce the *bandlimited Gauss transform and Weierstrass transform as integrals over a regular region* $\mathcal{K}$ involving the kernels G and W, respectively.

Definition 14.1. *Let $\mathcal{K}$ be a regular region. For $\tau > 0$, the Gauss transform $\mathcal{G}_\tau^{\mathcal{K}}(F)$ of $F \in \mathrm{C}^{(0)}(\overline{\mathcal{K}})$ at $x \in \mathbb{R}^q$ is defined by*

$$\mathcal{G}_\tau^{\mathcal{K}}(F)(x) = \left(\frac{1}{\tau}\right)^{q/2} \int_{\mathcal{K}} e^{-\frac{\pi}{\tau}(x-y)^2} F(y)\, dV_{(q)}(y), \tag{14.5}$$

while the Weierstrass transform $\mathcal{W}_\tau^{\mathcal{K}}(F)$ of $F \in \mathrm{C}^{(0)}(\overline{\mathcal{K}})$ at $x \in \mathbb{R}^q$ is defined by

$$\mathcal{W}_\tau^{\mathcal{K}}(F)(x) = \int_{\mathcal{K}} e^{-\pi\tau y^2} F(y)\, e^{2\pi i x\cdot y}\, dV_{(q)}(y). \tag{14.6}$$

The relation between the Gauss transform and the Weierstrass transform is documented by the following identities.

Lemma 14.2. *For $\tau > 0$, and $z \in \mathbb{R}^q$,*

$$\begin{aligned} \mathcal{W}_\tau(F_{\mathcal{K}}^{\wedge})(z) &= \mathcal{G}_\tau^{\mathcal{K}}(F)(z), && (14.7)\\ \mathcal{G}_\tau(F_{\mathcal{K}}^{\wedge})(z) &= \mathcal{W}_\tau^{\mathcal{K}}(F)(-z). && (14.8) \end{aligned}$$

In fact, $\mathcal{W}_\tau(F_{\mathcal{K}}^{\wedge})(z)$ written out is given in the form

$$\mathcal{W}_\tau(F_{\mathcal{K}}^{\wedge})(z) = \int_{\mathbb{R}^q} e^{-\pi\tau y^2} \int_{\mathcal{K}} F(x) e^{-2\pi i x\cdot y}\, dV_{(q)}(x)\, e^{2\pi i z\cdot y}\, dV_{(q)}(y). \tag{14.9}$$

The order of the integrations may be reversed such that

$$\begin{aligned} \mathcal{W}_\tau(F_{\mathcal{K}}^{\wedge})(z) &= \int_{\mathcal{K}} F(x) \left(\int_{\mathbb{R}^q} e^{-\pi\tau y^2} e^{2\pi i (z-x)\cdot y}\, dV_{(q)}(y) \right) dV_{(q)}(x)\\ &= \left(\frac{1}{\tau}\right)^{q/2} \int_{\mathcal{K}} e^{-\frac{\pi}{\tau}(x-z)^2} F(x)\, dV_{(q)}(x)\\ &= \mathcal{G}_\tau^{\mathcal{K}}(F)(z). && (14.10) \end{aligned}$$

Moreover, $\mathcal{G}_\tau(F_{\mathcal{K}}^{\wedge})(z)$ can be represented in the form

$$\mathcal{G}_\tau(F_{\mathcal{K}}^{\wedge})(z) = \left(\frac{1}{\tau}\right)^{q/2} \int_{\mathbb{R}^q} e^{-\frac{\pi}{\tau}(z-y)^2} \int_{\mathcal{K}} F(x)\, e^{-2\pi i x\cdot y}\, dV_{(q)}(x)\, dV_{(q)}(y). \tag{14.11}$$

By interchanging the order of integration we have

$$\begin{aligned} \mathcal{G}_\tau(F_{\mathcal{K}}^{\wedge})(z) &= \int_{\mathcal{K}} F(x) \int_{\mathbb{R}^q} \left(\frac{1}{\tau}\right)^{q/2} e^{-\frac{\pi}{\tau}(z-y)^2}\, e^{-2\pi i x\cdot y}\, dV_{(q)}(y)\, dV_{(q)}(x)\\ &= \int_{\mathcal{K}} e^{-\pi\tau x^2} F(x)\, e^{-2\pi i z\cdot x}\, dV_{(q)}(x)\\ &= \mathcal{W}_\tau^{\mathcal{K}}(F)(-z). && (14.12) \end{aligned}$$

Next our purpose is to show that the "integral mean" $\mathcal{W}_\tau(F_{\mathcal{K}}^{\wedge})(w)$ tends to $\alpha_{\mathcal{K}}(w)\ F(w)$ as $\tau \to 0$, where $\alpha_{\mathcal{K}}(w)$ is the solid angle at $w \in \mathbb{R}^q$ subtended by the surface $\partial\mathcal{K}$.

Lemma 14.3. *Suppose that F is of class $F \in \mathrm{C}^{(0)}(\overline{\mathcal{K}})$. Then*

$$\lim_{\substack{\tau\to 0\\ \tau>0}} \mathcal{G}_\tau^{\mathcal{K}}(F)(w) = \lim_{\substack{\tau\to 0\\ \tau>0}} \mathcal{W}_\tau(F_{\mathcal{K}}^{\wedge})(w) = \alpha_{\mathcal{K}}(w)\ F(w), \quad w \in \overline{\mathcal{K}}. \tag{14.13}$$

Proof. We start from

$$\begin{aligned} \mathcal{G}_\tau^{\mathcal{K}}(F)(w) &= \left(\frac{1}{\tau}\right)^{\frac{q}{2}} \int_{\mathcal{K}} e^{-\frac{\pi}{\tau}(w-x)^2} F(x)\ dV_{(q)}(x) \\ &= \left(\frac{1}{\tau}\right)^{\frac{q}{2}} \int_{\mathbb{R}^q} e^{-\frac{\pi}{\tau}(w-x)^2} \tilde{F}(x)\ dV_{(q)}(x), \end{aligned} \tag{14.14}$$

where

$$\tilde{F}(x) = \begin{cases} F(x) & , \quad x \in \overline{\mathcal{K}} \\ 0 & , \quad x \notin \overline{\mathcal{K}}. \end{cases} \tag{14.15}$$

By introducing polar coordinates $x = w + r\xi$, $r > 0$, $\xi \in \mathbb{S}^{q-1}$, we obtain

$$\mathcal{G}_\tau^{\mathcal{K}}(F)(w) = \left(\frac{1}{\tau}\right)^{\frac{q}{2}} \int_0^\infty e^{-\frac{\pi}{\tau}r^2} \Phi(r;w)\ r^{q-1}\ dr, \tag{14.16}$$

where

$$\Phi(r;w) = \int_{\mathbb{S}^{q-1}} \tilde{F}(w + r\xi)\ dS_{(q-1)}(\xi). \tag{14.17}$$

Observing that $\tilde{F} \in \mathrm{L}^1(\mathbb{R}^q)$ we find that

$$\Phi(0;z) = \|\mathbb{S}^{q-1}\|\ \alpha_{\mathcal{K}}(w)\ F(w). \tag{14.18}$$

We therefore obtain from (14.16) and (14.18) by use of a well-known 1D-limit relation (see W. Freeden [2011], Chapter 3, Lemma 3.20)

$$\lim_{\substack{\tau\to 0\\ \tau>0}} \left(\frac{1}{\tau}\right)^{\frac{q}{2}} \int_0^\infty e^{-\frac{\pi}{\tau}r^2} \Phi(r;w)\ r^{q-1}\ dr = \frac{\Phi(0;w)}{\|\mathbb{S}^{q-1}\|} = \alpha_{\mathcal{K}}(w)\ F(w), \tag{14.19}$$

as desired. □

14.2 Gauss–Weierstrass Mean Euler–Green Function

The Λ-*lattice τ-Gauss–Weierstrass mean Euler–Green function with respect to the operator* $L + \lambda$ is defined as a convolution integral.

Definition 14.4. *For sufficiently small $\tau > 0$*

$$\mathcal{G}_\tau^W(L+\lambda; x) = \int_{\mathbb{R}^q} G(L+\lambda; x-y)\, W_\tau(y)\, dV(y), \quad x \in \mathbb{R}^q, \tag{14.20}$$

is called the Λ-lattice τ-Gauss–Weierstrass mean Euler–Green function with respect to the operator $L+\lambda$ (note that the superscript W denotes the Gauss–Weierstrass means obtained via the Weierstrass convolution (14.20)).

The Λ-lattice τ-Gauss–Weierstrass Euler–Green mean $G_\tau^W(L+\lambda;\cdot)$ is continuous in $\mathbb{R}^q$. Furthermore, $G_\tau^W(L+\lambda,\cdot)$ has the absolutely and uniformly convergent bilinear expansion

$$G_\tau^W(L+\lambda; x-y) = \frac{1}{\|\mathcal{F}_\Lambda\|} \sum_{\substack{(L+\lambda)^\wedge(h)\neq 0 \\ h\in\Lambda^{-1}}} e^{-\tau\pi^2 h^2}\, \frac{e^{2\pi i h\cdot x}\, e^{-2\pi i h\cdot y}}{-(L+\lambda)^\wedge(h)} \tag{14.21}$$

(note that the Gauss–Weierstrass means (14.21) does not only provide absolute and uniform convergence, but also an exponentially accelerated summation).

14.3 Gauss–Weierstrass Mean Euler Summation over Regular Regions

The Euler-type summation formula in Gauss–Weierstrass nomenclature can be obtained in a straightforward way: Suppose that $\mathcal{G} \subset \mathbb{R}^q$ is a regular region. For $a, y \in \mathbb{R}^q$ and $F \in \mathrm{C}^{(2)}(\overline{\mathcal{G}})$, we have the limit relation

$$\begin{aligned}
&\lim_{\substack{\tau\to 0\\ \tau>0}} \Bigg((\tau\pi)^{-\frac{q}{2}} \sum_{\substack{a+g\in\overline{\mathcal{G}} \\ g\in\Lambda}} \int_{\mathcal{G}} e^{-\frac{|x(a+g)|^2}{\tau}}\, e^{2\pi i y\cdot x} F(x)\, dV(x) \qquad (14.22)\\
&\quad - \frac{1}{\|\mathcal{F}_\Lambda\|} \sum_{\substack{(L+\lambda)^\wedge(h)=0 \\ h\in\Lambda^{-1}}} e^{-\tau\pi^2 h^2} e^{2\pi i h\cdot a} \int_{\mathcal{G}} F(x) e^{-2\pi i x\cdot(h-y)}\, dV(x) \Bigg) \\
&= \lim_{\substack{\tau\to 0\\ \tau>0}} \Bigg(\int_{\substack{x-y\in\mathcal{G}\\ x\in\mathbb{R}^q}} G_\tau^W(L+\lambda; x) e^{2\pi i a\cdot(x-y)} (\Delta_x + \lambda + 4\pi^2 a^2) F(x-y)\, dV(x) \\
&\quad - \int_{\substack{x-y\in\partial\mathcal{G}\\ x\in\mathbb{R}}} e^{2\pi i a\cdot(x-y)} G_\tau^W(L+\lambda; x) \frac{\partial}{\partial\nu_x} F(x-y)\, dS(x) \\
&\quad - \int_{\substack{x-y\in\partial\mathcal{G}\\ x\in\mathbb{R}^q}} e^{2\pi i a\cdot(x-y)} F(x-y) \left(\frac{\partial}{\partial\nu_x} + 4\pi i a\cdot\nu(x) \right) G_\tau^W(L+\lambda; x)\, dS(x) \Bigg).
\end{aligned}$$

For $\tau \to 0$, the Fourier inversion formula in the Gauss–Weierstrass nomenclature shows that

$$\lim_{\substack{\tau\to 0\\ \tau>0}} (\tau\pi)^{-\frac{q}{2}} \int_{\mathcal{G}} e^{-\frac{|x-(a+g)|^2}{\tau}} e^{2\pi i y\cdot x} F(x)\, dV(x) = \alpha_{\mathcal{G}}(a+g)\, e^{2\pi i y\cdot(a+g)} F(a+g). \tag{14.23}$$

Moreover, integration by parts enables us to show that the right side of (14.22) is equal to

$$\lim_{\substack{\tau\to 0\\ \tau>0}} \frac{1}{\|\mathcal{F}_\Lambda\|} \sum_{\substack{(L+\lambda)^\wedge(h)\neq 0\\ h\in\Lambda^{-1}}} e^{-\tau\pi^2 h^2} e^{2\pi i h\cdot a} \int_{\mathcal{G}} F(x) e^{-2\pi i x\cdot(h-y)}\, dV(x). \tag{14.24}$$

Summarizing our considerations we obtain the *Poisson-type summation formula in Gauss–Weierstrass summability.*

Theorem 14.5. *Let $\mathcal{G} \in \mathbb{R}^q$ be a regular region. Let Λ be a lattice in $\mathbb{R}^q$. Suppose that F is a member of class $C^{(2)}(\overline{\mathcal{G}})$. Then, for all $\tau > 0$ and all $a, y \in \mathbb{R}^q$, the series*

$$\sum_{h\in\Lambda^{-1}} e^{-\tau\pi^2 h^2} e^{2\pi i h\cdot a} \int_{\mathcal{G}} F(x) e^{-2\pi i x\cdot(h-y)}\, dV(x) \tag{14.25}$$

is absolutely and uniformly convergent. Moreover, for all $a, y \in \mathbb{R}^q$ we have the lattice point identity

$$\sum_{\substack{a+g\in\overline{\mathcal{G}}\\ g\in\Lambda}}{}' e^{2\pi i y\cdot(a+g)} F(a+g) \tag{14.26}$$

$$= \lim_{\substack{\tau\to 0\\ \tau>0}} \frac{1}{\|\mathcal{F}_\Lambda\|} \sum_{h\in\Lambda^{-1}} e^{-\tau\pi^2 h^2}\, e^{2\pi i h\cdot a} \int_{\mathcal{G}} F(x) e^{-2\pi i x\cdot(h-y)}\, dV(x), \tag{14.27}$$

where

$$\sum_{\substack{a+g\in\overline{\mathcal{G}}\\ g\in\Lambda}}{}' e^{2\pi i y(a+g)} F(a+g) = \sum_{\substack{a+g\in\overline{\mathcal{G}}\\ g\in\Lambda}} \alpha_{\mathcal{G}}(a+g) e^{2\pi i y\cdot(a+g)}. \tag{14.28}$$

14.4 Bandlimited Gauss–Weierstrass Shannon-Type Sampling

Let $\mathcal{G}$ be a regular region in $\mathbb{R}^q$ such that $\mathcal{G} \subset \mathcal{F}_\Lambda$. Suppose that Λ is a lattice in $\mathbb{R}^q$ generated by the basis vectors $g_1, \ldots, g_q \in \mathbb{R}^q$. Assume that F is of class $C^{(2)}(\overline{\mathcal{G}})$. Then it follows that

$$\int_{\mathcal{F}_\Lambda} \sum_{\substack{a+g\in\overline{\mathcal{G}}\\ g\in\Lambda}}{}' F(a+g) e^{-2\pi i y\cdot(a+g)}\, dV(a) = \int_{\mathcal{G}} F(x)\, e^{-2\pi i x\cdot y} dV(x) = F_{\mathcal{G}}^\wedge(y). \tag{14.29}$$

Together with Theorem 14.5 this leads to the exponentially accelerated Shannon-type sampling identity

$$F_{\mathcal{G}}^{\wedge}(y) = \lim_{\substack{\tau\to 0\\ \tau>0}} \sum_{h\in\Lambda^{-1}} e^{-\tau\pi^2 h^2} F_{\mathcal{G}}^{\wedge}(h) \frac{1}{\|\mathcal{F}_\Lambda\|} \int_{\mathcal{F}_\Lambda} e^{2\pi i a\cdot(h-y)}\, dV(a), \tag{14.30}$$

holds true for all $y \in \mathbb{R}^q$, where

$$\frac{1}{\|\mathcal{F}_\Lambda\|} \int_{\mathcal{F}_\Lambda} e^{2\pi i a\cdot(h-y)}\, dV(a) = \frac{\sin(\pi(g_1\cdot(y-h)))}{\pi(g_1\cdot(y-h))} \cdot \ldots \cdot \frac{\sin(\pi(g_q\cdot(y-h)))}{\pi(g_q\cdot(y-h))}. \tag{14.31}$$

For given regular regions $\mathcal{G}, \mathcal{H} \subset \mathbb{R}^q$ with $\overline{\mathcal{G}}, \overline{\mathcal{H}} \subset \overline{\mathcal{F}_\Lambda}$ we obtain from Theorem 14.5

$$\begin{aligned}
&\int_{\mathcal{H}} \sum_{\substack{a+g\in\overline{\mathcal{G}}\\ g\in\Lambda^{-1}}}{}^{'} e^{-2\pi i y\cdot(a+g)} F(a+g)\, dV(a) \qquad\qquad (14.32)\\
&= \int_{\mathcal{G}\cap\mathcal{H}} F(a) e^{-2\pi i y\cdot a} dV(a)\\
&= F_{\mathcal{G}\cap\mathcal{H}}^{\wedge}(y)\\
&= \lim_{\substack{\tau\to 0\\ \tau>0}} \sum_{h\in\Lambda^{-1}} e^{-\tau\pi^2 h^2} \underbrace{\int_{\mathcal{G}} F(x)\, e^{-2\pi i x\cdot h}\, dV(x)}_{=F_{\mathcal{G}}^{\wedge}(h)} \frac{1}{\|\mathcal{F}_\Lambda\|} \underbrace{\int_{\mathcal{H}} e^{2\pi i a\cdot(h-y)}\, dV(a)}_{=K_{\mathcal{H}}(h-y)},
\end{aligned}$$

where, of course, the series on the right side is absolutely and uniformly convergent on $\mathbb{R}^q$.

In other words, by using Gauss–Weierstrass means, general geometries in Shannon–type sampling become discussible (without any number-theoretic restriction), however, under the assumption that the function F under consideration is twice continuously differentiable. It should be remarked that the cardinal series shows an exponentially accelerated convergence.

Remark 14.6. *Based on a more subtle estimation procedure (under the special choice of $a = y = 0$), W. Freeden [2011] succeeded in generalizing Theorem 14.5 to functions $F \in \mathrm{C}^{(0)}(\overline{\mathcal{G}})$ (instead of $F \in \mathrm{C}^{(2)}(\overline{\mathcal{G}})$). Essential tool is the functional equation for the Theta function. The details (for general choices of $a, y \in \mathbb{R}^q$) will be presented in Section 15.2.*

Chapter 15

From Gauss–Weierstrass to Ordinary Lattice Point Poisson–Type Summation

First, our intention for the use of Theta functions is to formulate multivariate Hardy–Landau-type lattice point formulas in Gauss–Weierstrass summability, thereby guaranteeing successfully the validity of both mere continuity for the weight functions F and general geometry of $\mathcal{G}$.

Second, the lattice point identities expressed in Gauss–Weierstrass summability serve as initial tools for developing bandlimited Shannon–type sampling formulas. Actually, we are able to guarantee that the resulting bandlimited, i.e., $\mathcal{G}$-Fourier transformed values–based Shannon-type sampling formulas obtained by a "smoothing" integration process are also valid under non-summability assumptions. This is a remarkable theoretical feature, even though Gauss–Weierstrass summability in Shannon-type sampling formulas is senseful and extremely helpful for fast numerical calculations.

Third, our approach enables us to handle explicitly all situations resulting in over- and undersampling, where the aliasing error becomes concretely derivable within the multivariate lattice point context.

15.1 Theta Function and Functional Equation

We start with some information about multivariate Theta functions (for the univariate theory and some more multivariate details see, e.g., W. Freeden [2011]).

Definition 15.1. *For arbitrary points $x, y \in \mathbb{R}^q$, and arbitrary lattices $\Lambda \subset \mathbb{R}^q$ $\vartheta_n^{(q)}(\cdot\,; x, y; \Lambda)$ given by*

$$\vartheta_n^{(q)}(\sigma; x, y; \Lambda) = \sum_{g\in\Lambda} e^{-\pi\sigma|g-x|^2} H_n(q; g-x)\, e^{2\pi i g\cdot y}, \quad \sigma \in \mathbb{C},\ \Re(\sigma) > 0, \quad (15.1)$$

is called the Theta function of degree n and dimension q (note that, in the (general) formulation (15.1), $\vartheta_n^{(q)}(\cdot\,; x, y; \Lambda)$ is dependent on the choice of a homogeneous harmonic polynomial $H_n : x \mapsto H_n(q; x) = |x|^n Y_n(q; \xi), x \in \mathbb{R}^q$, $x = |x|\ \xi,\ \xi \in \mathbb{S}^{q-1}$, of degree n and dimension q, where $Y_n(q; \cdot)$ is a member of class Harm$_n$ *of spherical harmonics of degree n and dimension q).*

Functional Equation. For the function $\Xi : \mathbb{R}^q \to \mathbb{C}$ with exponential decay given by

$$\Xi(z) = e^{-\pi\sigma|z-x|^2} H_n(q; z - x)\ e^{2\pi i z \cdot y}, \quad z \in \mathbb{R}^q, \tag{15.2}$$

all classical variants of the Poisson summation formula are certainly true, and the series on both sides are absolutely convergent (as required by the approach, e.g., due to E.M. Stein, G. Weiss [1971]). This yields

$$\sum_{g\in\Lambda} \Xi(g) = \frac{1}{\|\mathcal{F}_\Lambda\|} \sum_{h\in\Lambda^{-1}} \int_{\mathbb{R}^q} \Xi(z)\ e^{2\pi i h \cdot z}\ dV(z). \tag{15.3}$$

When the formula (15.3) is explicitly written out for the function (15.2), we obtain the following identity:

$$\begin{aligned} &\vartheta_n^{(q)}(\sigma; x, y; \Lambda) \\ &\quad= \sum_{g\in\Lambda} e^{-\pi\sigma|g-x|^2} H_n(q; g - x)\ e^{2\pi i g\cdot y} \\ &\quad= \frac{1}{\|\mathcal{F}_\Lambda\|} \sum_{h\in\Lambda^{-1}} e^{2\pi i x\cdot(y+h)} \int_{\mathbb{R}^q} e^{-\pi\sigma|z-x|^2} H_n(q; z - x) e^{2\pi i (z-x)\cdot(y+h)}\ dV(z). \end{aligned} \tag{15.4}$$

Introducing polar coordinates $z - x = r\eta$ and $y + h = \rho\ \xi$ with $\eta, \xi \in \mathbb{S}^{q-1}$, we have

$$\begin{aligned} &\vartheta_n^{(q)}(\sigma; x, y; \Lambda) \\ &\quad= \frac{1}{\|\mathcal{F}_\Lambda\|} \sum_{h\in\Lambda^{-1}} e^{2\pi i x\cdot(\rho\xi)} \\ &\qquad\qquad \times \int_0^\infty \left(\int_{\mathbb{S}^{q-1}} Y_n(q; \eta) e^{2\pi i r\rho\xi\cdot\eta}\ dS(\eta) \right) e^{-\pi\sigma r^2}\ r^{n+q-1}\ dr. \end{aligned} \tag{15.5}$$

The formula of Funk and Hecke of the theory of spherical harmonics in connection with the integral representation (6.54) of the Bessel function gives

$$\int_{\mathbb{S}^{q-1}} Y_n(q; \eta) e^{2\pi i r\rho\xi\cdot\eta}\ dS(\eta) = i^n\ \|\mathbb{S}^{q-1}\|\ J_n(q; 2\pi r\rho)\ Y_n(q; \xi). \tag{15.6}$$

Thus we obtain

$$\begin{aligned} \vartheta_n^{(q)}(\sigma; x, y; \Lambda) \;=\; & \frac{i^n \|\mathbb{S}^{q-1}\|}{\|\mathcal{F}_\Lambda\|} \sum_{h\in\Lambda^{-1}} e^{2\pi i x\cdot(y+h)}\ Y_n\left(q; \frac{y+h}{|y+h|}\right) \\ & \times \int_0^\infty e^{-\pi\sigma r^2} J_n(q; 2\pi r\rho)\ r^{n+q-1}\ dr. \end{aligned} \tag{15.7}$$

The integral on the right side of (15.7) may be handled with specific technicalities known from the theory of Bessel functions. In fact, we are able to borrow from W. Freeden [2011] the following lemma.

Lemma 15.2. *For $\rho > 0, n \in \mathbb{N}_0$, and $\sigma \in \mathbb{C}$ with $\Re(\sigma) > 0$ we have*

$$\int_0^\infty e^{-\pi\sigma r^2} J_n(q; 2\pi r\rho)\, r^{n+q-1}\, dr \;=\; \tfrac{1}{2}\Gamma\left(\tfrac{q}{2}\right) \frac{\rho^n}{\pi^{\frac{q}{2}}\sigma^{n+\frac{q}{2}}}\, e^{-\frac{\pi\rho^2}{\sigma}}.$$

In particular, with $x, y \in \mathbb{R}^q$ and $\rho = |x-y|$ we therefore obtain the identity (cf. Lemma 8.14)

$$\|\mathbb{S}^{q-1}\| \int_0^\infty e^{-\pi\sigma r^2} J_0(q; 2\pi|x-y|r) r^{q-1}\, dr = \left(\frac{1}{t}\right)^{\frac{q}{2}} e^{-\frac{\pi}{\sigma}(x-y)^2}. \tag{15.8}$$

Now, by virtue of Lemma 15.2 it follows from (15.7) that

$$\begin{aligned} &\vartheta_n^{(q)}(\sigma; x, y; \Lambda) \qquad\qquad (15.9)\\ &= \sum_{g\in\Lambda} e^{-\pi\sigma|g-x|^2} H_n(q; g-x)\, e^{2\pi i g\cdot y} \\ &= \frac{i^n \left\|\mathbb{S}^{q-1}\right\|}{2\|\mathcal{F}_\Lambda\|} \frac{\Gamma\left(\frac{q}{2}\right)}{\pi^{\frac{q}{2}}\sigma^{n+\frac{q}{2}}}\, e^{2\pi i x\cdot y} \sum_{h\in\Lambda^{-1}} e^{-\frac{\pi}{\sigma}|h+y|^2} H_n(q; h+y)\, e^{2\pi i h\cdot x}, \end{aligned}$$

so that the explicit value for $\|\mathbb{S}^{q-1}\|$, i.e., $\|\mathbb{S}^{q-1}\| = 2\pi^{\frac{q}{2}}\left(\Gamma\left(\frac{q}{2}\right)\right)^{-1}$, finally implies that

$$\begin{aligned} &\vartheta_n^{(q)}(\sigma; x, y; \Lambda) \qquad\qquad (15.10)\\ &= \sum_{g\in\Lambda} e^{-\pi\sigma|g-x|^2} H_n(q; g-x)\, e^{2\pi i g\cdot y} \\ &= \frac{i^n}{\|\mathcal{F}_\Lambda\|} e^{2\pi i x\cdot y}\sigma^{-n-\frac{q}{2}} \underbrace{\sum_{h\in\Lambda^{-1}} e^{-\frac{\pi}{\sigma}|h+y|^2} H_n(q; h+y)\, e^{2\pi i h\cdot x}}_{=\,\vartheta_n^{(q)}\left(\frac{1}{\sigma}; -y, x; \Lambda^{-1}\right)}. \end{aligned}$$

Consequently we obtain the *functional equation of the Theta function of degree n and dimension q.*

Theorem 15.3. *For $\sigma \in \mathbb{C}$ with $\Re(\sigma) > 0$, the Theta function $\vartheta_n^{(q)}(\cdot; x, y; \Lambda)$ is holomorphic, and we have*

$$\vartheta_n^{(q)}(\sigma; x, y; \Lambda) = \frac{i^n}{\|\mathcal{F}_\Lambda\|} e^{2\pi i x\cdot y}\sigma^{-n-\frac{q}{2}}\, \vartheta_n^{(q)}\left(\frac{1}{\sigma}; -y, x; \Lambda^{-1}\right). \tag{15.11}$$

15.2 Poisson-Type Summation over Regular Regions (Gauss–Weierstrass Approach)

Now, let $\mathcal{G}$ be a regular region in $\mathbb{R}^q$. Suppose that x is an arbitrary but fixed point of the Euclidean space $\mathbb{R}^q$. Let F be a continuous function on $\overline{\mathcal{G}}$.

An interesting question that arises from our previous Gauss–Weierstrass context is in what respect the infinite series

$$\lim_{N\to\infty} \frac{1}{\|\mathcal{F}_\Lambda\|} \sum_{\substack{|h-y|\leq N\\ h\in\Lambda^{-1}}} e^{2\pi a\cdot h} \int_{\mathcal{G}} F(x)\; e^{-2\pi i x\cdot(h-y)}\; dV(x) \tag{15.12}$$

for a continuous function on $\overline{\mathcal{G}}$ can be "summed" to the finite sum

$$\sum_{\substack{a+g\in\overline{\mathcal{G}}\\ g\in\Lambda}}{}' e^{2\pi i y\cdot(a+g)} F(a+g) = \sum_{\substack{a+g\in\overline{\mathcal{G}}\\ g\in\Lambda}} \alpha(a+g)\; e^{2\pi i y\cdot(a+g)} F(a+g), \tag{15.13}$$

where $\alpha(g+a)$ (more accurately, $\alpha_{\mathcal{G}}(g+a)$) denotes the solid angle subtended by the boundary $\partial\mathcal{G}$ at $g+a$.

As a matter of fact, the answer cannot be given directly because of serious convergence problems of the series (15.12) and the discontinuous behavior of (15.13) as a function of the variable a. Thus, our basic idea is to identify (15.12) and (15.13) in the sense of the Gauss–Weierstrass means. Essential tools are the Fourier inversion formula and the functional equation of the Theta function in $\mathbb{R}^q$.

To be more concrete, for all values $\tau\in\mathbb{R}$ with $\tau>0$, $x\in\mathbb{R}^q$, and $F\in \mathrm{C}^{(0)}(\overline{\mathcal{G}})$ we readily see from Theorem 15.3 that

$$\begin{aligned} &\frac{1}{\|\mathcal{F}_\Lambda\|} \int_{\mathcal{G}} \vartheta_0^{(q)}\left(\pi\tau; 0, x-a; \Lambda^{-1}\right)\; e^{2\pi i y\cdot x} F(x)\; dV(x) \\ &= \quad (\pi\tau)^{\frac{q}{2}} \int_{\mathcal{G}} \vartheta_0^{(q)}\left(\frac{1}{\pi\tau}; a-x; 0; \Lambda\right)\; e^{2\pi i y\cdot x}\; F(x)\; dV(x), \end{aligned} \tag{15.14}$$

So, in equivalent series notation, we have

$$\begin{aligned} &\frac{1}{\|\mathcal{F}_\Lambda\|} \int_{\mathcal{G}} \sum_{h\in\Lambda^{-1}} e^{-\tau\pi^2 h^2} e^{2\pi i y\cdot x} F(x) e^{2\pi i h\cdot(a-x)}\; dV(x) \\ &= \quad (\tau\pi)^{-\frac{q}{2}} \int_{\mathcal{G}} \sum_{g\in\Lambda} e^{-\frac{|x-(g+a)|^2}{\tau}} e^{2\pi i y\cdot x} F(x)\; dV(x) \end{aligned} \tag{15.15}$$

for all $\tau > 0$ and all $x \in \mathbb{R}^q$, for arbitrary but fixed $a, y \in \mathbb{R}^q$, and all regular regions $\mathcal{G}$ as well as for all arbitrary (weight) functions F of class $\mathrm{C}^{(0)}(\overline{\mathcal{G}})$.

Interchanging sums and integrals we immediately obtain

$$(\tau\pi)^{-\frac{q}{2}} \sum_{g\in\Lambda} \int_{\mathcal{G}} e^{-\frac{|x-(g+a)|^2}{\tau}} e^{2\pi i y\cdot x} F(x)\, dV(x) \tag{15.16}$$
$$= \frac{1}{\|\mathcal{F}_\Lambda\|} \sum_{h\in\Lambda^{-1}} e^{-\tau\pi^2 h^2} e^{2\pi i h\cdot a} \int_{\mathcal{G}} F(x) e^{-2\pi i x\cdot(h-y)}\, dV(x).$$

After these preparations we are led to the main result of this chapter (cf. W. Freeden [2011]).

Theorem 15.4. *(Poisson Summation-Type Formula over Regular Regions $\mathcal{G}$ in Gauss–Weierstrass Summability). Let $\mathcal{G} \subset \mathbb{R}^q$ be a regular region. Suppose that F is a member of class $\mathrm{C}^{(0)}(\overline{\mathcal{G}})$. Then, for all $\tau > 0$ and all $a, y \in \mathbb{R}^q$, the series*

$$\sum_{h\in\Lambda^{-1}} e^{-\tau\pi^2 h^2} e^{2\pi i h\cdot a} \int_{\mathcal{G}} F(x) e^{-2\pi i x\cdot(h-y)}\, dV(x) \tag{15.17}$$

is absolutely convergent. Moreover, for all $a \in \mathbb{R}^q$, we have the lattice point identity

$$\sum_{\substack{a+g\in\overline{\mathcal{G}}\\ g\in\Lambda}}{}' e^{2\pi i y\cdot(a+g)} F(a+g) \tag{15.18}$$
$$= \lim_{\substack{\tau\to 0\\ \tau>0}} \frac{1}{\|\mathcal{F}_\Lambda\|} \sum_{h\in\Lambda^{-1}} e^{-\tau\pi^2 h^2} e^{2\pi i h\cdot a} \int_{\mathcal{G}} F(x) e^{-2\pi i x\cdot(h-y)}\, dV(x).$$

Proof. The key idea (cf. Lemma 14.3) is the already known *Fourier inversion formula in Gauss–Weierstrass nomenclature.* It states

$$\lim_{\substack{\tau\to 0\\ \tau>0}} (\tau\pi)^{-\frac{q}{2}} \int_{\mathcal{G}} e^{-\frac{|x-w|^2}{\tau}} e^{2\pi i y\cdot x} F(x)\, dV(x) = \alpha_{\mathcal{G}}(w)\, e^{2\pi i y\cdot w}\, F(w), \tag{15.19}$$

where $\alpha_{\mathcal{G}}(w)$ is the solid angle subtended at $w \in \overline{\mathcal{G}}$ by the surface $\partial\mathcal{G}$.

For given $a, y \in \mathbb{R}^q$, take a sufficiently large positive constant R such that $|a + x| \leq R$ for all $x \in \overline{\mathcal{G}}$ and $|x - (g + a)|^2 \geq \frac{1}{2}|g + a|^2$ for all $g \in \Lambda$ with

$|g+a| > R$. Then, because of the continuity of F, we see that

$$\begin{aligned}
&(\tau\pi)^{-\frac{q}{2}} \sum_{\substack{|g+a|>R \\ g\in\Lambda}} \int_{\mathcal{G}} e^{-\frac{|x-(g+a)|^2}{\tau}} e^{2\pi i y\cdot x} F(x)\, dV(x) \qquad (15.20)\\
&= O\left((\tau\pi)^{-\frac{q}{2}} \sum_{\substack{|g+a|\geq R \\ g\in\Lambda}} e^{\frac{-|g+a|^2}{2\tau}} \right)\\
&= O\left((\tau\pi)^{-\frac{q}{2}} \int_{\substack{|a+x|\geq R \\ x\in\mathbb{R}^q}} e^{\frac{-|a+x|^2}{2\tau}}\, dV(x) \right)
\end{aligned}$$

for $\tau \to 0$. Moreover, there exists a positive constant $T > 0$ such that $\operatorname{dist}(g, \overline{\mathcal{G}}) \geq T$ holds for all $g \notin \overline{\mathcal{G}}$. Thus, we are able to verify that

$$\begin{aligned}
&\sum_{\substack{g+a\notin\overline{\mathcal{G}} \\ |g+a|\leq R \\ g\in\Lambda}} (\tau\pi)^{-\frac{q}{2}} \int_{\mathcal{G}} e^{-\frac{|x-(g+a)|^2}{\tau}} e^{2\pi i y\cdot x} F(x)\, dV(x) \qquad (15.21)\\
&= O\left((\tau\pi)^{-\frac{q}{2}} \sum_{\substack{g+a\notin\overline{\mathcal{G}} \\ |g+a|\leq R \\ g\in\Lambda}} \int_{\mathcal{G}} e^{-\frac{|x-(g+a)|^2}{\tau}}\, dV(x) \right)\\
&= O\left((\tau\pi)^{-\frac{q}{2}} \left(\int_{\substack{|x|\geq T \\ x\in\mathbb{R}^q}} e^{-\frac{|x|^2}{\tau}}\, dV(y) \right) \sum_{\substack{g+a\notin\overline{\mathcal{G}} \\ |g+a|\leq R \\ g\in\Lambda}} 1 \right)\\
&= O\left((\tau\pi)^{-\frac{q}{2}} \int_{\substack{|x|\geq T \\ x\in\mathbb{R}^q}} e^{-\frac{|x|^2}{\tau}}\, dV(x) \right)
\end{aligned}$$

for $\tau \to 0$. Collecting our results we therefore obtain the asymptotic result

$$\begin{aligned}
&\sum_{\substack{g+a\notin\overline{\mathcal{G}} \\ g\in\Lambda}} (\tau\pi)^{-\frac{q}{2}} \int_{\mathcal{G}} e^{-\frac{|x-(g+a)|^2}{\tau}} e^{2\pi i y\cdot x} F(x)\, dV(x) \qquad (15.22)\\
&= O\left(\int_{\frac{R^2}{2\tau}}^{\infty} e^{-r} r^{\frac{q}{2}-1} dr + \int_{\frac{T^2}{\tau}}^{\infty} e^{-r} r^{\frac{q}{2}-1} dr \right)\\
&= o(1).
\end{aligned}$$

As is well-known, for every $\rho > 0$, we have

$$(\tau\pi)^{-\frac{q}{2}} \int_{\mathbb{B}_\rho^q} e^{-\frac{|x|^2}{\tau}} \, dV(x) = \|\mathbb{S}^{q-1}\| \, \pi^{-\frac{q}{2}} \int_0^\rho e^{-\frac{r^2}{\tau}} r^{q-1} \, dr \tag{15.23}$$

$$= \frac{1}{2} \|\mathbb{S}^{q-1}\| \, \pi^{-\frac{q}{2}} \int_0^{\frac{\rho^2}{\tau}} e^{-s} s^{\frac{q}{2}-1} \, ds.$$

Furthermore, it is known that

$$\frac{1}{2} \|\mathbb{S}^{q-1}\| \, \pi^{-\frac{q}{2}} = \left(\Gamma(\frac{q}{2})\right)^{-1}, \tag{15.24}$$

such that

$$\lim_{\substack{\tau\to 0 \\ \tau>0}} (\tau\pi)^{-\frac{q}{2}} \int_{\mathbb{B}_\rho^q} e^{-\frac{|x|^2}{\tau}} \, dV(x) = 1. \tag{15.25}$$

Now, it follows from the Fourier inversion formula in Gauss–Weierstrass nomenclature (15.19) that

$$\lim_{\substack{\tau\to 0 \\ \tau>0}} (\tau\pi)^{-\frac{q}{2}} \sum_{\substack{g+a\in\overline{\mathcal{G}} \\ g\in\Lambda}} \int_{\mathcal{G}} e^{-\frac{|x-(g+a)|^2}{\tau}} e^{2\pi i y\cdot x} F(x) \, dV(x) \tag{15.26}$$

$$= \sum_{\substack{g+a\in\overline{\mathcal{G}} \\ g\in\Lambda}} \alpha_{\mathcal{G}}(g+a) \, e^{2\pi i y\cdot(a+g)} F(g+a).$$

Combining the relations (15.16), (15.19), and (15.26) we finally obtain the assertion of Theorem 15.4

$$\sum_{\substack{a+g\in\overline{\mathcal{G}} \\ g\in\Lambda}}{}' \, e^{2\pi i y\cdot(a+g)} F(a+g) \tag{15.27}$$

$$= \lim_{\substack{\tau\to 0 \\ \tau>0}} \frac{1}{\|\mathcal{F}_\Lambda\|} \sum_{h\in\Lambda^{-1}} e^{-\tau\pi^2 h^2} e^{2\pi i h\cdot a} \int_{\mathcal{G}} F(x) e^{-2\pi i x\cdot(h-y)} \, dV(x).$$

□

15.3 Poisson-Type Summation over Regular Regions (Ordinary Approach)

It is interesting to investigate under which assumptions Gauss–Weierstrass convergence allows the transition to ordinary convergence, i.e., the limit $\lim_{\tau\to 0} \ldots$ can be interchanged with the sum $\sum_{h\in\Lambda^{-1}} \ldots$. We already know some conditions under which Gauss–Weierstrass summability can be avoided (for more details see W. Freeden [2011]). Here, we only present some significant realizations.

Theorem 15.5. *Let Λ be an arbitrary lattice in $\mathbb{R}^q$. Let $\mathcal{G} \subset \mathbb{R}^q$ be a regular region. Suppose that F is of class $\mathrm{C}^{(0)}(\overline{\mathcal{G}})$. Furthermore, assume that*

$$\lim_{N\to\infty} \frac{1}{\|\mathcal{F}_\Lambda\|} \sum_{\substack{|h-y|\le N \\ h\in\Lambda^{-1}}} e^{2\pi i a\cdot h} \int_{\mathcal{G}} F(x) e^{-2\pi i x\cdot(h-y)} dV(x) \tag{15.28}$$

is convergent. Then we have

$$\begin{aligned} &\sum_{\substack{a+g\in\overline{\mathcal{G}} \\ g\in\Lambda}}{}' e^{2\pi i y\cdot(a+g)} F(a+g) \\ &\quad = \lim_{N\to\infty} \frac{1}{\|\mathcal{F}_\Lambda\|} \sum_{\substack{|h-y|\le N \\ h\in\Lambda^{-1}}} e^{2\pi i a\cdot h} \int_{\mathcal{G}} F(x) e^{-2\pi i x\cdot(h-y)} dV(x). \end{aligned} \tag{15.29}$$

Proof. Under the assumption of Theorem 15.5, the series

$$\lim_{N\to\infty} \frac{1}{\|\mathcal{F}_\Lambda\|} \sum_{\substack{|h-y|\le N \\ h\in\Lambda^{-1}}} e^{-\tau\pi h^2} e^{2\pi i a\cdot h} \int_{\mathcal{G}} F(x)\, e^{-2\pi i x\cdot(h-y)} dV(x) \tag{15.30}$$

is uniformly convergent with respect to τ. Hence, we are allowed to interchange the limits $\tau \to 0$ and $N \to \infty$. This proves the assertion of Theorem 15.5. □

Next we are concerned with homogeneous boundary weights. A rough condition to establish sufficient criteria for the validity of the Poisson summation formula (15.28) on regular regions $\mathcal{G} \subset \mathbb{R}^q$ is as follows:

Lemma 15.6. *Let Λ be an arbitrary lattice in $\mathbb{R}^q$. Let $\mathcal{G}$ be a regular region in $\mathbb{R}^q$. Suppose that $F \in \mathrm{C}^{(2m)}(\overline{\mathcal{G}})$, $m > \frac{q}{2}$, satisfies the homogeneous boundary conditions*

$$\Delta^k F \mid_{\partial\mathcal{G}} = 0, \quad 0 \le k \le m-1, \tag{15.31}$$

and

$$\nabla\Delta^k F \mid_{\partial\mathcal{G}} = 0, \quad 0 \le k \le m-2. \tag{15.32}$$

Then the series

$$\sum_{h\in\Lambda^{-1}} e^{2\pi i a\cdot x} \int_{\mathcal{G}} F(x) e^{-2\pi i x\cdot(h-y)}\; dV(x) \tag{15.33}$$

is convergent, and (15.28) is valid.

Proof. From the extended second Green theorem we obtain

$$
\begin{aligned}
&\int_{\mathcal{G}} F(x)\, e^{-2\pi i a\cdot x}\, dV(x) \qquad (15.34)\\
&= \frac{1}{(-4\pi^2 a^2)^m} \int_{\mathcal{G}} (\Delta^m F(y))\, e^{-2\pi i a\cdot x}\, dV(x)\\
&\quad - \sum_{k=0}^{m-1} \frac{1}{(-4\pi^2 a^2)^{k+1}} \int_{\partial\mathcal{G}} \left(\frac{\partial}{\partial\nu}\Delta^k F(y)\right) e^{-2\pi i a\cdot x}\, dS(x)\\
&\quad + \sum_{k=0}^{m-1} \frac{1}{(-4\pi^2 a^2)^{k+1}} \int_{\partial\mathcal{G}} (\Delta^k F(x))\, \frac{\partial}{\partial\nu} e^{-2\pi i a\cdot x}\, dS(x)
\end{aligned}
$$

for all $a \in \mathbb{R}^q$, $a \neq 0$, such that under the homogeneous boundary assumptions imposed on the function F under consideration we find that

$$
\begin{aligned}
&\int_{\mathcal{G}} F(x)\, e^{-2\pi i a\cdot y}\, dV(x) \qquad (15.35)\\
&= \frac{1}{(-4\pi^2 a^2)^m} \int_{\mathcal{G}} (\Delta^m F(x))\ e^{-2\pi i a\cdot x}\, dV(x)\\
&\quad - \frac{1}{(-4\pi^2 a^2)^m} \int_{\partial\mathcal{G}} \left(\frac{\partial}{\partial\nu}\Delta^{m-1} F(x)\right)\ e^{-2\pi i a\cdot x}\, dS(x).
\end{aligned}
$$

Because of the continuity properties of the weight function F and its derivatives $\Delta^m F$, $\nabla\Delta^{m-1}F$ on $\overline{\mathcal{G}}$, it follows that

$$
\int_{\mathcal{G}} F(x)\, e^{-2\pi i a\cdot x}\, dV(x) = O\left(\frac{1}{|a|^{2m}}\right), \quad |a| \to \infty. \qquad (15.36)
$$

Consequently, under the assumption $2m > q$, we are able to guarantee the absolute and uniform convergence of the series (15.33), such that Theorem 15.5 guarantees the required result. □

Lemma 15.6 opens a way out of Gauss–Weierstrass summability in higher dimensions for lattice point sums under the assumption of certain vanishing weights on the boundary. Furthermore, it should be noted that metaharmonic theory (cf. W. Freeden [2011]) is able to reduce the number of homogeneous conditions in Lemma 15.6 by far, when the spherical case is under specific consideration.

Part VII

Multivariate Shannon-Type Sampling Formulas over Regular Regions

Chapter 16

Shannon-Type Sampling Based on Poisson-Type Summation Formulas

For a function $F \in \mathrm{C}^{(0)}(\overline{\mathcal{G}})$, the lattice point identities

$$\sum_{\substack{a+g\in\overline{\mathcal{G}}\\ g\in\Lambda}}{}' F(a+g)\, e^{-2\pi i y\cdot(a+g)} \tag{16.1}$$

$$= \lim_{\substack{\tau\to 0\\ \tau>0}} \sum_{h\in\Lambda^{-1}} e^{-\tau\pi^2 h^2} \frac{1}{\|\mathcal{F}_\Lambda\|}\, e^{2\pi i h\cdot a} \int_{\mathcal{G}} F(x) e^{-2\pi i x\cdot(h+y)}\, dV(x)$$

and

$$\lim_{\substack{\tau\to 0\\ \tau>0}} \sum_{g\in\Lambda} e^{-\tau\pi^2 g^2} e^{2\pi i a\cdot(g+y)} \int_{\mathcal{G}} F(x) e^{-2\pi i x\cdot(g+y)}\, dV(x) \tag{16.2}$$

$$= \sum_{\substack{a+h\in\overline{\mathcal{G}}\\ h\in\Lambda^{-1}}}{}' F(a+h)\, \frac{1}{\|\mathcal{F}_\Lambda\|}\, e^{-2\pi i y\cdot h}$$

can be viewed as two equivalent formulations of Theorem 15.4. Nevertheless, it is the goal of this chapter to document that the identities (16.1) and (16.2), respectively, offer completely different entry points for purposes of Shannon-type sampling, namely

- a cardinal series resulting from the right side of (16.1) and summing up functional values of the Fourier transform of F over $\mathcal{G}$ in y-shifted lattice points (this case has been already studied by W. Freeden, M.Z. Nashed [2015]),
- a cardinal series resulting from the left side of (16.2) and summing up functional values of F over $\mathcal{G}$ in a-shifted lattice points.

Clearly, the role of functional values and Fourier transformed functional values of F in (16.1) and (16.2) is exchangeable, since the Fourier transformed values of the Fourier transform canonically lead back to the functional values via the Fourier inversion formula. Nevertheless, it will be shown in this chapter that the different entry points of the same Poisson-type summation formula provided by (16.1) and (16.2), respectively, may be also regarded as different points of departure to Shannon-type sampling. As a result, we indeed obtain two different variants of Shannon-type sampling over regular regions in the sense of Gauss–Weierstrass summability, and later in ordinary sense.

16.1 Fourier-Transformed Values–Based Shannon-Type Sampling (Gauss–Weierstrass Approach)

As already mentioned, this part of Shannon-type sampling is essentially based on the publication W. Freeden, M.Z. Nashed [2015]:

Integrating the identity (16.1) over a regular region $\mathcal{H} \subset \mathbb{R}^q$, $q \geq 2$, we are led to a *Fourier-transformed values–based Shannon-type sampling approach.*

Lemma 16.1. *If $\mathcal{G}, \mathcal{H} \subset \mathbb{R}^q$ are arbitrary regular regions, then*

$$\begin{aligned} &\int_{\mathcal{H}} \sum_{\substack{a+g\in\overline{\mathcal{G}} \\ g\in\Lambda}}{}' F(a+g)\, e^{-2\pi i y\cdot(a+g)}\, dV(a) \qquad (16.3) \\ &= \lim_{\substack{\tau\to 0 \\ \tau>0}} \sum_{h\in\Lambda^{-1}} e^{-\tau\pi^2 h^2} \underbrace{\int_{\mathcal{G}} F(x) e^{-2\pi i h\cdot x}\, dV(x)}_{=F^{\wedge}_{\mathcal{G}}(h)} \frac{1}{\|\mathcal{F}_\Lambda\|} \underbrace{\int_{\mathcal{H}} e^{2\pi i a\cdot(h-y)}\, dV(a)}_{=K_{\mathcal{H}}(h-y)} \end{aligned}$$

is valid for all points $y \in \mathbb{R}^q$, where the lattice constant $\frac{1}{\|\mathcal{F}_\Lambda\|}$ can be regarded as an indicator of the sampling density.

We denote by $\Lambda(\mathcal{H})$ the finite set of lattice points $g' \in \Lambda \subset \mathbb{R}^q$, such that $\mathcal{F}_\Lambda + \{g'\}$ contains points of $\mathcal{H}$, i.e., in mathematical formalism

$$\Lambda(\mathcal{H}) = \{g' \in \Lambda : (\mathcal{F}_\Lambda + \{g'\}) \cap \mathcal{H} \neq \emptyset\}. \qquad (16.4)$$

Keeping this abbreviation in mind, we are able to see that

$$\begin{aligned} &\int_{\mathcal{H}} \sum_{\substack{a+g\in\overline{\mathcal{G}} \\ g\in\Lambda}}{}' F(a+g)\, e^{-2\pi i y\cdot(a+g)}\, dV(a) \\ &= \sum_{g'\in\Lambda(\mathcal{H})} \int_{\mathcal{H}_{g'}} \sum_{\substack{a+g\in\overline{\mathcal{G}} \\ g\in\Lambda}}{}' F(a+g)\, e^{-2\pi i y\cdot(a+g)}\, dV(a), \qquad (16.5) \end{aligned}$$

where the set $\mathcal{H}_{g'}, g' \in \Lambda(\mathcal{H})$, is given by

$$\mathcal{H}_{g'} = (\mathcal{F}_\Lambda + \{g'\}) \cap \mathcal{H}, \quad g' \in \Lambda(\mathcal{H}). \qquad (16.6)$$

Since the function

$$a \mapsto \sum_{\substack{a+g\in\overline{\mathcal{G}} \\ g\in\Lambda}}{}' F(a+g)\, e^{-2\pi i y\cdot(a+g)}, \quad a \in \mathbb{R}^q, \qquad (16.7)$$

is Λ–periodic, the right side of (16.5) can be rewritten in the form

$$\begin{aligned}
&\sum_{g'\in\Lambda(\mathcal{H})} \int_{\mathcal{H}_{g'}} \sum_{\substack{g+a\in\overline{\mathcal{G}}\\ g\in\Lambda}}{}' e^{-2\pi i y\cdot(a+g)} F(a+g)\, dV(a) \qquad (16.8)\\
&= \sum_{g'\in\Lambda(\mathcal{H})} \int_{\mathcal{H}^0_{g'}} \sum_{\substack{a+g\in\overline{\mathcal{G}}\\ g\in\Lambda}}{}' F(a+g)\, e^{-2\pi i y\cdot(a+g)}\, dV(a)\\
&= \sum_{g'\in\Lambda(\mathcal{H})} \int_{\mathcal{G}} \chi_{\mathcal{H}^0_{g'}+\Lambda}(x)\, F(x)\, e^{-2\pi i y\cdot x}\, dV(x),
\end{aligned}$$

where $\mathcal{H}^0_{g'}, g'\in\Lambda(\mathcal{H})$, is the shift of $\mathcal{H}_{g'}$ to the fundamental cell, i.e.,

$$\mathcal{H}^0_{g'} = \mathcal{H}_{g'} - \{g'\},\ g'\in\Lambda(\mathcal{H}) \qquad (16.9)$$

is a subset of $\mathcal{F}_\Lambda$, and the characteristic function $x\mapsto \chi_{\mathcal{H}^0_{g'}+\Lambda}(x)$, $x\in\mathbb{R}^q$, is given by

$$\chi_{\mathcal{H}^0_{g'}+\Lambda}(x) = \begin{cases} 1, & x\in \bigcup_{g\in\Lambda}\left(\mathcal{H}^0_{g'}+\{g\}\right),\\ 0, & \text{otherwise}.\end{cases} \qquad (16.10)$$

Consequently, we obtain the identity

$$\begin{aligned}
&\int_{\mathcal{H}} \sum_{\substack{a+g\in\overline{\mathcal{G}}\\ g\in\Lambda}}{}' F(a+g)\, e^{-2\pi i y\cdot(a+g)}\, dV(a) \qquad (16.11)\\
&= \sum_{g'\in\Lambda(\mathcal{H})} \int_{\mathcal{G}} \chi_{\mathcal{H}^0_{g'}+\Lambda}(x)\, F(x)\, e^{-2\pi i y\cdot x}\, dV(x),\\
&= \sum_{\substack{(\mathcal{F}_\Lambda+\{g'\})\cap\mathcal{H}\neq\emptyset\\ g'\in\Lambda}} \int_{\mathcal{G}\cap\bigcup_{g\in\Lambda}(((\mathcal{H}\cap(\mathcal{F}_\Lambda+\{g'\}))-\{g'\})+\{g\})} F(x)\, e^{-2\pi i y\cdot x}\, dV(x).
\end{aligned}$$

Collecting our results we obtain the following Shannon-type sampling theorem in Gauss–Weierstrass summability.

Theorem 16.2. *Let $\mathcal{G},\mathcal{H}$ be regular regions in $\mathbb{R}^q$. Suppose that F is a member of the class $\mathrm{C}^{(0)}(\overline{\mathcal{G}})$. Then*

$$\begin{aligned}
&\sum_{\substack{(\mathcal{F}_\Lambda+\{g'\})\cap\mathcal{H}\neq\emptyset\\ g'\in\Lambda}} F^\wedge_{\mathcal{G}\cap\bigcup_{g\in\Lambda}(((\mathcal{H}\cap(\mathcal{F}_\Lambda+\{g'\}))-\{g'\})+\{g\})}(y) \qquad (16.12)\\
&= \sum_{\substack{(\mathcal{F}_\Lambda+\{g'\})\cap\mathcal{H}\neq\emptyset\\ g'\in\Lambda}} \int_{\mathcal{G}\cap\bigcup_{g\in\Lambda}(((\mathcal{H}\cap(\mathcal{F}_\Lambda+\{g'\}))-\{g'\})+\{g\})} F(x)\, e^{-2\pi i y\cdot x}\, dV(x)\\
&= \lim_{\substack{\tau\to 0\\ \tau>0}} \sum_{h\in\Lambda^{-1}} e^{-\tau\pi^2 h^2} \underbrace{\int_{\mathcal{G}} F(x)\, e^{-2\pi i h\cdot x}\, dV(x)}_{=F^\wedge_{\mathcal{G}}(h)} \frac{1}{\|\mathcal{F}_\Lambda\|} \underbrace{\int_{\mathcal{H}} e^{2\pi i a\cdot(h-y)}\, dV(a)}_{=K_{\mathcal{H}}(h-y)}.
\end{aligned}$$

In fact, Theorem 16.2 includes all manifestations of over- and undersampling involving regular regions $\mathcal{G}, \mathcal{H} \subset \mathbb{R}^q$.

Some geometric configurations of regular regions $\mathcal{G}, \mathcal{H}$ should be studied in more detail (cf. W. Freeden, M.Z. Nashed [2015]):

We begin with $\overline{\mathcal{H}} \subset \overline{\mathcal{F}_\Lambda}$ and $\mathcal{G}$ arbitrary. In this case $\Lambda(\overline{\mathcal{H}}) = \{0\}$ and $\mathcal{H}_0^0$ is in the interior of $\mathcal{F}_\Lambda$ such that

$$\begin{aligned}
&\sum_{\substack{(\mathcal{F}_\Lambda+\{g'\})\cap\mathcal{H}\neq\emptyset \\ g'\in\Lambda}} F^\wedge_{\mathcal{G}\cap\bigcup\limits_{g\in\Lambda}(((\mathcal{H}\cap(\mathcal{F}_\Lambda+\{g'\}))-\{g'\})+\{g\})}(y) \qquad (16.13)\\
&= F^\wedge_{\mathcal{G}\cap\bigcup\limits_{g\in\Lambda}(\mathcal{H}+\{g\})}(y)\\
&= \lim_{\substack{\tau\to 0\\ \tau>0}} \sum_{h\in\Lambda^{-1}} e^{-\tau\pi^2 h^2} \underbrace{\int_{\mathcal{G}} F(x) e^{-2\pi i y\cdot x}\, dV(x)}_{=F^\wedge_{\mathcal{G}}(h)} \frac{1}{\|\mathcal{F}_\Lambda\|} \underbrace{\int_{\mathcal{H}} e^{2\pi i a\cdot(h-y)}\, dV(a)}_{=K_{\mathcal{H}}(h-y)}.
\end{aligned}$$

For $\overline{\mathcal{H}} = \overline{\mathcal{F}_\Lambda}$ we therefore obtain

$$\begin{aligned}
&\sum_{\substack{(\mathcal{F}_\Lambda+\{g'\})\cap\mathcal{H}\neq\emptyset \\ g'\in\Lambda}} F^\wedge_{\mathcal{G}\cap\bigcup\limits_{g\in\Lambda}(((\mathcal{H}\cap(\mathcal{F}_\Lambda+\{g'\}))-\{g'\})+\{g\})}(y) \qquad (16.14)\\
&= F^\wedge_{\mathcal{G}}(y)\\
&= \lim_{\substack{\tau\to 0\\ \tau>0}} \sum_{h\in\Lambda^{-1}} e^{-\tau\pi^2 h^2} \underbrace{\int_{\mathcal{G}} F(x)\, e^{-2\pi i h\cdot x}\, dV(x)}_{=F^\wedge_{\mathcal{G}}(h)} \frac{1}{\|\mathcal{F}_\Lambda\|} \underbrace{\int_{\mathcal{F}_\Lambda} e^{2\pi i a\cdot(h-y)}\, dV(a)}_{=K_{\mathcal{F}_\Lambda}(h-y)}.
\end{aligned}$$

We continue with $\overline{\mathcal{G}} \subset \overline{\mathcal{F}_\Lambda}$ and $\mathcal{H}$ arbitrary. This yields the identity

$$\begin{aligned}
&\sum_{\substack{(\mathcal{F}_\Lambda+\{g'\})\cap\mathcal{H}\neq\emptyset \\ g'\in\Lambda}} F^\wedge_{\mathcal{G}\cap\bigcup\limits_{g\in\Lambda}(((\mathcal{H}\cap(\mathcal{F}_\Lambda+\{g'\}))-\{g'\})+\{g\})}(y) \qquad (16.15)\\
&= \sum_{\substack{(\mathcal{F}_\Lambda+\{g'\}\cap\mathcal{H}\neq\emptyset) \\ g'\in\Lambda}} F^\wedge_{\mathcal{G}\cap((\mathcal{H}\cap(\mathcal{F}_\Lambda+\{g'\}))-\{g'\})}(y)\\
&= \lim_{\substack{\tau\to 0\\ \tau>0}} \sum_{h\in\Lambda^{-1}} e^{-\tau\pi^2 h^2} \underbrace{\int_{\mathcal{G}} F(x)\, e^{-2\pi i h\cdot x}\, dV(x)}_{=F^\wedge_{\mathcal{G}}(h)} \frac{1}{\|\mathcal{F}_\Lambda\|} \underbrace{\int_{\mathcal{H}} e^{2\pi i a\cdot(h-y)}\, dV(a)}_{=K_{\mathcal{H}}(h-y)}.
\end{aligned}$$

For $\overline{\mathcal{G}}, \overline{\mathcal{H}} \subset \overline{\mathcal{F}_\Lambda}$ we have

$$\sum_{\substack{(\mathcal{F}_\Lambda+\{g'\})\cap\mathcal{H}\neq\emptyset \\ g'\in\Lambda}} F^{\wedge}_{\mathcal{G}\cap\bigcup_{g\in\Lambda}(((\mathcal{H}\cap(\mathcal{F}_\Lambda+\{g'\}))-\{g'\})+\{g\})}(y) \tag{16.16}$$
$$= F^{\wedge}_{\mathcal{G}\cap\mathcal{H}}(y)$$
$$= \lim_{\substack{\tau\to 0\\ \tau>0}} \sum_{h\in\Lambda^{-1}} e^{-\tau\pi^2 h^2} F^{\wedge}_{\mathcal{G}}(h) \frac{1}{\|\mathcal{F}_\Lambda\|} K_{\mathcal{H}}(h-y).$$

For $\overline{\mathcal{G}} \subset \overline{\mathcal{H}} \subset \overline{\mathcal{F}_\Lambda}$ we have a sampling process without aliasing

$$F^{\wedge}_{\mathcal{G}}(y) = \lim_{\substack{\tau\to 0\\ \tau>0}} \sum_{h\in\Lambda^{-1}} e^{-\tau\pi^2 h^2} F^{\wedge}_{\mathcal{G}}(h) \frac{1}{\|\mathcal{F}_\Lambda\|} K_{\mathcal{H}}(h-y), \tag{16.17}$$

whereas, for $\overline{\mathcal{H}} \subset \overline{\mathcal{G}} \subset \overline{\mathcal{F}}_\Lambda$,

$$F^{\wedge}_{\mathcal{H}}(y) = \lim_{\substack{\tau\to 0\\ \tau>0}} \sum_{h\in\Lambda^{-1}} e^{-\tau\pi^2 h^2} F^{\wedge}_{\mathcal{G}}(h) \frac{1}{\|\mathcal{F}_\Lambda\|} K_{\mathcal{H}}(h-y), \tag{16.18}$$

so that aliasing occurs as far as $\overline{\mathcal{H}} \subsetneq \overline{\mathcal{G}}$. In particular, for $\overline{\mathcal{G}} = \overline{\mathcal{H}} \subset \overline{\mathcal{F}_\Lambda}$, we are able to formulate the following identity

$$F^{\wedge}_{\mathcal{G}}(y) = \lim_{\substack{\tau\to 0\\ \tau>0}} \sum_{h\in\Lambda^{-1}} e^{-\tau\pi^2 h^2} F^{\wedge}_{\mathcal{G}}(h) \frac{1}{\|\mathcal{F}_\Lambda\|} K_{\mathcal{G}}(h-y),\ y\in\mathbb{R}^q. \tag{16.19}$$

Obviously, by choosing a sampling density such that $\overline{\mathcal{F}_\Lambda}$ covers the compact support $\overline{\mathcal{G}}$ of the original signal $F^{\wedge}_{\mathcal{G}}$, the number of samples $F^{\wedge}_{\mathcal{G}}(h), h \in \Lambda^{-1}$, for the reconstruction is high and vice versa. In practice, we are therefore required to find a compromise between sampling density and total number of samples. This can be achieved by a choice of Λ such that $\overline{\mathcal{F}_\Lambda}$ covers $\overline{\mathcal{G}}$ tightly.

For $\overline{\mathcal{G}} \subset \overline{\mathcal{F}_\Lambda} = \overline{\mathcal{H}}$ and the lattice Λ generated by the vectors $g_1, \ldots, g_q \in \mathbb{R}^q$ we have in explicitly written form

$$F^{\wedge}_{\mathcal{G}}(y) = \lim_{\substack{\tau\to 0\\ \tau>0}} \sum_{h\in\Lambda^{-1}} e^{-\tau\pi^2 h^2} F^{\wedge}_{\mathcal{G}}(h) \frac{\sin(\pi(g_1\cdot(h-y)))}{\pi(g_1\cdot(h-y))} \cdot \ldots \cdot \frac{\sin(\pi g_q\cdot(h-y))}{\pi(g_q\cdot(h-y))}. \tag{16.20}$$

In other words, for sufficiently small $\tau > 0$, $F^{\wedge}_{\mathcal{G}}$ can be expressed by the series on the right side of (16.20) in exponential convergence, i.e.,

$$(F^{(\tau)})^{\wedge}_{\mathcal{G}}(y) \approx \sum_{h\in\Lambda^{-1}} e^{-\tau\pi^2 h^2} F^{\wedge}_{\mathcal{G}}(h) \frac{\sin(\pi(g_1\cdot(h-y)))}{\pi(g_1\cdot(h-y))} \cdot \ldots \cdot \frac{\sin(\pi g_q\cdot(h-y))}{\pi(g_q\cdot(h-y))}. \tag{16.21}$$

Replacing the lattice Λ by its dilated lattice $\sigma\Lambda$, $\sigma > 1$, we find that

$$F^{\wedge}_{\mathcal{G}}(y) = \lim_{\substack{\tau\to 0\\ \tau>0}} \sigma^q \sum_{h\in\Lambda^{-1}} e^{-\tau\pi^2 (\frac{h}{\sigma})^2} F^{\wedge}_{\mathcal{G}}\left(\frac{h}{\sigma}\right) \frac{\sin(\pi(\sigma g_1\cdot(\frac{h}{\sigma}-y)))}{\pi(\sigma g_1\cdot(\frac{h}{\sigma}-y))} \cdot \ldots \cdot \frac{\sin(\pi(\sigma g_q\cdot(\frac{h}{\sigma}-y)))}{\pi(\sigma g_q\cdot(\frac{h}{\sigma}-y))}. \tag{16.22}$$

The standard form of sampling in Gauss–Weierstrass summability is provided by taking $\overline{\mathcal{G}} = \overline{\mathcal{H}} = \overline{\mathcal{F}_\Lambda}$, i.e.,

$$F^\wedge_{\mathcal{F}_\Lambda}(y) = \lim_{\substack{\tau \to 0 \\ \tau > 0}} \sum_{h \in \Lambda^{-1}} e^{-\tau \pi^2 h^2} F^\wedge_{\mathcal{F}_\Lambda}(h) \, \frac{1}{\|\mathcal{F}_\Lambda\|} \, K_{\mathcal{F}_\Lambda}(h-y), \; y \in \mathbb{R}^q. \tag{16.23}$$

Next we discuss some interesting examples (in Gauss–Weierstrass summability), which again demonstrate the close relationship of sampling and lattice point theory.

Example 16.3. *Suppose that $\overline{\mathcal{H}}$ is the closure of the union of all periodic cells corresponding to lattice points $g' \in \Lambda$ inside the boundary of the ball $\overline{\mathbb{B}^q_R}$ of radius $R > 0$ around the origin:*

$$\overline{\mathcal{H}} = \bigcup_{g' \in \overline{\mathbb{B}^q_R}} \left(\overline{\mathcal{F}_\Lambda} + \{g'\} \right). \tag{16.24}$$

Then, for all $y \in \mathbb{R}^q$,

$$\left(\sum_{\substack{|g| \le R \\ g \in \Lambda}} 1 \right) F^\wedge_{\mathcal{G}}(y) \quad = \quad \lim_{\substack{\tau \to 0 \\ \tau > 0}} \sum_{h \in \Lambda^{-1}} e^{-\tau \pi^2 h^2} F^\wedge_{\mathcal{G}}(h) \, \frac{1}{\|\mathcal{F}_\Lambda\|} \, K_{\mathcal{H}}(h-y) \tag{16.25}$$

with

$$K_{\mathcal{H}}(h-y) = \sum_{\substack{|g'| \le R \\ g' \in \Lambda}} K_{\mathcal{F}_\Lambda + \{g'\}}(h-y) = \underbrace{\left(\sum_{\substack{|g'| \le R \\ g' \in \Lambda}} e^{2\pi i g' \cdot y} \right)}_{\sim \sum_{h \in \Lambda^{-1}} e^{2\pi i y \cdot h} R^q \|\mathbb{S}^{q-1}\| \frac{J_1(q; 2\pi |h| R)}{2\pi |h| R}} K_{\mathcal{F}_\Lambda}(h-y) \tag{16.26}$$

holds true for an arbitrary regular region $\mathcal{G} \subset \mathbb{R}^q$, whereas the choice $\overline{\mathcal{H}} = \overline{\mathbb{B}^q_R}$ provides the identity

$$\begin{aligned} &\sum_{\substack{\mathcal{F}_\Lambda + \{g'\} \cap \overline{\mathbb{B}^q_R} \neq \emptyset \\ g' \in \Lambda}} F^\wedge_{\mathcal{G} \cap \bigcup_{g \in \Lambda} (((\overline{\mathbb{B}^q_R} \cap \mathcal{F}_\Lambda + \{g'\}) - \{g'\}) + \{g\})}(y) \\ &\quad = \frac{R^q}{\|\mathcal{F}_\Lambda\|} \lim_{\substack{\tau \to 0 \\ \tau > 0}} \sum_{h \in \Lambda^{-1}} e^{-\tau \pi^2 h^2} F^\wedge_{\mathcal{G}}(h) \, \frac{J_{\frac{q}{2}}(2\pi |h-y| R)}{(|h-y| R)^{\frac{q}{2}}}, \end{aligned} \tag{16.27}$$

for an arbitrary regular region $\mathcal{G} \subset \mathbb{R}^q$. *If* $\overline{\mathcal{H}} = \overline{\mathbb{B}^q_R}$ *is a subset of* $\overline{\mathcal{F}_\Lambda}$, *we especially have for an arbitrary regular* $\mathcal{G} \subset \mathbb{R}^q$

$$F^{\wedge}_{\mathcal{G}\cap \bigcup_{g\in\Lambda}(\mathbb{B}^q_R+\{g\})}(y) \tag{16.28}$$
$$= \quad R^q \lim_{\substack{\tau\to 0\\ \tau>0}} \sum_{h\in\Lambda^{-1}} e^{-\tau\pi^2h^2} F^{\wedge}_{\mathcal{G}}(h) \, \frac{1}{\|\mathcal{F}_\Lambda\|} \frac{J_{\frac{q}{2}}(2\pi|h-y|R)}{(|h-y|R)^{\frac{q}{2}}},$$

such that the additional condition $\overline{\mathcal{G}} \subset \overline{\mathcal{F}_\Lambda}$ *leads back to the case (already discussed in our 2D context, but now for general* $\mathcal{G} \subset \mathbb{R}^q$ *and in Gauss–Weierstrass nomenclature)*

$$F^{\wedge}_{\mathcal{G}\cap\overline{\mathbb{B}^q_R}}(y) = R^q \lim_{\substack{\tau\to 0\\ \tau>0}} \sum_{h\in\Lambda^{-1}} e^{-\tau\pi^2h^2} F^{\wedge}_{\mathcal{G}}(h) \frac{1}{\|\mathcal{F}_\Lambda\|} \frac{J_{\frac{q}{2}}(2\pi|h-y|R)}{(|h-y|R)^{\frac{q}{2}}}. \tag{16.29}$$

Remark 16.4. *Given an arbitrary lattice* $\Lambda \subset \mathbb{R}^q$ *and an arbitrary regular region* $\mathcal{H} \subset \mathbb{R}^q$, *we are able to find a constant* $\sigma \in \mathbb{R}$, *such that*

$$\overline{\mathcal{H}} \subset \overline{\mathcal{F}_{\sigma\Lambda}} \tag{16.30}$$

as tightly as possible. Under these circumstances we obtain

$$F^{\wedge}_{\mathcal{G}\cap \bigcup_{g\in\sigma\Lambda}(\mathcal{H}+\{g\})}(y) \tag{16.31}$$
$$= \quad \lim_{\substack{\tau\to 0\\ \tau>0}} \sum_{h\in(\sigma\Lambda)^{-1}} e^{-2\tau\pi^2h^2} F^{\wedge}_{\mathcal{G}}(h) \frac{1}{\|\mathcal{F}_{\sigma\Lambda}\|} K_{\mathcal{H}}(h-y)$$
$$= \quad \lim_{\substack{\tau\to 0\\ \tau>0}} \frac{1}{\sigma^q} \sum_{h\in\Lambda^{-1}} e^{-\tau\pi^2(\frac{h}{\sigma})^2} F^{\wedge}_{\mathcal{G}}\left(\frac{h}{\sigma}\right) \frac{1}{\|\mathcal{F}_\Lambda\|} K_{\mathcal{H}}\left(\frac{h}{\sigma}-y\right).$$

Compared to sampling with respect to the lattice Λ, *the identity (16.31) related to* $\sigma\Lambda$ *provides upsampling for* $\sigma > 1$ *or downsampling for* $\sigma < 1$, *respectively. Of particular significance are choices of* σ, *such that* Λ *is a sublattice of* $\sigma\Lambda$, $\sigma > 1$, *or* $\sigma\Lambda$, $\sigma < 1$, *is a sublattice of* Λ.

Example 16.5. *Let* $\mathcal{G}, \mathcal{H}$ *be regular regions. Suppose that* $F = 1$ *on* $\overline{\mathcal{G}}$. *Then, for all* $y \in \mathbb{R}^q$, *we obtain some formulas leading over to the Parseval-type identities*

$$\int_{\mathcal{H}} \sum_{\substack{a+g\in\overline{\mathcal{G}}\\ g\in\Lambda}}{}' e^{-2\pi i y\cdot(a+g)} \, dV(a) \tag{16.32}$$
$$= \quad \sum_{\substack{(\mathcal{F}_\Lambda+\{g'\})\cap\mathcal{H}\neq\emptyset\\ g\in\Lambda}} \int_{\mathcal{G}\cap\bigcup_{g\in\Lambda}(((\mathcal{H}\cap(\mathcal{F}_\Lambda+\{\gamma'\})-\{g'\})+\{g\}))} e^{-2\pi i y\cdot x} \, dV(x)$$
$$= \quad \lim_{\substack{\tau\to 0\\ \tau>0}} \sum_{h\in\Lambda^{-1}} e^{-\tau\pi^2h^2} \int_{\mathcal{G}} e^{-2\pi i h\cdot x} \, dV(x) \, \frac{1}{\|\mathcal{F}_\Lambda\|} \int_{\mathcal{H}} e^{2\pi i x\cdot(h-y)} \, dV(x).$$

such that, under the special choice $\mathcal{H} = \mathcal{G}$,

$$\int_{\mathcal{G}} \sum_{\substack{a+g\in\overline{\mathcal{G}} \\ g\in\Lambda}}{}' e^{-2\pi i y\cdot(a+g)} \, dV(a) \tag{16.33}$$

$$= \sum_{\substack{(\mathcal{F}_\Lambda+\{g'\})\cap\mathcal{G}\neq\emptyset \\ g\in\Lambda}} \int_{\mathcal{G}\cap\bigcup_{g\in\Lambda}(((\mathcal{G}\cap(\mathcal{F}_\Lambda+\{\gamma'\})-\{g'\})+\{g\}))} e^{-2\pi i y\cdot x} \, dV(x)$$

$$= \lim_{\substack{\tau\to 0 \\ \tau>0}} \sum_{h\in\Lambda^{-1}} e^{-\tau\pi^2 h^2} \int_{\mathcal{G}} e^{-2\pi i h\cdot x} \, dV(x) \frac{1}{\|\mathcal{F}_\Lambda\|} \int_{\mathcal{G}} e^{2\pi i x\cdot(h-y)} \, dV(x).$$

In particular, for $y = 0$, we obtain

$$\int_{\mathcal{H}} \sum_{\substack{a+g\in\overline{\mathcal{G}} \\ g\in\Lambda}}{}' 1 \, dV(a) \tag{16.34}$$

$$= \sum_{\substack{(\mathcal{F}_\Lambda+\{g'\})\cap\mathcal{H}\neq\emptyset \\ g\in\Lambda}} \int_{\mathcal{G}\cap\bigcup_{g\in\Lambda}(((\mathcal{H}\cap(\mathcal{F}_\Lambda+\{g'\})-\{g'\})+\{g\}))} 1 \, dV(x)$$

$$= \lim_{\substack{\tau\to 0 \\ \tau>0}} \sum_{h\in\Lambda^{-1}} e^{-\tau\pi^2 h^2} \int_{\mathcal{G}} e^{-2\pi i h\cdot x} \, dV(x) \frac{1}{\|\mathcal{F}_\Lambda\|} \int_{\mathcal{H}} e^{2\pi i x\cdot h} \, dV(x)$$

and, especially for $\mathcal{H} = \mathcal{G}$,

$$\int_{\mathcal{G}} \sum_{\substack{a+g\in\overline{\mathcal{G}} \\ g\in\Lambda}}{}' 1 \, dV(a) \tag{16.35}$$

$$= \sum_{\substack{(\mathcal{F}_\Lambda+\{g'\})\cap\mathcal{G}\neq\emptyset \\ g'\in\Lambda}} \int_{\mathcal{G}\cap\bigcup_{g\in\Lambda}(((\mathcal{G}\cap(\mathcal{F}_\Lambda+\{g'\})-\{g'\})+\{g\}))} 1 \, dV(x)$$

$$= \lim_{\substack{\tau\to 0 \\ \tau>0}} \frac{1}{\|\mathcal{F}_\Lambda\|} \sum_{h\in\Lambda^{-1}} e^{-\tau\pi^2 h^2} \left| \int_{\mathcal{G}} e^{-2\pi i h\cdot x} \, dV(x) \right|^2.$$

16.2 Parseval-Type Identity (Gaussian/Ordinary Approach)

As already announced, the identities (16.34) and (16.35) may be regarded as canonical preparations for a turn over to the Parseval-type identity involving regular regions $\mathcal{G}, \mathcal{H} \subset \mathbb{R}^q$ within the Gauss–Weierstrass framework.

Remark 16.6. *The formula (16.35) was already used in 2D Euclidean space $\mathbb{R}^2$. In fact, C. Müller [1956] noticed that the Parseval-type* *identity*

$$\int_{\mathcal{G}} \sum_{\substack{a+g\in\overline{\mathcal{G}} \\ g\in\mathbb{Z}^2}} {}' \, 1 \; dV(a) \;=\; \sum_{h\in\mathbb{Z}^2} \left| \int_{\mathcal{G}} e^{-2\pi i h\cdot x} \; dV(x) \right|^2 \tag{16.36}$$

holds true for all symmetric (with respect to the origin) and convex regions $\mathcal{G} \subset \mathbb{R}^2$. *An easy consequence of (16.36) is the inequality*

$$\int_{\mathcal{G}} \sum_{\substack{a+g\in\overline{\mathcal{G}} \\ g\in\mathbb{Z}^2}} {}' \, 1 \; dV(a) \;\geq\; \|\mathcal{G}\|^2 = \left(\int_{\mathcal{G}} dV(x) \right)^2 , \tag{16.37}$$

such that under the assumption $\|\mathcal{G}\| > 4$, *an essential step towards Minkowski's Theorem can be made by considering simultaneously the lattice* $2\mathbb{Z}^2$ *(see, e.g., H. Minkowski [1896], E. Hlawka [1943]). Indeed, the inequality*

$$\int_{\mathcal{G}} \sum_{\substack{a+g\in\overline{\mathcal{G}} \\ g\in 2\mathbb{Z}^2}} {}' \, 1 \; dV(a) \;\geq\; \frac{\|\mathcal{G}\|}{4} \, \|\mathcal{G}\| \;>\; \|\mathcal{G}\| \tag{16.38}$$

is the key to guarantee that a symmetric (with respect to the origin) and convex region $\mathcal{G} \subset \mathbb{R}^2$ *with* $\|\mathcal{G}\| > 4$ *contains lattice points of* $\mathbb{Z}^2$ *different from the origin.*

It should be noted that Müller's 2D approach (cf. C. Müller [1956]) will be generalized here in various ways. For example, the Parseval identity of type (16.36) can be extended to arbitrary lattices $\Lambda \subset \mathbb{R}^q$ and regular regions $\mathcal{G} \subset \mathbb{R}^q$. Moreover, arbitrary continuous weight functions can be included instead of constant weights.

Theorem 16.7. *(Parseval-Type Identity in Gauss–Weierstrass Summability). Let* Λ *be a lattice in* $\mathbb{R}^q$. *Let* $\mathcal{G}, \mathcal{H} \subset \mathbb{R}^q$ *be arbitrary regular regions. Suppose that* F *is a real function of class* $\mathrm{C}^{(0)}(\overline{\mathcal{G}})$ *and* G *is a real function of class* $\mathrm{C}^{(0)}(\overline{\mathcal{H}})$, *respectively. Then we have*

$$\begin{aligned} &\lim_{\substack{\tau\to 0 \\ \tau>0}} \frac{1}{\|\mathcal{F}_\Lambda\|} \sum_{h\in\Lambda^{-1}} e^{-\tau\pi^2 h^2} \int_{\mathcal{G}} F(x) e^{-2\pi i x\cdot h} dV(x) \overline{\int_{\mathcal{H}} G(x) e^{-2\pi i x\cdot h} \; dV(x)} \\ &\quad = \lim_{\substack{\tau\to 0 \\ \tau>0}} \frac{1}{\|\mathcal{F}_\Lambda\|} \sum_{h\in\Lambda^{-1}} e^{-\tau\pi^2 h^2} F_{\mathcal{G}}^{\wedge}(h) \, \overline{G_{\mathcal{H}}^{\wedge}(h)} \\ &\quad = \int_{\mathcal{H}} \sum_{\substack{a+g\in\overline{\mathcal{G}} \\ g\in\Lambda}} {}' F(a+g) \, \overline{G(a)} \; dV(a). \end{aligned} \tag{16.39}$$

Proof. First we introduce an auxiliary function H by letting

$$H(x) = F(a+x) \,, \qquad x \in \overline{\mathcal{G}} - \{a\}. \tag{16.40}$$

Then it is clear that H is of class $\mathrm{C}^{(0)}\left(\overline{\mathcal{G}-\{a\}}\right)$. From (15.15) we are able to deduce that

$$
\begin{aligned}
&(\pi\tau)^{-\frac{q}{2}} \sum_{g\in\Lambda} \int_{\mathcal{G}-\{a\}} e^{-\frac{|x-g|^2}{\tau}} \; H(x)\; dV(x) \qquad (16.41)\\
&= \frac{1}{\|\mathcal{F}_\Lambda\|} \sum_{h\in\Lambda^{-1}} e^{-\tau\pi^2 h^2} \int_{\mathcal{G}-\{a\}} H(x)\; e^{-2\pi i h\cdot x}\; dV(x)\\
&= \frac{1}{\|\mathcal{F}_\Lambda\|} \sum_{h\in\Lambda^{-1}} e^{-\tau\pi^2 h^2} e^{2\pi i a\cdot h} \int_{\mathcal{G}} F(x)\; e^{-2\pi i h\cdot x}\; dV(x).
\end{aligned}
$$

As a consequence we obtain

$$
\begin{aligned}
&(\pi\tau)^{-\frac{q}{2}} \int_{\mathcal{H}} G(a) \sum_{g\in\Lambda} \int_{\mathcal{G}-\{a\}} e^{-\frac{|x-g|^2}{\tau}} \; H(x)\; dV(x)\; dV(a) \qquad (16.42)\\
&= \frac{1}{\|\mathcal{F}_\Lambda\|} \sum_{h\in\Lambda^{-1}} e^{-\tau\pi^2 h^2} \int_{\mathcal{G}} F(x)\; e^{-2\pi i h\cdot x}\; dV(x) \int_{\mathcal{H}} G(a)\; e^{2\pi i a\cdot h}\; dV(a)\\
&= \frac{1}{\|\mathcal{F}_\Lambda\|} \sum_{h\in\Lambda^{-1}} e^{-\tau\pi^2 h^2} \int_{\mathcal{G}} F(x)\; e^{-2\pi i h\cdot x}\; dV(x)\; \overline{\int_{\mathcal{H}} G(a)\; e^{-2\pi i a\cdot h}\; dV(a)}.
\end{aligned}
$$

Set $\gamma = \sup_{a\in\overline{\mathcal{G}}} |a|$. For all $a \in \overline{\mathcal{G}}$ we have

$$
\begin{aligned}
&\left| (\pi\tau)^{-\frac{q}{2}} \sum_{g\in\Lambda} \int_{\mathcal{G}-\{a\}} e^{-\frac{|y-g|^2}{\tau}} F(a+x)\; dV(x) \right| \qquad (16.43)\\
&\leq\; (\pi\tau)^{-\frac{q}{2}}\; \|F\|_{\mathrm{C}^{(0)}(\overline{\mathcal{G}})} \sum_{g\in\Lambda} \int_{|x|\leq 2\gamma} e^{-\frac{|x-g|^2}{\tau}} \; dV(x).
\end{aligned}
$$

The series on the right side of (16.43) is convergent, hence, the expression on the left side of (16.43) is bounded. Thus there exists a constant C, such that

$$
\left| (\pi\tau)^{-\frac{q}{2}} \sum_{g\in\Lambda} \int_{\mathcal{G}-\{a\}} e^{-\frac{|x-g|^2}{\tau}} H(x)\; dV(x) \right| \;\leq\; C. \qquad (16.44)
$$

Because of (16.40) we are able to write

$$
\sum_{\substack{g+a\in\overline{\mathcal{G}}\\ g\in\Lambda}}{}' \; H(g)\; \overline{G(a)} = \sum_{\substack{g+a\in\overline{\mathcal{G}}\\ g\in\Lambda}}{}' \; F(a+g)\; \overline{G(a)}, \quad a\in\mathbb{R}^q. \qquad (16.45)
$$

The function

$$
a \mapsto \sum_{\substack{g+a\in\overline{\mathcal{G}}\\ g\in\Lambda}}{}' \; F(a+g)\; \overline{G(a)}, \quad a\in\mathbb{R}^q \qquad (16.46)
$$

constitutes (apart from the finite number of surfaces $\overline{\mathcal{H}} \cap (\partial\mathcal{G} - \{g\})$) a continuous function in $\overline{\mathcal{H}}$. The discontinuities of (16.46) are, in fact, finite jumps so that (16.46) is not only Lebesgue integrable, but also Riemann integrable). Hence, in accordance with Lebesgue's convergence theorem, we are therefore allowed to conclude that

$$\begin{aligned}
\lim_{\substack{\tau\to 0\\ \tau>0}} (\pi\tau)^{-\frac{q}{2}} \int_{\mathcal{H}} G(a) \sum_{g\in\Lambda} \int_{\mathcal{G}-\{a\}} e^{-\frac{|x-g|^2}{\tau}} F(a+x)\, dV(x)\, dV(a) \qquad (16.47)\\
= \int_{\mathcal{H}} \sum_{\substack{a+g\in\overline{\mathcal{G}}\\ g\in\Lambda}}{}' F(a+g)\, \overline{G(a)}\, dV(a)\\
= \lim_{\substack{\tau\to 0\\ \tau>0}} \frac{1}{\|\mathcal{F}_\Lambda\|} \sum_{h\in\Lambda^{-1}} e^{-\tau\pi^2h^2} \int_{\mathcal{G}} F(x)\, e^{-2\pi i h\cdot x}\, dV(x)\\
\times \overline{\int_{\mathcal{H}} G(a)\, e^{-2\pi i a\cdot h}\, dV(a)}.
\end{aligned}$$

This is the desired result of Theorem 16.7. □

Example 16.8. *We choose* $G = \mathbb{B}^q_\rho$, $\mathcal{H} = \mathbb{B}^q_R$, $\rho, R > 0$, *and* $F = 1$ *on* $\overline{\mathcal{G}}$ *as well as* $G = 1$ *on* $\overline{\mathcal{H}}$. *From Theorem 16.7 we obtain*

$$\begin{aligned}
\frac{1}{\|\mathcal{F}_\Lambda\|} \rho^q R^q \lim_{\substack{\tau\to 0\\ \tau>0}} \sum_{h\in\Lambda^{-1}} e^{-\tau\pi h^2} \frac{J_{\frac{q}{2}}(2\pi|h-y|\rho)}{(|h|\rho)^{\frac{q}{2}}} \frac{J_{\frac{q}{2}}(2\pi|h-y|R)}{(|h|R)^{\frac{q}{2}}}\\
= \int_{\overline{\mathbb{B}^q_R}} \sum_{\substack{a+g\in\overline{\mathbb{B}^q_\rho}\\ g\in\Lambda}}{}' 1\, dV(a), \qquad (16.48)
\end{aligned}$$

so that

$$\begin{aligned}
\frac{1}{\|\mathcal{F}_\Lambda\|} \rho^q R^q \sum_{h\in\Lambda^{-1}} \frac{J_{\frac{q}{2}}(2\pi|h-y|\rho)}{(|h|\rho)^{\frac{q}{2}}} \frac{J_{\frac{q}{2}}(2\pi|h-y|R)}{(|h|R)^{\frac{q}{2}}} \qquad (16.49)\\
= \int_{\overline{\mathbb{B}^q_R}} \sum_{\substack{a+g\in\overline{\mathbb{B}^q_\rho}\\ g\in\Lambda}}{}' 1\, dV(a)\\
= \int_{\overline{\mathbb{B}^q_R}} \chi_{\overline{\mathbb{B}^q_\rho}+\Lambda}(a)\, dV(a)\\
= \int_{\overline{\mathbb{B}^q_R}} \chi_{\overline{\mathbb{B}^q_\rho}+\Lambda}(a)\, \chi_{\overline{\mathbb{B}^q_R}}(a)\, dV(a).
\end{aligned}$$

Now, if $\overline{\mathcal{G}}, \overline{\mathcal{H}}$ are subsets of $\overline{\mathcal{F}_\Lambda}$, then we obtain from the extended Parseval identity (cf. Theorem 16.7)

$$\lim_{\substack{\tau\to 0\\ \tau>0}} \frac{1}{\|\mathcal{F}_\Lambda\|} \sum_{g\in\Lambda^{-1}} e^{-\tau\pi^2h^2} F^\wedge_{\mathcal{G}}(h)\, \overline{G^\wedge_{\mathcal{H}}(h)} = \int_{\mathcal{G}\cap\mathcal{H}} F(a)\, \overline{G(a)}\, dV(a). \qquad (16.50)$$

If $\overline{\mathcal{G}} \subset \overline{\mathcal{H}} \subset \overline{\mathcal{F}_\Lambda}$, then

$$\lim_{\substack{\tau\to 0\\ \tau>0}} \frac{1}{\|\mathcal{F}_\Lambda\|} \sum_{g\in\Lambda^{-1}} e^{-\tau\pi^2 h^2} F^\wedge_{\mathcal{G}}(h)\, \overline{G^\wedge_{\mathcal{H}}(h)} = \int_{\mathcal{G}} F(a)\, \overline{G(a)}\, dV(a). \tag{16.51}$$

Thus, the special configuration $\overline{\mathcal{G}} = \overline{\mathcal{H}} = \overline{\mathcal{F}_\Lambda}$ leads back to the conventional Parseval identity of Fourier theory (however, in Gauss–Weierstrass summability).

By virtue of Theorem 16.7 (with $\mathcal{H} = \mathcal{G}$) we arrive at the following corollary (in Gauss–Weierstrass summability).

Corollary 16.9. *(Parseval-Type Identity in Gauss–Weierstrass Summability). Let Λ be a lattice in $\mathbb{R}^q$. Let $\mathcal{G} \subset \mathbb{R}^q$ be a regular region. Suppose that F, G are of class $\mathrm{C}^{(0)}(\overline{\mathcal{G}})$. Then the following variant of the Parseval identity holds true:*

$$\lim_{\substack{\tau\to 0\\ \tau>0}} \frac{1}{\|\mathcal{F}_\Lambda\|} \sum_{h\in\Lambda^{-1}} e^{-\tau\pi^2 h^2} F^\wedge_{\mathcal{G}}(h)\, \overline{G^\wedge_{\mathcal{G}}(h)} = \int_{\mathcal{G}} \sum_{\substack{a+g\in\overline{\mathcal{G}}\\ g\in\Lambda}}{}' F(a+g)\, \overline{G(a)}\, dV(a). \tag{16.52}$$

In particular, we have

$$\lim_{\substack{\tau\to 0\\ \tau>0}} \frac{1}{\|\mathcal{F}_\Lambda\|} \sum_{h\in\Lambda^{-1}} e^{-\tau\pi^2 h^2} \left|F^\wedge_{\mathcal{G}}(h)\right|^2 = \int_{\mathcal{G}} \sum_{\substack{a+g\in\overline{\mathcal{G}}\\ g\in\Lambda}}{}' F(a+g)\overline{F(a)}\, dV(a). \tag{16.53}$$

Next we are interested in a way to avoid the Gauss–Weierstrass summability independently of the dimension of the Euclidean space $\mathbb{R}^q, q \geq 2$. In fact, the same arguments leading to Theorem 15.5 allow us to conclude that limit and sum on the right of (16.52) may be interchanged. As an immediate consequence we obtain the following result.

Corollary 16.10. *(Parseval-Type Identity for Regular Regions). Under the assumptions of Theorem 16.7 we have*

$$\frac{1}{\|\mathcal{F}_\Lambda\|} \sum_{h\in\Lambda^{-1}} F^\wedge_{\mathcal{G}}(h)\, \overline{G^\wedge_{\mathcal{H}}(h)} = \int_{\mathcal{H}} \sum_{\substack{a+g\in\overline{\mathcal{G}}\\ g\in\Lambda}}{}' F(a+g)\, \overline{G(a)}\, dV(a). \tag{16.54}$$

In particular, if $\overline{\mathcal{G}} = \overline{\mathcal{H}}$ and $F = G$, then

$$\frac{1}{\|\mathcal{F}_\Lambda\|} \sum_{h\in\Lambda^{-1}} \left|F^\wedge_{\mathcal{G}}(h)\right|^2 = \int_{\mathcal{G}} \sum_{\substack{a+g\in\overline{\mathcal{G}}\\ g\in\Lambda}}{}' F(a+g)\, \overline{F(a)}\, dV(a). \tag{16.55}$$

Under the choice $\overline{\mathcal{G}} = \overline{\mathcal{H}} \subset \overline{\mathcal{F}_\Lambda}$ and $F = G$ we find that

$$\frac{1}{\|\mathcal{F}_\Lambda\|} \sum_{h\in\Lambda^{-1}} \left|F^\wedge_{\mathcal{G}}(h)\right|^2 = \int_{\mathcal{G}} |F(a)|^2\, dV(a). \tag{16.56}$$

A special case is the classical Parseval identity (see (8.31)) with $\overline{\mathcal{G}} = \overline{\mathcal{H}} = \overline{\mathcal{F}_\Lambda}$ and $F = G$

$$\frac{1}{\|\mathcal{F}_\Lambda\|} \sum_{h \in \Lambda^{-1}} \left| F^\wedge_{\mathcal{F}_\Lambda}(h) \right|^2 = \int_{\mathcal{F}_\Lambda} |F(a)|^2 \, dV(a). \tag{16.57}$$

16.3 Fourier-Transformed Values–Based Shannon-Type Sampling (Ordinary Approach)

It is remarkable that Corollary 16.10 also enables us to formulate the Shannon sampling theorem in ordinary sense avoiding Gauss–Weierstrass nomenclature (by letting $G(x) = 1$ for $x \in \mathcal{H}$ and replacing $F(x)$ by $F(x)e^{-2\pi i y \cdot x}$ for $x \in \mathcal{G}$).

Theorem 16.11. *(Shannon-Type Sampling Theorem). Let $\mathcal{G}, \mathcal{H}$ be regular regions in $\mathbb{R}^q$. Suppose that F is a member of the class $\mathrm{C}^{(0)}(\overline{\mathcal{G}})$. Then*

$$\int_{\mathcal{H}} \sum_{\substack{a+g \in \overline{\mathcal{G}} \\ g \in \Lambda}}{}' e^{-2\pi i y \cdot (a+g)} F(a+g) \, dV(a) \tag{16.58}$$

$$= \sum_{\substack{(\mathcal{F}_\Lambda + \{g'\}) \cap \mathcal{H} \neq \emptyset \\ g' \in \Lambda}} F^\wedge_{\mathcal{G} \cap \bigcup_{g \in \Lambda} (((\mathcal{H} \cap (\mathcal{F}_\Lambda + \{g'\})) - \{g'\}) + \{g\})}(y)$$

$$= \sum_{h \in \Lambda^{-1}} F^\wedge_{\mathcal{G}}(h) \, \frac{1}{\|\mathcal{F}_\Lambda\|} \, K_{\mathcal{H}}(h-y) \tag{16.59}$$

is valid for all $y \in \mathbb{R}^q$. For $\overline{\mathcal{G}} \subset \overline{\mathcal{H}} \subset \overline{\mathcal{F}_\Lambda}$, we have

$$F^\wedge_{\mathcal{G}}(y) = \sum_{h \in \Lambda^{-1}} F^\wedge_{\mathcal{G}}(h) \, \frac{1}{\|\mathcal{F}_\Lambda\|} \, K_{\mathcal{H}}(h-y), \ y \in \mathbb{R}^q. \tag{16.60}$$

Finally, under the assumption $\overline{\mathcal{G}} = \overline{\mathcal{H}} \subset \overline{\mathcal{F}_\Lambda}$, the identity (16.60) implies that

$$F^\wedge_{\mathcal{G}}(y) = \sum_{h \in \Lambda^{-1}} F^\wedge_{\mathcal{G}}(h) \, \frac{1}{\|\mathcal{F}_\Lambda\|} \, K_{\mathcal{G}}(h-y), \ y \in \mathbb{R}^q. \tag{16.61}$$

Furthermore, for $\overline{\mathcal{G}} \subset \overline{\mathcal{H}} = \overline{\mathcal{F}_\Lambda}$, we obtain

$$F^\wedge_{\mathcal{G}}(y) = \sum_{h \in \Lambda^{-1}} F^\wedge_{\mathcal{G}}(h) \, \frac{1}{\|\mathcal{F}_\Lambda\|} \, K_{\mathcal{F}_\Lambda}(h-y), \ y \in \mathbb{R}^q. \tag{16.62}$$

Explicitly written out (cf. E. Parzen [1956]) we obtain for the lattice Λ generated by the vectors $g_1, \ldots, g_q \in \mathbb{R}^q$

$$F^\wedge_{\mathcal{G}}(y) = \sum_{h \in \Lambda^{-1}} F^\wedge_{\mathcal{G}}(h) \frac{\sin(\pi(g_1 \cdot (h-y)))}{\pi(g_1 \cdot (h-y))} \cdot \ldots \cdot \frac{\sin(\pi g_q \cdot (h-y))}{\pi(g_q \cdot (h-y))}, \ y \in \mathbb{R}^q. \tag{16.63}$$

Restoring Lost Samples. Let $\mathcal{G}, \mathcal{H}$ be regular regions with $\overline{\mathcal{G}} \subset \overline{\mathcal{H}} \subset \overline{\mathcal{F}_\Lambda}$. Suppose that $\mathfrak{Q} \subset \mathbb{R}^q$ is a regular region such that the total number $\#_{\Lambda^{-1}}(\mathfrak{Q})$ of lattice points $h' \in \Lambda^{-1}$ in $\mathfrak{Q}$ is positive.

The formula (16.60) enables us to restore a finite set of lost samples (cf. R.J. Marks II [1991]) from those remaining for a bandlimited signal $F_{\mathcal{G}}^{\wedge}$. In fact, for all $y \in \mathbb{R}^q$, we obtain from Theorem 16.11

$$\frac{1}{\|\mathcal{F}_\Lambda\|} \sum_{\substack{h \in \mathfrak{Q} \\ h \in \Lambda^{-1}}} F_{\mathcal{G}}^{\wedge}(h)\, K_{\mathcal{H}}(h-y) \;-\; F_{\mathcal{G}}^{\wedge}(y) = \underbrace{-\frac{1}{\|\mathcal{F}_\Lambda\|} \sum_{\substack{h \notin \mathfrak{Q} \\ h \in \Lambda^{-1}}} F_{\mathcal{G}}^{\wedge}(h)\, K_{\mathcal{H}}(h-y)}_{=Q(y)}. \tag{16.64}$$

We are interested in determining the unknown finite set of coefficients $F_{\mathcal{G}}^{\wedge}(h)$, $h \in \mathfrak{Q} \cap \Lambda^{-1}$, from the known values $Q(h)$, $h \in \mathfrak{Q}$. In fact, the coefficients $F_{\mathcal{G}}^{\wedge}(h)$, $h \in \mathfrak{Q}$, can be calculated via the linear equations

$$\sum_{h \in \mathfrak{Q}} F_{\mathcal{G}}^{\wedge}(h) \left(\frac{1}{\|\mathcal{F}_\Lambda\|} K_{\mathcal{H}}(h-h') - \delta_{h,h'} \right) = Q(h'), \quad h' \in \mathfrak{Q}, \tag{16.65}$$

thereby assuming that the solution is not singular. We have

$$K_{\mathcal{H}}(h-h') = \int_{\mathcal{H}} e^{2\pi i a \cdot h}\, \overline{e^{2\pi i a \cdot h'}}\, dV(a), \quad h, h' \in \mathfrak{Q}, \tag{16.66}$$

such that $(K_{\mathcal{H}}(h-h'))_{h.h' \in \mathfrak{Q}}$ is regular as Gram matrix of linearly independent functions. Note that a sufficient condition for (16.65) to be singular can be realized by letting $\overline{\mathcal{H}} = \overline{\mathcal{F}_\Lambda}$ because of the fact

$$\frac{1}{\|\mathcal{F}_\Lambda\|} K_{\mathcal{F}_\Lambda}(h-h') = \frac{1}{\|\mathcal{F}_\Lambda\|} \int_{\mathcal{F}_\Lambda} e^{2\pi i a \cdot (h-h')}\, dV(a) = \delta_{h,h'}. \tag{16.67}$$

This procedure explains the *restoring of a finite set of lost samples.*

16.4 Functional Values–Based Shannon-Type Sampling (Gaussian Approach)

Integrating the identity (16.2) over a regular region $\mathcal{H} \subset \mathbb{R}^q$, $q \geq 2$, we obtain the following lattice point formula, that serves as point of departure for our *functional values–based Shannon-type sampling approach.*

Lemma 16.12. *If $\mathcal{G}, \mathcal{H} \subset \mathbb{R}^q$ are arbitrary regular regions, then*

$$\sum_{\substack{a+h\in\overline{\mathcal{G}} \\ h\in\Lambda^{-1}}}{}' \frac{1}{\|\mathcal{F}_\Lambda\|} \int_{\mathcal{H}} e^{-2\pi i y\cdot h}\, dV(y)\, F(a+h) = \int_{\mathcal{H}} \sum_{g\in\Lambda} e^{2\pi i(g+y)\cdot a} \int_{\mathcal{G}} F(w) e^{-2\pi i w\cdot(g+y)}\, dV(w)\, dV(y) \tag{16.68}$$

is valid for all points $a \in \mathbb{R}^q$, where the lattice constant $\frac{1}{\|\mathcal{F}_\Lambda\|}$ can be regarded as an indicator of the sampling density.

The left side of (16.68) may be understood as a cardinal series in a Shannon-type sampling process

$$\sum_{\substack{a+h\in\overline{\mathcal{G}} \\ h\in\Lambda^{-1}}}{}' \frac{1}{\|\mathcal{F}_\Lambda\|} \int_{\mathcal{H}} e^{-2\pi i y\cdot h}\, dV(y)\, F(a+h) = \sum_{\substack{a+h\in\overline{\mathcal{G}} \\ h\in\Lambda^{-1}}}{}' \frac{1}{\|\mathcal{F}_\Lambda\|} K_{\mathcal{H}}(-h)\, F(a+h).$$

In what follows, we are interested in a reformulation of the right side of (16.68): Again, we denote by $\Lambda(\mathcal{H})$ the set of lattice points $g' \in \Lambda \subset \mathbb{R}^q$, such that $\mathcal{F}_\Lambda + \{g'\}$ contains points of $\mathcal{H}$, i.e.,

$$\Lambda(\mathcal{H}) = \{g' \in \Lambda : (\mathcal{F}_\Lambda + \{g'\}) \cap \mathcal{H} \neq \emptyset\}. \tag{16.69}$$

Using this abbreviation we are able to write

$$\int_{\mathcal{H}} \sum_{g\in\Lambda} \int_{\mathcal{G}} F(w) e^{2\pi i(a-w)\cdot(g+y)}\, dV(w)\, dV(y) = \sum_{g'\in\Lambda(\mathcal{H})} \int_{\mathcal{H}_{g'}} \sum_{g\in\Lambda} \int_{\mathcal{G}} F(w) e^{2\pi i(a-w)\cdot(g+y)}\, dV(w)\, dV(y),$$

where the set $\mathcal{H}_{g'}, g' \in \Lambda(\mathcal{H})$, is given by

$$\mathcal{H}_{g'} = (\mathcal{F}_\Lambda + \{g'\}) \cap \mathcal{H}, \quad g' \in \Lambda(\mathcal{H}). \tag{16.70}$$

Since the function

$$y \mapsto \sum_{g\in\Lambda} \int_{\mathcal{G}} F(w) e^{2\pi i(a-w)\cdot(g+y)}\, dV(w), \quad y \in \mathbb{R}^q, \tag{16.71}$$

is Λ-periodic, we are able to reformulate (16.70) as follows:

$$\sum_{g'\in\Lambda(\mathcal{H})} \int_{\mathcal{H}_{g'}} \sum_{g\in\Lambda} \int_{\mathcal{G}} F(w) e^{2\pi i(a-w)\cdot(g+y)}\, dV(w)\, dV(y) = \sum_{g'\in\Lambda(\mathcal{H})} \int_{\mathcal{H}^0_{g'}} \sum_{g\in\Lambda} \int_{\mathcal{G}} F(w) e^{2\pi i(a-w)\cdot(g+y)}\, dV(w)\, dV(y),$$

where $\mathcal{H}^0_{g'}, g' \in \Lambda(\mathcal{H})$, is the shift of $\mathcal{H}_{g'}$ to the fundamental cell, i.e.,

$$\mathcal{H}^0_{g'} = \mathcal{H}_{g'} - \{g'\},\ g' \in \Lambda(\mathcal{H}) \tag{16.72}$$

forms a subset of $\mathcal{F}_\Lambda$. Now, we arrive at the identity

$$\begin{aligned} &\sum_{g' \in \Lambda(\mathcal{H})} \int_{\mathcal{H}^0_{g'}} \sum_{g \in \Lambda} \int_{\mathcal{G}} F(w) e^{2\pi i (a-w)\cdot(g+y)}\ dV(w)\ dV(y) \qquad (16.73) \\ &= \sum_{g' \in \Lambda(\mathcal{H})} \int_{\mathbb{R}^q} \chi_{\mathcal{H}^0_{g'}+\Lambda}(x) \int_{\mathcal{G}} F(w) e^{2\pi i (a-w)\cdot x}\ dV(w)\ dV(x), \end{aligned}$$

where the characteristic function $x \mapsto \chi_{\mathcal{H}^0_{g'}+\Lambda}(x)$, $x \in \mathbb{R}^q$, is, as usual, given by

$$\chi_{\mathcal{H}^0_{g'}+\Lambda}(x) = \begin{cases} 1. & x \in \bigcup\limits_{g \in \Lambda} \left(\mathcal{H}^0_{g'} + \{g\}\right) \\ 0, & \text{otherwise.} \end{cases} \tag{16.74}$$

Summarizing our results we obtain the following lattice point Shannon-type sampling formula.

Theorem 16.13. *(Shannon-Type Sampling Theorem). Let $\mathcal{G}, \mathcal{H}$ be regular regions in $\mathbb{R}^q$. Assume that Λ is an arbitrary lattice in $\mathbb{R}^q$. Suppose that F is a member of the class $\mathrm{C}^{(0)}(\overline{\mathcal{G}})$. Then, for $a \in \mathbb{R}^q$, we have*

$$\begin{aligned} &\sum_{g' \in \Lambda(\mathcal{H})} \int_{\mathbb{R}^q} \chi_{\mathcal{H}^0_{g'}+\Lambda}(x) \int_{\mathcal{G}} F(w) e^{2\pi i (a \quad w)\ x}\ dV(w)\ dV(x) \\ &= \sum_{\substack{(\mathcal{F}_\Lambda + \{g'\}) \cap \mathcal{H} \neq \emptyset \\ g' \in \Lambda}} \int_{\bigcup\limits_{g \in \Lambda} (((\mathcal{H} \cap (\mathcal{F}_\Lambda + \{g'\})) - \{g'\}) + \{g\})} e^{2\pi i a \cdot x}\ F^\wedge_{\mathcal{G}}(x)\ \ dV(x) \\ &= \sum_{\substack{a+h \in \overline{\mathcal{G}} \\ h \in \Lambda^{-1}}}{}' F(a+h) \frac{1}{\|\mathcal{F}_\Lambda\|} K_{\mathcal{H}}(-h), \end{aligned} \tag{16.75}$$

where the cardinal series is a finite sum.

We only discuss the following three cases: $\overline{\mathcal{H}} \subsetneq \overline{\mathcal{F}_\Lambda}$, $\overline{\mathcal{H}} = \overline{\mathcal{F}_\Lambda}$, and $\overline{\mathcal{H}} \supsetneq \overline{\mathcal{F}_\Lambda}$:

For $\overline{\mathcal{H}} \subsetneq \overline{\mathcal{F}_\Lambda}$ we find that

$$\begin{aligned} &\int_{\mathbb{R}^q} \chi_{\mathcal{H}+\Lambda}(x)\ e^{2\pi i a \cdot x}\ F^\wedge_{\mathcal{G}}(x)\ \ dV(x) \\ &\quad = \sum_{\substack{a+h \in \overline{\mathcal{G}} \\ h \in \Lambda^{-1}}}{}' F(a+h) \frac{1}{\|\mathcal{F}_\Lambda\|} K_{\mathcal{H}}(-h). \end{aligned} \tag{16.76}$$

As a consequence we are able to see by application of the Fourier inversion formula that

$$\alpha_{\mathcal{G}}(a)F(a) - \int_{\mathbb{R}^q} \chi_{(\mathcal{F}_\Lambda \setminus \{\mathcal{H}\})+\Lambda}(x)\ e^{2\pi i a\cdot x}\ F_{\mathcal{G}}^{\wedge}(x)\ \ dV(x)$$

$$= \sum_{\substack{a+h\in\overline{\mathcal{G}} \\ h\in\Lambda^{-1}}}{}' \ F(a+h)\frac{1}{\|\mathcal{F}_\Lambda\|}K_{\mathcal{H}}(-h), \tag{16.77}$$

so that the aliasing error is given by the integral on the left side.

As an example, by choosing $\overline{\mathcal{H}} = \overline{\mathbb{B}^q_\tau} \subset \overline{\mathcal{F}_\Lambda}$, we particularly obtain

$$\int_{\mathbb{R}^q} \chi_{\mathbb{B}^q_\tau+\Lambda}(x)\ e^{2\pi i a\cdot x}\ F_{\mathcal{G}}^{\wedge}(x)\ \ dV(x)$$

$$= \sum_{\substack{a+h\in\overline{\mathcal{G}} \\ h\in\Lambda^{-1}}}{}' \ F(a+h)\frac{\|\mathbb{S}^{q-1}\|}{\|\mathcal{F}_\Lambda\|}\ \ \tau^q\ \frac{J_1(q;2\pi|h|\tau)}{2\pi|h|\tau}. \tag{16.78}$$

For $\overline{\mathcal{H}} = \overline{\mathcal{F}_\Lambda}$, we obtain by application of the Fourier inversion formula

$$\int_{\mathbb{R}^q} \chi_{\mathcal{F}_\Lambda+\Lambda}(x)\ \ e^{2\pi i a\cdot x}\ F_{\mathcal{G}}^{\wedge}(x)\ \ dV(x) \tag{16.79}$$

$$\begin{aligned} &= \int_{\mathbb{R}^q} e^{2\pi i a\cdot x}\ F_{\mathcal{G}}^{\wedge}(x)\ \ dV(x) \\ &= \alpha_{\mathcal{G}}(a)\ F(a) \\ &= \sum_{\substack{a+h\in\overline{\mathcal{G}} \\ h\in\Lambda^{-1}}}{}' \ F(a+h)\frac{1}{\|\mathcal{F}_\Lambda\|}K_{\mathcal{F}_\Lambda}(-h). \end{aligned}$$

For $\overline{\mathcal{H}} \supsetneq \overline{\mathcal{F}_\Lambda}$ we find that

$$\sum_{\substack{(\mathcal{F}_\Lambda+\{g'\})\cap\mathcal{H}\neq\emptyset \\ g'\in\Lambda}} \int_{\mathbb{R}^q} \chi_{\mathcal{H}^0_{g'}+\Lambda}(x)\ e^{2\pi i a\cdot x}\ F_{\mathcal{G}}^{\wedge}(x)\ \ dV(x)$$

$$= \sum_{\substack{a+h\in\overline{\mathcal{G}} \\ h\in\Lambda^{-1}}}{}' \ F(a+h)\frac{1}{\|\mathcal{F}_\Lambda\|}K_{\mathcal{H}}(-h). \tag{16.80}$$

In our examples, we are confronted with the following situation: The closer $\overline{\mathcal{H}}$ is to a finite (closed) union of lattice cells of Λ, the smaller is the aliasing error by this variant of Shannon-type sampling.

In conclusion, Theorem 15.4 opens two different manifestations of Shannon-type sampling. The Fourier integral–based context provides a rich cornucopia of sampling capabilities. The functional based Shannon-type sampling, however, does not provide results comparable to the Fourier integral–based case. Nevertheless, also in this framework, a way is found to detect an explicit representation of the aliasing error.

Chapter 17

Paley–Wiener Space Framework and Spline Approximation

Many extensions of the Shannon-type sampling theorems can be studied in more detail. An important aspect is the Paley–Wiener reproducing kernel Hilbert space theory, which we only investigate for the special case that $\overline{\mathcal{G}}$ is a subset of $\overline{\mathcal{F}_\Lambda}$. Our approach opens the perspective to introduce Paley–Wiener splines and to discuss some of their significant applications, such as the multivariate antenna theory.

17.1 Paley–Wiener Reproducing Kernel Structure

Restricting ourselves to regular regions $\mathcal{G} \subset \mathbb{R}^q$ with $\overline{\mathcal{G}} \subset \overline{\mathcal{F}_\Lambda}$, we already know that we are able to recover the continuous signal

$$F^\wedge_{\mathcal{G}}(y) = \int_{\mathcal{G}} F(a)\, e^{-2\pi i a\cdot y}\, dV(a),\ y \in \mathbb{R}^q \tag{17.1}$$

from the signal values $F^\wedge_{\mathcal{G}}(h), h \in \Lambda^{-1}$, so that

$$F^\wedge_{\mathcal{G}}(y) = \sum_{h\in\Lambda^{-1}} F^\wedge_{\mathcal{G}}(h)\, \frac{1}{\|\mathcal{F}_\Lambda\|}\, K_{\mathcal{G}}(h-y),\ y \in \mathbb{R}^q \tag{17.2}$$

with

$$K_{\mathcal{G}}(x-y) = \int_{\mathcal{G}} e^{2\pi i a\cdot(x-y)}\, dV(a),\ x, y \in \mathbb{R}^q. \tag{17.3}$$

This enables us to deduce some interesting results in the area of *multivariate integration* (cf. W. Freeden, M. Gutting [2017]): The bandlimited function $F^\wedge_{\mathcal{G}}$ allows it to express its integral over the Euclidean space $\mathbb{R}^q$ by the product of the lattice density and the sum over all samples in points of the inverse lattice

$$\frac{1}{\|\mathcal{F}_\Lambda\|} \sum_{h\in\Lambda^{-1}} F^\wedge_{\mathcal{G}}(h) = \lim_{N\to\infty} \int_{\substack{|x|\le N\\ x\in\mathbb{R}^q}} F^\wedge_{\mathcal{G}}(x)\, dV(x). \tag{17.4}$$

Furthermore, the Parseval identity is valid:

$$\frac{1}{\|\mathcal{F}_\Lambda\|} \sum_{h\in\Lambda^{-1}} |F_{\mathcal{G}}^{\wedge}(h)|^2 = \int_{\mathcal{G}} |F(a)|^2 \, dV(a). \tag{17.5}$$

From the Fourier theory, it follows that

$$\int_{\mathcal{G}} |F(a)|^2 \, dV(a) = \int_{\mathbb{R}^q} \left|F_{\mathcal{G}}^{\wedge}(y)\right|^2 \, dV(x). \tag{17.6}$$

In other words, if $F_{\mathcal{G}}^{\wedge}$, $\overline{\mathcal{G}} \subset \overline{\mathcal{F}_\Lambda}$, belongs to the inner product space

$$\mathrm{B}^{(0)}(\overline{\mathcal{G}}) = \left\{ y \mapsto \underbrace{\int_{\mathcal{G}} e^{-2\pi i a \dot{y}} F(a) \, dV(a)}_{=F_{\mathcal{G}}^{\wedge}(y)}, \; y \in \mathbb{R}^q : F \in \mathrm{C}^{(0)}(\overline{\mathcal{G}}) \right\}, \tag{17.7}$$

equipped with the standard $\mathrm{L}^2(\mathbb{R}^q)$-topology, then

$$\int_{\mathbb{R}^q} |F_{\mathcal{G}}^{\wedge}(y)|^2 \, dV(y) = \frac{1}{\|\mathcal{F}_\Lambda\|} \sum_{h\in\Lambda^{-1}} |F_{\mathcal{G}}^{\wedge}(h)|^2. \tag{17.8}$$

Replacing Λ^{-1} by its inverse lattice, i.e., Λ, we obtain

$$\frac{1}{\|\mathcal{F}_\Lambda\|} \int_{\mathbb{R}^q} |F_{\mathcal{G}}^{\wedge}(y)|^2 \, dV(y) = \sum_{g\in\Lambda} |F_{\mathcal{G}}^{\wedge}(g)|^2. \tag{17.9}$$

Looking critically at our approach we notice that the sampling theory is formulated on the reference set $\mathrm{C}^{(0)}_{\overline{\mathcal{G}}}$, which is a strict subset of the *Paley–Wiener space*

$$\mathrm{B}(\overline{\mathcal{G}}) = \left\{ y \mapsto \int_{\mathcal{G}} F(a) \, e^{-2\pi i a\cdot y} \, dV(a), \; y \in \mathbb{R}^q : F \in \mathrm{L}^2(\mathcal{G}) \right\}. \tag{17.10}$$

This observation, however, does not affect us very much, since every function $F \in \mathrm{L}^2(\mathcal{G})$ can be approximated (in $\mathrm{L}^2(\mathcal{G})$-sense) by a function $F_\varepsilon \in \mathrm{C}^{(0)}(\overline{\mathcal{G}})$ in ε-accuracy such that

$$\begin{aligned}
&\sup_{y\in\mathbb{R}^q} \left|F_{\mathcal{G}}^{\wedge}(y) - (F_\varepsilon)_{\mathcal{G}}^{\wedge}(y)\right| \qquad (17.11)\\
&\leq \sup_{y\in\mathbb{R}^q} \left| \int_{\mathcal{G}} e^{-2\pi i a\cdot y} (F(a) - F_\varepsilon(a)) \, dV(a) \right| \\
&\leq \left(\int_{\mathcal{G}} |F(a) - F_\varepsilon(a)|^2 \, dV(a) \right)^{1/2} \left(\int_{\mathcal{G}} |e^{-2\pi i a\cdot y}|^2 \, dV(a) \right)^{1/2} \\
&\leq \left(\int_{\mathbb{R}^q} \left|F_{\mathcal{G}}^{\wedge}(a) - (F_\varepsilon)_{\mathcal{G}}^{\wedge}(a)\right|^2 \, dV(a) \right)^{1/2} \left(\int_{\mathcal{G}} |e^{-2\pi i a\cdot y}|^2 \, dV(a) \right)^{1/2} \\
&= \sqrt{\|\mathcal{G}\|} \;\; \varepsilon.
\end{aligned}$$

As a consequence, if $\mathcal{G} \subset \mathbb{R}^q$ is a regular region with $\overline{\mathcal{G}} \subset \overline{\mathcal{F}_\Lambda}$, then the Paley–Wiener space $\mathrm{B}(\overline{\mathcal{G}})$ is the completion of the space $\mathrm{B}^{(0)}(\overline{\mathcal{G}})$ under the $\mathrm{L}^2(\mathbb{R}^q)$–topology:

$$\mathrm{B}(\overline{\mathcal{G}}) = \overline{\mathrm{B}^{(0)}(\overline{\mathcal{G}})}^{\|\cdot\|_{\mathrm{L}^2(\mathbb{R}^q)}}. \tag{17.12}$$

The set $\mathrm{B}(\overline{\mathcal{G}})$ forms a *reproducing kernel Hilbert space* with the uniquely determined kernel

$$K_{\mathcal{G}}(x-y) = \int_{\mathcal{G}} e^{-2\pi i a\cdot(y-x)}\, dV(a). \tag{17.13}$$

In fact, by virtue of (17.6), we see that

$$|F_{\mathcal{G}}^{\wedge}(y)| \leq \sqrt{\|\mathcal{G}\|}\sqrt{\int_{\mathcal{G}} |F(a)|^2\, dV(a)} = \sqrt{\|\mathcal{G}\|}\sqrt{\int_{\mathbb{R}^q} |F_{\mathcal{G}}^{\wedge}(w)|^2\, dV(w)}. \tag{17.14}$$

Moreover, standard Fourier inversion yields the identity

$$F_{\mathcal{G}}^{\wedge}(y) = \int_{\mathbb{R}^q} F_{\mathcal{G}}^{\wedge}(x) \left(\int_{\mathcal{G}} e^{2\pi i a\cdot(x-y)}\, dV(a)\right) dV(x), \tag{17.15}$$

so that the *reproducing kernel property*

$$F_{\mathcal{G}}^{\wedge}(y) = \int_{\mathbb{R}^q} F_{\mathcal{G}}^{\wedge}(x)\, K_{\mathcal{G}}(x-y)\, dV(x) \tag{17.16}$$

holds for all $y \in \mathbb{R}^q$, where

$$\int_{\mathbb{R}^q} \cdots \quad = \quad \lim_{N\to\infty} \int_{\substack{|x|\leq N \\ x\in\mathbb{R}^q}} \cdots \quad . \tag{17.17}$$

Summarizing we are able to conclude the following result:

Theorem 17.1. $\mathrm{B}(\overline{\mathcal{G}})$ *as given by*

$$\mathrm{B}(\overline{\mathcal{G}}) = \left\{ y \mapsto \int_{\mathcal{G}} F(a)\, e^{-2\pi i a\cdot y}\, dV(a),\ y \in \mathbb{R}^q : F \in \mathrm{L}^2(\mathcal{G}) \right\} \tag{17.18}$$

is a reproducing kernel Hilbert space with reproducing kernel (17.19)

$$K_{\mathcal{G}}(x-y) = \int_{\mathcal{G}} e^{2\pi i a\cdot(x-y)}\, dV(a). \tag{17.19}$$

An immediate consequence is the following lemma.

Lemma 17.2. $\{K_{\mathcal{H}}(\cdot - h)\}$ *constitutes a closed and complete system in* $\mathrm{B}(\overline{\mathcal{G}})$, *i.e.,*

$$\overline{\mathrm{span}_{h\in\Lambda^{-1}} K_{\mathcal{H}}(\cdot\ - h)} = \mathrm{B}(\overline{\mathcal{G}}), \tag{17.20}$$

where the completion is understood in the sense of $\|\cdot\|_{\mathrm{L}^2(\mathbb{R}^q)}$.

Proof. $F_{\mathcal{G}}^{\wedge} \in \mathrm{B}(\overline{\mathcal{G}})$ and the conditions

$$\int_{\mathbb{R}^q} F_{\mathcal{G}}^{\wedge}(x)\ K_{\mathcal{G}}(x-h)\ dV(x) \ = \ 0 \tag{17.21}$$

for all $h \in \Lambda^{-1}$ imply that $F_{\mathcal{G}}^{\wedge}(h) = 0$ for all $h \in \Lambda^{-1}$. Now, we have

$$F_{\mathcal{G}}^{\wedge}(y) = \frac{1}{\|\mathcal{F}_\Lambda\|} \sum_{h \in \Lambda^{-1}} \underbrace{F_{\mathcal{G}}^{\wedge}(h)}_{=0}\ K_{\mathcal{G}}(h-y) \tag{17.22}$$

for all $y \in \mathbb{R}^q$. Consequently, $(F_{\mathcal{G}}^{\wedge}, K_{\mathcal{H}}(h-\cdot))_{\mathrm{L}^2(\mathbb{R}^q)} = F_{\mathcal{G}}^{\wedge}(h) = 0$ is equivalent to $F_{\mathcal{G}}^{\wedge} = 0$. This guarantees the completeness in $\mathrm{B}(\overline{\mathcal{G}})$. Since closure and completenes are equivalent in $\mathrm{B}(\overline{\mathcal{G}})$, this also proves the desired result of Lemma 17.2. □

Going over to the Paley–Wiener space $\mathrm{B}(\overline{\mathcal{F}_\Lambda}), \Lambda = \mathbb{Z}^q$, we are able to guarantee, in addition, that the reproducing kernels form an orthonormal closed and complete system with discrete orthogonality property (as explained in the univariate case by, e.g., M.Z. Nashed [2010]).

17.2 Spline Interpolation in Paley–Wiener Spaces

There are diverse important spline concepts in the context of lattice point sampling, for example, A. Aldroubi, K. Gröchenig [2001], C. de Boor et al. [1985], P.L. Butzer et al. [1986], C.K. Chui et al. [1987], P. Devaraj et al. [2019], W. Sun, X. Zhou [2002a], mostly based on (iterated) univariate approaches. In what follows, in the line of W. Freeden [1981, 1988], we propose a specific spline understanding resulting from the reproducing kernel structure of the Paley–Wiener space.

Again, let $\mathcal{G}$ be a regular region in $\mathbb{R}^q$ with $\overline{\mathcal{G}} \subset \overline{\mathcal{F}_\Lambda}$. Suppose that $\mathfrak{Q} \subset \mathbb{R}^q$ is a regular region such that the total number $\#_{\Lambda^{-1}}(\mathfrak{Q})$ of lattice points $h' \in \Lambda^{-1}$ in $\mathfrak{Q}$ is positive. Assume that the set of samples $\{F_{\mathcal{G}}^{\wedge}(h')\}_{h' \in \mathfrak{Q}}$ corresponding to the finite set of lattice points $h' \in \mathfrak{Q}$ is known.

Our purpose is to deal with spline interpolation, i.e., minimum norm interpolation in the Paley–Wiener space $\mathrm{B}(\overline{\mathcal{G}})$. We determine the spline interpolant of $F_{\mathcal{G}}^{\wedge}$ from the known values $F_{\mathcal{G}}^{\wedge}(h'), h' \in \mathfrak{Q}$ as follows: Suppose that the matrix

$$\mathbf{k}_{\mathfrak{Q}} \ = \ (K_{\mathcal{G}}(h'-h))_{h',h \in \mathfrak{Q}} \tag{17.23}$$

is non-singular (note that $\mathbf{k}_{\mathfrak{Q}}$ is a Gram matrix). Then, any function $S_{\mathfrak{Q}}$ of the form

$$S_{\mathfrak{Q}}(y) = \sum_{h' \in \mathfrak{Q}} a_{h'} \int_{\mathcal{G}} e^{2\pi i a\cdot(h'-y)} dV(a), \quad y \in \mathbb{R}^q, \tag{17.24}$$

is called a $B(\overline{\mathcal{G}})$-*spline function relative to* $\mathfrak{Q}$. The class of all $B(\overline{\mathcal{G}})$-splines relative to $\mathfrak{Q}$ is denoted by $\mathrm{Spline}_{B(\overline{\mathcal{G}})}(\mathfrak{Q})$.

Obviously, $\mathrm{Spline}_{B(\overline{\mathcal{G}})}(\mathfrak{Q})$ is a finite subset of $B(\overline{\mathcal{G}})$.

It is easy to verify the so-called *spline interpolation formula*

$$\sum_{h'\in\mathfrak{Q}} a_{h'} P(h') = \sum_{h'\in\mathfrak{Q}} a_{h'} \int_{\mathbb{R}^q} K_{\mathcal{G}}(h'-y)\, \overline{P(y)}\, dV(y) = \int_{\mathbb{R}^q} S_{\mathfrak{Q}}(y)\, \overline{P(y)}\, dV(y) \tag{17.25}$$

for all $P \in B(\overline{\mathcal{G}})$.

Under the regularity assumption (17.23), there exists one and only one spline function $\hat{S_{\mathfrak{Q}}^{F}} \in \mathrm{Spline}_{B(\overline{\mathcal{G}})}(\mathfrak{Q})$ of the form

$$\hat{S_{\mathfrak{Q}}^{F}}(y) = \sum_{h'\in\mathfrak{Q}} \hat{a}_{h'}^{F} \int_{\mathcal{G}} e^{2\pi i a\cdot(h'-y)} dV(a), \quad y \in \mathbb{R}^q, \tag{17.26}$$

satisfying the linear equations

$$\hat{S_{\mathfrak{Q}}^{F}}(h) = F_{\mathcal{G}}^{\wedge}(h), \quad h \in \mathfrak{Q}, \tag{17.27}$$

i.e.,

$$\sum_{h'\in\mathfrak{Q}} \hat{a}_{h'}^{F} \int_{\mathcal{G}} e^{2\pi i a\cdot(h'-h)} dV(a) \;=\; F_{\mathcal{G}}^{\wedge}(h), \quad h \in \mathfrak{Q}. \tag{17.28}$$

From (17.28) it follows that

$$\int_{\mathbb{R}^q} \hat{S_{\mathfrak{Q}}^{F}}(y)\, \overline{(\hat{S_{\mathfrak{Q}}^{F}}(y) - P(y))}\, dV(y) = \sum_{h'\in\mathfrak{Q}} \hat{a}_{h'}^{F} \hat{S_{\mathfrak{Q}}^{F}}(h) - \sum_{h'\in\mathfrak{Q}} \hat{a}_{h'}^{F} P(h') \;=\; 0 \tag{17.29}$$

holds for all $P \in B(\overline{\mathcal{G}})$ with $P(h) = F_{\mathcal{G}}^{\wedge}(h)$, $h \in \mathfrak{Q}$. This implies that

$$\begin{aligned} \int_{\mathbb{R}^q} P(y)\, \overline{P(y)}\, dV(y) \;&=\; \int_{\mathbb{R}^q} \hat{S_{\mathfrak{Q}}^{F}}(y)\, \overline{\hat{S_{\mathfrak{Q}}^{F}}(y)}\, dV(y) \\ &+\; \int_{\mathbb{R}^q} \left(\hat{S_{\mathfrak{Q}}^{F}}(y) - P(y)\right) \overline{(\hat{S_{\mathfrak{Q}}^{F}}(y) - P(y))}\, dV(y). \end{aligned} \tag{17.30}$$

Summarizing our results we obtain the following *minimum norm characterization of an interpolating spline function in the Paley–Wiener space* $B(\overline{\mathcal{G}})$-*framework.*

Theorem 17.3. *(Variational Formulation of Paley–Wiener Spline Interpolation). Suppose that a finite discrete dataset* $\{F_{\mathcal{G}}^{\wedge}(h),\ h \in \mathfrak{Q}\}$ *is known. Then, the spline function given by*

$$\hat{S_{\mathfrak{Q}}^{F}}(y) = \sum_{h'\in\mathfrak{Q}} \hat{a}_{h'}^{F} \int_{\mathcal{G}} e^{2\pi i a\cdot(h'-y)}\, dV(a),\ y \in \mathbb{R}^q, \tag{17.31}$$

with coefficients $\hat{a}^F_{h'}$ *determined by the linear equations (17.25)*

$$\sum_{h' \in \mathfrak{Q}} \hat{a}^F_{h'} \int_{\mathcal{G}} e^{2\pi i a\cdot(h'-h)} dV(a) \;=\; F^{\wedge}_{\mathcal{G}}(h), \quad h \in \mathfrak{Q} \tag{17.32}$$

fulfills the minimum norm property

$$\left(\int_{\mathcal{G}} \left|\hat{S^F_{\mathfrak{Q}}}(y)\right|^2 dV(y)\right)^{1/2} = \min_{\substack{P\in \mathrm{B}(\overline{\mathcal{G}})\\ P(h)=F^{\wedge}_{\mathcal{G}}(h)\\ h\in\Lambda^{-1}\cap\mathfrak{Q}}} \left(\int_{\mathcal{G}} |P(y)|^2 \, dV(y)\right)^{1/2}. \tag{17.33}$$

Additional spline methods (such as spline smoothing of noisy data, best approximation of linear functionals, Sard's Theorem, Schönberg's Theorem, etc.) can be realized in a canonical way (see, e.g., W. Freeden [1981, 1988] for the methodology in the multi-periodic case) by use of the reproducing property of the kernel (17.13). The details are omitted here. Moreover, combined spline interpolation/smoothing (as proposed by W. Freeden, B. Witte [1982]) can be used instead of spline interpolation if the data are only (partially) noisy.

Paley–Wiener Spline Interpolatory Error Estimates. Suppose that the finite dataset

$$\left\{F^{\wedge}_{\mathcal{G}}(h) = \int_{\mathcal{G}} F(a)\, e^{-2\pi i a\cdot h}\, dV(a) : h \in \mathfrak{Q}\right\} \tag{17.34}$$

is given. We introduce the $\mathfrak{Q}$-*width* $\Theta^{\Lambda^{-1}}_{\mathfrak{Q}}$ by setting

$$\Theta^{\Lambda^{-1}}_{\mathfrak{Q}} = \sup_{y\in\mathbb{R}^q} \left(\min_{\substack{h\in\mathfrak{Q}\\ h\in\Lambda^{-1}}} |y-h| \right). \tag{17.35}$$

For any given point $y \in \mathfrak{Q}$, there exists a lattice point $h \in \mathfrak{Q}$ satisfying the condition

$$|\, y-h \,| \leq \Theta^{\Lambda^{-1}}_{\mathfrak{Q}}. \tag{17.36}$$

Observing the fact that $\hat{S^F_{\mathfrak{Q}}}(h) = F^{\wedge}_{\mathcal{G}}(h)$, $h \in \mathfrak{Q}$, we obtain

$$\hat{S^F_{\mathfrak{Q}}}(y) - F^{\wedge}_{\mathcal{G}}(y) \;=\; \hat{S^F_{\mathfrak{Q}}}(y) - \hat{S^F_{\mathfrak{Q}}}(h) + F^{\wedge}_{\mathcal{G}}(h) - F^{\wedge}_{\mathcal{G}}(y). \tag{17.37}$$

Thus, by the triangle inequality, we have

$$|\, \hat{S^F_{\mathfrak{Q}}}(y) - F^{\wedge}_{\mathcal{G}}(y) \,|\leq|\, \hat{S^F_{\mathfrak{Q}}}(y) - \hat{S^F_{\mathfrak{Q}}}(h) \,| + |\, F^{\wedge}_{\mathcal{G}}(y) - F^{\wedge}_{\mathcal{G}}(h) \,|\,. \tag{17.38}$$

From the reproducing kernel property (17.16), it can be deduced that

$$\hat{S^F_{\mathfrak{Q}}}(y) - \hat{S^F_{\mathfrak{Q}}}(h) \;=\; \int_{\mathbb{R}^q} \hat{S^F_{\mathfrak{Q}}}(a)\, (K_{\mathcal{G}}(y-a) - K_{\mathcal{G}}(h-a))\, dV(a) \tag{17.39}$$

and

$$F_{\mathcal{G}}^{\wedge}(y) - F_{\mathcal{G}}^{\wedge}(h) = \int_{\mathbb{R}^q} F_{\mathcal{G}}^{\wedge}(a) \left(K_{\mathcal{G}}(y-a) - K_{\mathcal{G}}(h-a)\right) dV(a) \tag{17.40}$$

for all $y \in \mathfrak{Q}$. Therefore, the Cauchy–Schwarz inequality yields the estimates

$$| \hat{S_{\mathfrak{Q}}^F}(y) - \hat{S_{\mathfrak{Q}}^F}(h) | \leq (\kappa_{\mathfrak{Q}}(y,h))^{1/2} \left(\int_{\mathcal{G}} \left| \hat{S_{\mathfrak{Q}}^F}(y) \right|^2 dV(y) \right)^{1/2} \tag{17.41}$$

and

$$| F_{\mathcal{G}}^{\wedge}(y) - F_{\mathcal{G}}^{\wedge}(h) | \leq (\kappa_{\Omega}(y,h))^{1/2} \left(\int_{\mathcal{G}} \left| F_{\mathcal{G}}^{\wedge}(y) \right|^2 dV(y) \right)^{1/2}, \tag{17.42}$$

where we have used the abbreviation

$$\kappa_{\mathfrak{Q}}(y,h) = \int_{\mathbb{R}^q} \left| K_{\mathcal{G}}(y-a) - K_{\mathcal{G}}(h-a) \right|^2 dV(a). \tag{17.43}$$

The interpolating spline, $\hat{S_{\mathfrak{Q}}}$, is the smoothest interpolant. Therefore we have

$$\left(\int_{\mathcal{G}} \left| \hat{S_{\mathfrak{Q}}^F}(a) \right|^2 dV(a) \right)^{1/2} \leq \left(\int_{\mathcal{G}} \left| F_{\mathcal{G}}^{\wedge}(y) \right|^2 dV(y) \right)^{1/2}. \tag{17.44}$$

Consequently we obtain from (17.38), (17.41), and (17.42)

$$| \hat{S_{\mathfrak{Q}}^F}(y) - F_{\mathcal{G}}^{\wedge}(y) | \leq 2\, (\kappa_{\mathfrak{Q}}(y,h))^{1/2} \left(\int_{\mathcal{G}} \left| F_{\mathcal{G}}^{\wedge}(a) \right|^2 dV(a) \right)^{1/2}. \tag{17.45}$$

Elementary calculations yield

$$\kappa_{\mathfrak{Q}}(y,h) = \overline{K_{\mathcal{G}}(y-y) - K_{\mathcal{G}}(h-y)} - \left(K_{\mathcal{G}}(y-h) - K_{\mathcal{G}}(h-h)\right). \tag{17.46}$$

Observing the well-known estimate for the exponential function

$$e^{\alpha} - 1 = \alpha + R(a) \tag{17.47}$$

with

$$|R(a)| \leq e^{|\alpha|} \frac{|\alpha|^2}{2!} \tag{17.48}$$

we obtain

$$\overline{|K_{\mathcal{G}}(y-y) - K_{\mathcal{G}}(h-y)|} = 2\pi\, |y-h| \underbrace{\int_{\mathcal{G}} |a|\, dV(a)}_{=\beta_{\mathcal{G}}} + O(|y-h|^2), \tag{17.49}$$

so that

$$| \kappa_{\mathfrak{Q}}(y,h) | = 4\pi\, \beta_{\mathcal{G}}\, |y-h| + O(|y-h|^2). \tag{17.50}$$

Collecting our results we find

$$\begin{aligned} &\sup_{y\in\mathfrak{Q}} \mid \hat{S_{\mathfrak{Q}}^F}(y) - F_{\mathcal{G}}^{\wedge}(y) \mid & (17.51)\\ &= 2\left(\sqrt{4\pi}\,\sqrt{\beta_{\mathcal{G}}}\,(\Theta_{\mathfrak{Q}}^{\Lambda^{-1}})^{1/2} + O(\Theta_{\mathfrak{Q}}^{\Lambda^{-1}})\right)\left(\int_{\mathcal{G}} \left|F_{\mathcal{G}}^{\wedge}(a)\right|^2 \, dV(a)\right)^{1/2}. \end{aligned}$$

Paley–Wiener Spline Interpolatory Convergence. As an immediate consequence of (17.51) we are led to the following statement:

Let $F_{\mathcal{G}}^{\wedge}$ be of class $\mathrm{B}(\overline{\mathcal{G}})$. Suppose that $\{\Lambda_N^{-1}\}$ is a sequence of lattices Λ_N^{-1}, such that $\Theta_{\mathfrak{Q}}^{\Lambda_N^{-1}} \to 0$ as $N \to \infty$. Then the sequence $\{(\hat{S}_{\mathfrak{Q}}^F)^{\Lambda_N^{-1}}\}$ of the uniquely determined solutions $(\hat{S}_{\mathfrak{Q}}^F)^{\Lambda_N^{-1}}$ of the variational problems

$$\left(\int_{\mathcal{G}} \left|(\hat{S}_{\mathfrak{Q}}^F)^{\Lambda_N^{-1}}(y)\right|^2 dV(y)\right)^{1/2} = \min_{\substack{P\in \\ P(h)=F_{\mathcal{G}}^{\wedge}(h) \\ h\in\Lambda_N^{-1}\cap\mathfrak{Q}}} \left(\int_{\mathcal{G}} |P(y)|^2 \, dV(y)\right)^{1/2} \quad (17.52)$$

satisfies the limit relation

$$\lim_{N\to\infty} \sup_{y\in\mathfrak{Q}} \mid (\hat{S}_{\mathfrak{Q}}^F)^{\Lambda_N^{-1}}(y) - F_{\mathcal{G}}^{\wedge}(y) \mid = 0. \quad (17.53)$$

This means that we are able to approximate any function $F_{\mathcal{G}}^{\wedge} \in \mathrm{B}(\overline{\mathcal{G}})$ in a constructive way using Paley–Wiener spline interpolation provided that $\Theta_{\mathfrak{Q}}^{\Lambda_N^{-1}} \to 0$ as $N \to \infty$.

17.3 Paley–Wiener Spline Interpolatory Sampling

The space $\mathrm{Spline}_{\mathrm{B}(\overline{\mathcal{G}})}(\mathfrak{Q}) \subset \mathrm{B}(\overline{\mathcal{G}})$ spanned by the column vector

$$k_y = (K_{\mathcal{G}}(h' - y))_{h'\in\mathfrak{Q}} \quad (17.54)$$

is, of course, a finite-dimensional reproducing kernel Hilbert space. The reproducing kernel

$$\mathbb{R}^q \times \mathbb{R}^q \ni (x,y) \mapsto K_{\mathcal{G}}^{\mathfrak{Q}}(x,y) \quad (17.55)$$

of the set $\mathrm{Spline}_{\mathrm{B}(\overline{\mathcal{G}})}(\mathfrak{Q})$ is given by

$$K_{\mathcal{G}}^{\mathfrak{Q}}(x,y) = k_x \cdot \mathbf{k}_{\mathfrak{Q}}^{-1} k_y, \quad x,y \in \mathbb{R}^q. \quad (17.56)$$

We set

$$q_y = (Q_{\mathcal{G}}(h',y))_{h'\in\mathfrak{Q}} = \mathbf{k}_{\mathfrak{Q}}^{-1} k_y \quad (17.57)$$

and

$$\mathbf{q} \;=\; (Q_{\mathcal{G}}(h',h))_{h',h\in\mathfrak{Q}}\,. \tag{17.58}$$

Then it is not difficult to see that

$$\mathbf{q} = \mathbf{k}_{\mathfrak{Q}}^{-1}\;\mathbf{k}_{\mathfrak{Q}} \;=\; \mathbf{i}, \tag{17.59}$$

where $\mathbf{i}$ is the unit matrix. Thus it it follows from (17.59) that the sampling property

$$S(y) \;=\; \sum_{h\in\mathfrak{Q}} \; S(h)\;Q_{\mathcal{G}}(h-y) \tag{17.60}$$

holds true for all $S \in \mathrm{Spline}_{\mathrm{B}(\overline{\mathcal{G}})}(\mathfrak{Q})$.

In particular, we have

$$\hat{S_{\mathfrak{Q}}^{F}}(y) \;=\; \sum_{h\in\mathfrak{Q}} \; \underbrace{\hat{S_{\mathfrak{Q}}^{F}}(h)}_{=F_{\mathcal{G}}^{\wedge}(h)}\;Q_{\mathcal{G}}(h-y), \tag{17.61}$$

so that, by virtue of (17.57), we obtain

$$\begin{aligned}\hat{S_{\mathfrak{Q}}^{F}}(y) \;&=\; \sum_{h\in\mathfrak{Q}}\left(\sum_{h'\in\mathfrak{Q}} \;\hat{a}_{h'}^{F}\;K_{\mathcal{G}}(h'-h)\right)\;Q_{\mathcal{G}}(h-y) \\ &=\; \sum_{h'\in\mathfrak{Q}} \;\hat{a}_{h'}^{F}\;\sum_{h\in\mathfrak{Q}}\;K_{\mathcal{G}}(h'-h)\;Q_{\mathcal{G}}(h-y) \\ &=\; \sum_{h'\in\mathfrak{Q}} \;\hat{a}_{h'}^{F}\;K_{\mathcal{G}}(h'-y).\end{aligned} \tag{17.62}$$

For more detailed information about sampling in reproducing kernel Hilbert spaces the reader is referred to M.Z. Nashed [2010], M.Z. Nashed, G.G. Walter [1995], W. Freeden, M.Z. Nashed [2015].

17.4 Paley–Wiener Spline Interpolatory Cubature

Suppose that the dataset

$$\left\{F_{\mathcal{G}}^{\wedge}(h) = \int_{\mathcal{G}} F(a)\;e^{-2\pi i a\cdot h}\;da : h \in \mathcal{G}\cap\Lambda^{-1}\right\} \tag{17.63}$$

is given such that

$$\hat{S_{\mathfrak{Q}}^{F}}(y) = \sum_{h'\in\mathfrak{Q}} a_{h'}^{F}\int_{\mathcal{G}} e^{2\pi i a\cdot(h'-y)}dV(y), \quad y\in\mathbb{R}^q, \tag{17.64}$$

is "close" to $F_{\mathcal{G}}^{\wedge}$ in $\overline{\mathcal{G}}$ (for example, if $\overline{\mathcal{G}} \subset \mathfrak{Q}$ and the nodal width of Λ^{-1} is "small enough"). Then, for $x \in \overline{\mathcal{G}}$, the Fourier inversion formula yields the identity

$$\begin{aligned} \alpha_{\mathcal{G}}(x)\; F(x) &\approx \int_{\mathbb{R}^q} e^{2\pi i x\cdot y}\; \hat{S}_{\mathfrak{Q}}^{F}(y)\; dV(y) \\ &= \sum_{h'\in\mathfrak{Q}} \hat{a}_{h'}^{F} \int_{\mathbb{R}^q} e^{2\pi i x\cdot y} \int_{\mathcal{G}} e^{2\pi i a\cdot(h'-y)}\; dV(a)\; dV(y) \\ &= \sum_{h'\in\mathfrak{Q}} \hat{a}_{h'}^{F}\; e^{2\pi i x\cdot h'}, \end{aligned} \tag{17.65}$$

with coefficients $\hat{a}_{h'}^{F}$ determined from the already known linear (spline) equations (17.28). In other words, our spline interpolation technique approximately solves the multivariate inversion problem of determining the function $\alpha_{\mathcal{G}}\; F \in \mathrm{L}^2(\mathcal{G})$ from a finite set of discrete values of $F_{\mathcal{G}}^{\wedge}$. Integrating (17.65) over $\mathcal{G}$ we finally arrive at a *Paley–Wiener spline interpolation cubature rule* of the form

$$\int_{\mathcal{G}} F(x)\; dV(x) \approx \sum_{h'\in\mathfrak{Q}} \hat{a}_{h'}^{F}\; K_{\mathcal{G}}(h'), \tag{17.66}$$

that is important in practice. Again, it demonstrates the central role of the reproducing kernel in $\mathrm{B}(\overline{\mathcal{G}})$ (for more details on the theory of reproducing Hilbert spaces see. e.g., N Aronszajn (1950)).

17.5 Multivariate Antenna Problem

Finally, a generalization of the *multivariate antenna problem* should be explained briefly: In antenna theory, the identification problem is the one in which the far-field radiation pattern is known, either exactly or approximately, by means of physical measurements. The problem is to find the aperture distribution producing this given pattern. The synthesis problems are ones in which we are given a desired far-field pattern, specified through samples of the far field at a finite number of points, and we wish to determine a source (a constrained or unconstrained aperture distribution) whose far-field radiation pattern approximates the desired pattern in some acceptable manner.

In terms of the operator equation,

$$A_{\mathcal{G}}F(y) = F_{\mathcal{G}}^{\wedge}(y) = \int_{\mathcal{G}} F(a)\; e^{-2\pi i a\cdot y}\; dV(a), \tag{17.67}$$

F represents the aperture distribution, $F_{\mathcal{G}}^{\wedge}$ represents the far-field pattern, and $A_{\mathcal{G}}$ the operator which relates these two functions, characterizing the antenna structure (for regularization techniques of integral equations of the type (17.67) the reader is referred, e.g., W. Freeden, M.Z. Nashed [2018a]).

We basically recast the operator-theoretic setting of a finite area source antenna in our Paley–Wiener spline framework (for a detailed description of the 1D antenna problem in the context of inverse problems the reader is referred to M.Z. Nashed [1981]):

Consider the problem (17.67) of a linear aperture on $\mathcal{G}$, so that the aperture distribution F is related to the far field $F_{\mathcal{G}}^{\wedge} \in \mathrm{B}(\overline{\mathcal{G}})$ by means of an integral equation, the pattern being limited to a visible range $\mathfrak{Q}$ containing the lattice points $h \in \mathfrak{Q} \cap \Lambda^{-1}$. In fact, assuming that $F_{\mathcal{G}}^{\wedge}(h)$ is known for the lattice points $h \subset \mathfrak{Q} \cap \Lambda^{-1}$ we are led back to the aforementioned spline problem, and an approximation to the aperture distribution is known from the formula (17.65)

$$F(x) \approx \sum_{h' \in \mathfrak{Q}} \hat{a}_{h'}^{F} \, e^{2\pi i x \cdot h'}, \quad x \in \mathcal{G}. \tag{17.68}$$

In other words, our Shannon-type sampling procedure enables Paley–Wiener spline interpolation of Fourier transforms from discrete data, located in lattice points.

All in all, the extensions of the Shannon-type sampling theorem as presented here allow many applications in engineering and physics, for example, in signal processing, data transmission, cryptography, constructive approximation, and inverse problems such as, e.g., the antenna problem.

Part VIII

Multivariate Poisson-Type Summation Formulas over Euclidean Spaces

Chapter 18

Poisson-Type Summation Formulas over Euclidean Spaces

Next we apply the multi-dimensional Euler-type summation formulas with respect to the operator $L + \lambda$, especially to sums of spherical nature, i.e., to balls $\overline{\mathbb{B}^q_N(y)}$ with $y \in \mathbb{R}^q$ arbitrary, but fixed. Our interest is in investigating the asymptotic behavior with increasing radii N to derive sufficient criteria for the validity of Poisson-type summation formulas over the whole Euclidean space $\mathbb{R}^q$, which do not necessarily imply the absolute convergence of the infinite multivariate sums on both sides (as it is the case in the approach, e.g., proposed by E.M. Stein, G. Weiss [1971]) and, therefore, are of particular importance in the context of multivariate alternating sum convergence.

As essential tools we need some integral mean asmptotics for iterated Euler–Green functions, which are formulated in similarity to the bivariate deductions (see Section 10.1).

Finally, the resulting variants of the Poisson-type summation formulas over $\mathbb{R}^q$ act as transitional identities to provide Shannon-type sampling realizations.

18.1 Integral Means for Iterated Euler–Green Functions

The following integral estimates (see W. Freeden [2011] for the the Helmholtz operator $\Delta + \lambda$) play an important part in the discussion of series over alternating functions in $\mathbb{R}^q, q \geq 2$.

Theorem 18.1. *For all lattices $\Lambda \subset \mathbb{R}^q, q \geq 2$, and for all values $\lambda \in \mathbb{R}$ and vectors $a, y \in \mathbb{R}^q$, the estimates*

$$\int_{\substack{|x-y|=N \\ x \in \mathbb{R}^q}} |G(L+\lambda; x)| \, dS_{(q-1)}(x) = O(N^{q-1}) \tag{18.1}$$

and

$$\int_{\substack{|x-y|=N \\ x\in\mathbb{R}^q}} \left| \frac{\partial G(L+\lambda;x)}{\partial \nu} \right| \, dS_{(q-1)}(x) = O(N^{q-1}) \tag{18.2}$$

hold true for $N \to \infty$.

Proof. The bivariate case of Theorem 18.1 (i.e., $q = 2$) is already known from Section 10.1. Therefore, in what follows we restrict ourselves to dimensions $q \geq 3$.

There is a positive constant D dependent on the lattice Λ, such that $|g - g'| \geq D$ holds for all points $g, g' \in \Lambda$ with $g \neq g'$. We set

$$\delta = \frac{1}{2}\min(1, \mathrm{D}). \tag{18.3}$$

Then we notice that, for $\lambda \in \mathbb{R}$ and $a, y \in \mathbb{R}^q$ fixed but arbitrary, there exists a constant C such that the estimates

$$|G(L+\lambda; x+y)| \leq C|x+y-g|^{2-q} \tag{18.4}$$

and

$$\left| \nabla_x G(L+\lambda; x+y) - \frac{1}{(q-2)\|\mathbb{S}^{q-1}\|} \frac{x+y-g}{|x+y-g|^q} \right| \leq C \frac{1}{|x+y-g|^{q-2}} \tag{18.5}$$

hold uniformly with respect to $x + y \in \overline{\mathbb{B}^q_\delta(g)}, g \in \Lambda$.

For $x + y \in \mathbb{R}^q$ we denote the distance of $x + y \in \mathbb{R}^q$ to the lattice Λ by

$$D(x+y;\Lambda) = \mathrm{dist}(x+y;\Lambda) = \min_{g\in\Lambda} |x+y-g|. \tag{18.6}$$

It is clear that there is a constant B (depending on $\lambda \in \mathbb{R}$, $a, y \in \mathbb{R}^q$, and δ), such that the estimates

$$\begin{aligned} |G(\dot{L}+\lambda; x+y)| &\leq B, & (18.7)\\ |\nabla_x G(L+\lambda; x+y)| &\leq B & (18.8) \end{aligned}$$

are valid for all $x \in \mathbb{R}^q$ with

$$D(x+y;\Lambda) = \mathrm{dist}(x+y;\Lambda) \geq \delta. \tag{18.9}$$

Moreover, for all $w \in \mathbb{R}^q$, we are able to deduce from the well-known Gaussian estimate (cf. Theorem 7.10) that

$$\#_\Lambda \left(\overline{\mathbb{B}^q_{N+\delta}(w)} \setminus \mathbb{B}^q_{N-\delta}(w) \right) = \sum_{\substack{N-\delta\leq|g-w|\leq N+\delta \\ g\in\Lambda}} 1 = O\left(N^{q-1}\right) \tag{18.10}$$

for δ fixed, suitably chosen and $N \to \infty$. Thus, for all $w \in \mathbb{R}^q$, it follows that

$$\sum_{\substack{|g-w|=N \\ g\in\Lambda}} 1 = O\left(N^{q-1}\right), \quad N \to \infty. \tag{18.11}$$

Now, we use the different results for the cases $D(x+y;\Lambda) > \delta$ and $D(x+y;\Lambda) \leq \delta$ to obtain an estimate of the form

$$\begin{aligned} &\int_{\substack{|x|=N \\ x\in\mathbb{R}^q}} |G(L+\lambda; x+y)| \, dS_{(q-1)}(x) \\ &= \int_{\substack{|x|=N \\ D(x+y;\Lambda)>\delta \\ x\in\mathbb{R}^q}} |G(L+\lambda; x+y)| \, dS_{(q-1)}(x) \\ &\quad + \int_{\substack{|x|=N \\ D(x+y;\Lambda)\leq\delta \\ x\in\mathbb{R}^q}} |G(L+\lambda; x+y)| \, dS_{(q-1)}(x). \end{aligned} \tag{18.12}$$

From (18.7) and (18.8) it follows that

$$\int_{\substack{|x|=N \\ D(x+y;\Lambda)>\delta \\ x\in\mathbb{R}^q}} |G(L+\lambda; x+y)| \, dS_{(q-1)}(x) \leq B \, \|\mathbb{S}^{q-1}\| \, N^{q-1}. \tag{18.13}$$

Because of the characteristic singularity of $G(L+\lambda; x+y)$ we obtain the estimate

$$\begin{aligned} &\int_{\substack{|x|=N \\ D(x+y;\Lambda)\leq\delta \\ x\in\mathbb{R}^q}} |G(L+\lambda; x+y)| \, dS_{(q-1)}(x) \\ &= O\left(\left(\#_\Lambda \left(\overline{\mathbb{B}^q_{N+\delta}} \setminus \mathbb{B}^q_{N-\delta} \right) \right) \int_{\substack{|x|=N \\ |x+y-g|\leq\delta \\ x\in\mathbb{R}^q}} \frac{1}{|x+y-g|^{q-2}} \, dS_{(q-1)}(x) \right) \end{aligned} \tag{18.14}$$

for $N \to \infty$. In connection with (18.10) and Theorem 7.10, we therefore find that

$$\int_{\substack{|x|=N \\ D(x+y;\Lambda)\leq\delta \\ x\in\mathbb{R}^q}} |G(L+\lambda; x+y)| \, dS_{(q-1)}(x) = O\left(N^{q-1}\right) \tag{18.15}$$

for $N \to \infty$. This proves the first part of Theorem 18.1.

Concerning the second part of Theorem 18.1, we again split the integral such that

$$\begin{aligned} &\int_{\substack{|x|=N \\ x\in\mathbb{R}^q}} \left| \frac{\partial}{\partial\nu} G(L+\lambda; x+y) \right| \, dS_{(q-1)}(x) \\ &= \int_{\substack{|x|=N \\ D(x+y;\Lambda)>\delta \\ x\in\mathbb{R}^q}} \left| \frac{\partial}{\partial\nu} G(L+\lambda; x+y) \right| \, dS_{(q-1)}(x) \\ &\quad + \int_{\substack{|x|=N \\ D(x+y;\Lambda)\leq\delta \\ x\in\mathbb{R}^q}} \left| \frac{\partial}{\partial\nu} G(L+\lambda; x+y) \right| \, dS_{(q-1)}(x). \end{aligned} \tag{18.16}$$

Then, by virtue of estimates known from the concept of solid angles (see Section 6.3 for details), we obtain from the Gaussian estimate, i.e., Theorem 7.10,

$$\begin{aligned} & \int_{\substack{|x|=N \\ x\in\mathbb{R}^q}} \left| \frac{\partial}{\partial\nu} G(L+\lambda; x+y) \right| \, dS_{(q-1)}(x) \\ = \;& O\left(N^{q-1}\right) + O\left(\#_\Lambda\left(\overline{\mathbb{B}^q_{N+\delta}} \setminus \mathbb{B}^q_{N-\delta}\right)\right) \\ = \;& O\left(N^{q-1}\right) \end{aligned} \tag{18.17}$$

for $N \to \infty$. Altogether, this is the desired result of Theorem 18.1. □

Since each iteration reduces the order of the singularity by two, $G\left((L+\lambda)^m; \cdot\right)$ is continuous for $m > \frac{q}{2}$ and continuously differentiable for $m > \frac{q}{2} + 1$. The estimates

$$\int_{\substack{|x-y|=N \\ x\in\mathbb{R}^q}} \left| G\left((L+\lambda)^m; x\right) \right| \, dS_{(q-1)}(x) = O\left(N^{q-1}\right) \tag{18.18}$$

and

$$\int_{\substack{|x-y|=N \\ x\in\mathbb{R}^q}} \left| \frac{\partial}{\partial\nu} G\left((L+\lambda)^m; x\right) \right| \, dS_{(q-1)}(x) = O\left(N^{q-1}\right) \tag{18.19}$$

for $N \to \infty$, therefore, are obvious for all $m > \frac{q}{2} + 1$.

It remains to study the intermediate cases $m \subset (1, \frac{q}{2} + 1]$, where the integrals

$$\int_{\substack{|x-y|=N \\ x\in\mathbb{R}^q}} \cdots = \int_{\substack{|x-y|=N \\ D(x;\Lambda)>\delta \\ x\in\mathbb{R}^q}} \cdots + \int_{\substack{|x-y|=N \\ D(x;\Lambda)\le\delta \\ x\in\mathbb{R}^q}} \cdots \tag{18.20}$$

can be estimated in the same way as described above. As a consequence, analogously to the formulation of Theorem 18.1, we obtain the following asymptotic estimates.

Theorem 18.2. *For all $m \in \mathbb{N}$, $q \geq 3$, all lattices $\Lambda \subset \mathbb{R}^q$, and for all $\lambda \in \mathbb{R}$ and $a, y \in \mathbb{R}^q$, the asymptotic expansions*

$$\int_{\substack{|x-y|=N \\ x\in\mathbb{R}^q}} \left| G\left((L+\lambda)^m; x\right) \right| \, dS_{(q-1)}(x) = O(N^{q-1}) \tag{18.21}$$

and

$$\int_{\substack{|x-y|=N \\ x\in\mathbb{R}^q}} \left| \frac{\partial G\left((L+\lambda)^m; x\right)}{\partial\nu} \right| \, dS_{(q-1)}(x) = O(N^{q-1}) \tag{18.22}$$

hold true for $N \to \infty$.

18.2 Euler-Type Summation Formula over Increasing Balls Involving Euler–Green Functions

Let F be a twice continuously differentiable function in $\mathbb{R}^q$. Then, for $\lambda \in \mathbb{R}$, $a \in \mathbb{R}^q$, and all $N > 0$, the Euler-type summation formula (cf. Corollary 9.4) yields

$$\begin{aligned} &\sum_{\substack{|g-y|\le N \\ g\in\Lambda}} e^{2\pi i a\cdot g} F(g-y) \qquad (18.23)\\ &= \frac{1}{\|\mathcal{F}_\Lambda\|} e^{2\pi i a\cdot y} \sum_{\substack{4\pi^2\left((h+a)^2-a^2\right)=\lambda \\ h\in\Lambda^{-1}}} e^{2\pi i h\, y} \int_{\substack{|x|\le N \\ x\in\mathbb{R}^q}} F(x) e^{2\pi i x\cdot(a+h)} dV(x) \\ &+ \int_{\substack{|x-y|\le N \\ x\in\mathbb{R}^q}} G(L+\lambda;x)\, e^{2\pi i a\cdot x} \left(\Delta_x + \lambda + 4\pi^2 a^2\right) F(x-y)\, dV(x) \\ &+ \; R(N). \end{aligned}$$

In accordance with the Euler-type summation formula the "remainder term" abbreviated by $R(N)$ in (18.23) is explicitly given by

$$\begin{aligned} &R(N) \qquad (18.24)\\ &= \frac{1}{2} \sum_{\substack{|g-y|=N \\ g\in\Lambda}} e^{2\pi i a\cdot g} F(g-y) \\ &\quad + \int_{\substack{|x-y|=N \\ x\in\mathbb{R}^q}} e^{2\pi i a\cdot x} G(L+\lambda;x) \frac{\partial}{\partial\nu_x} F(x-y)\, dS(x) \\ &\quad - \int_{\substack{|x-y|=N \\ x\in\mathbb{R}^q}} e^{2\pi i a\cdot x} F(x-y) \left(\frac{\partial}{\partial\nu_x} + 4\pi i a\cdot\nu(x)\right) G(L+\lambda;x)\, dS(x). \end{aligned}$$

18.3 Spherically-Reflected Convergence Criteria (Basic Differentiability Order)

Questions of the spherical convergence as $N \to \infty$ require estimates of the constituting terms of the remainder term (18.24).

In the sequel, the properties developed for Λ-Euler–Green functions with respect to $L + \lambda$ (Theorem 18.1) are used to formulate convergence theorems for multi-dimensional (alternating) spherically summed series. We begin with the discussion of the remainder term (18.24) in $\mathbb{R}^q, q \geq 2$ (see W. Freeden [2011] for analogous results involving operators of the form $\Delta + \lambda$).

Proposition 18.3. *A function F is called of class $\mathrm{X}^{(2)}(\mathbb{R}^q)$, if F is of class $\mathrm{C}^{(2)}(\mathbb{R}^q)$ satisfying*

$$F(x) = o\left(|x|^{1-q}\right), \quad |x| \to \infty \tag{18.25}$$

and

$$|\nabla_x F(x)| = o\left(|x|^{1-q}\right), \quad |x| \to \infty. \tag{18.26}$$

Lemma 18.4. *Suppose that F is of class $\mathrm{X}^{(2)}(\mathbb{R}^q)$. Then, for $N \to \infty$, we have*

$$\begin{aligned} R(N) &= \frac{1}{2} \sum_{\substack{|g-y|=N \\ g\in\Lambda}} e^{2\pi i a\cdot g} F(g-y) \\ &\quad + \int_{\substack{|x-y|=N \\ x\in\mathbb{R}^q}} e^{2\pi i a\cdot x} G(L+\lambda;x) \frac{\partial}{\partial\nu_x} F(x-y)\, dS(x) \\ &\quad - \int_{\substack{|x-y|=N \\ x\in\mathbb{R}^q}} e^{2\pi i a\cdot x} F(x-y) \left(\frac{\partial}{\partial\nu_x} + 4\pi i a\cdot\nu(x)\right) G(L+\lambda;x)\, dS(x) \\ &= o(1). \end{aligned} \tag{18.27}$$

Proof. By virtue of Theorem 7.10 it follows that

$$\begin{aligned} \frac{1}{2} \sum_{\substack{|g-y|=N \\ g\in\Lambda}} F(g-y) &= o\left(N^{1-q} \sum_{\substack{|g-y|=N \\ g\in\Lambda}} 1\right) \\ &= o\left(N^{1-q} O(N^{q-1})\right) \\ &= o(1) \end{aligned} \tag{18.28}$$

for $N \to \infty$. Furthermore, by application of (18.1) and (18.2) we see that

$$\begin{aligned} &\int_{\substack{|x-y|=N \\ x\in\mathbb{R}^q}} e^{2\pi i a\cdot x} G(L+\lambda;x) \frac{\partial F(x-y)}{\partial\nu_x}\, dS(x) \\ &\quad - \int_{\substack{|x-y|=N \\ x\in\mathbb{R}^q}} e^{2\pi i a\cdot x} F(x-y) \left(\frac{\partial}{\partial\nu} + 4\pi i a\cdot\nu(x)\right) G(L+\lambda;x)\, dS(x) \\ &= o\left(N^{1-q} \int_{\substack{|x-y|=N \\ x\in\mathbb{R}^q}} |G((L+\lambda);x)| + \left|\frac{\partial}{\partial\nu} G(L+\lambda;x)\right|\, dS(x)\right) \end{aligned} \tag{18.29}$$

for $N \to \infty$, so that

$$\begin{aligned} &\int_{\substack{|x-y|=N \\ x\in\mathbb{R}^q}} e^{2\pi i a\cdot x} G(L+\lambda;x) \frac{\partial F(x-y)}{\partial\nu_x}\, dS(x) \\ &\quad - \int_{\substack{|x-y|=N \\ x\in\mathbb{R}^q}} e^{2\pi i a\cdot x} F(x-y) \left(\frac{\partial}{\partial\nu} + 4\pi i a\cdot\nu(x)\right) G(L+\lambda;x)\, dS(x) \\ &= o(1) \end{aligned} \tag{18.30}$$

for $N \to \infty$.

Collecting our results we obtain the desired result as stated in Lemma 18.4. □

From (18.23) we consequently obtain the asymptotic identity

$$\begin{aligned}
\sum_{\substack{|g-y|\le N\\ g\in\Lambda}} & e^{2\pi i a\cdot g} F(g-y) \qquad (18.31)\\
= & \frac{1}{\|\mathcal{F}_\Lambda\|} e^{2\pi i a\cdot y} \sum_{\substack{4\pi^2((h+a)^2-a^2)=\lambda\\ h\in\Lambda^{-1}}} e^{2\pi i h\cdot y} \int_{\substack{|x|\le N\\ x\in\mathbb{R}^q}} F(x)\, e^{2\pi i x\cdot(a+h)} dV(x)\\
+ & \int_{\substack{|x-y|\le N\\ x\in\mathbb{R}^q}} G(L+\lambda;x)\, e^{2\pi i a\cdot x} \left(\Delta_x+\lambda+4\pi^2\, a^2\right) F(x-y)\, dV(x)\\
& + \ o(1), \quad N\to\infty.
\end{aligned}$$

Next we deal with the volume integral in (18.31) involving the $(\Delta_x+\lambda+4\pi^2a^2)$-derivative of F.

Proposition 18.5. *For an appropriately specified $\lambda\in\mathbb{R}$, a function F is said to be of class $\mathrm{Y}^{(2)}_\lambda(\mathbb{R}^q)$, if F is of class $\mathrm{C}^{(2)}(\mathbb{R}^q)$ satisfying*

$$(\Delta_x+\lambda+4\pi^2a^2)F(x) = O\left(|x|^{-(q+\varepsilon)}\right), \quad \varepsilon>0,\ |x|\to\infty. \qquad (18.32)$$

Lemma 18.6. *Under the condition (18.32), the volume integral*

$$\int_{\mathbb{R}^q} G(L+\lambda;x)\, e^{2\pi i a\cdot x} \left(\Delta_x+\lambda+4\pi^2a^2\right) F(x-y)\, dV(x) \qquad (18.33)$$

is absolutely convergent.

Proof. By virtue of (18.32) we are immediately able to guarantee that, for suitable positive constants M, N,

$$\begin{aligned}
& \left| \int_{\substack{M\le|x-y|\le N\\ x\in\mathbb{R}^q}} G(L+\lambda;x)\, e^{2\pi i a\cdot x} \left(\Delta_x+\lambda+4\pi^2a^2\right) F(x-y)\, dV(x)\right| \qquad (18.34)\\
& = O\left(\int_M^N \frac{1}{(1+r)^{q+\varepsilon}} \left(\int_{\substack{|x-y|=r\\ x\in\mathbb{R}^q}} |G((L+\lambda);x)|\, dS(x)\right) dr\right)\\
& = O\left(\int_M^N \frac{r^{q-1}}{(1+r)^{q+\varepsilon}}\, dr\right).
\end{aligned}$$

Consequently, the absolute convergence of the integral (18.34) is guaranteed. □

Combining Lemma 18.4 and Lemma 18.6 we are able to formulate the following limit relation.

Theorem 18.7. *Suppose that, for a preassigned $\lambda \in \mathbb{R}$, the function F is of class $\mathrm{X}^{(2)}(\mathbb{R}^q) \cap \mathrm{Y}_\lambda^{(2)}(\mathbb{R}^q)$.*

Then, the limit

$$\lim_{N\to\infty}\left(\sum_{\substack{|g-y|\le N\\ g\in\Lambda}} e^{2\pi i a\cdot g}F(g-y) - \frac{1}{\|\mathcal{F}_\Lambda\|}\, e^{2\pi i a\cdot y} \sum_{\substack{4\pi^2\left((h+a)^2-a^2\right)=\lambda\\ h\in\Lambda^{-1}}} e^{2\pi i h\cdot y}\int_{\substack{|x|\le N\\ x\in\mathbb{R}^q}} F(x)\, e^{2\pi i x\cdot(a+h)} dV(x)\right) \tag{18.35}$$

exists, and we have

$$\begin{aligned}\lim_{N\to\infty}&\left(\sum_{\substack{|g-y|\le N\\ g\in\Lambda}} e^{2\pi i a\cdot g}F(g-y) - \frac{1}{\|\mathcal{F}_\Lambda\|}\, e^{2\pi i a\cdot y} \sum_{\substack{4\pi^2\left((h+a)^2-a^2\right)=\lambda\\ h\in\Lambda^{-1}}} e^{2\pi i h\cdot y}\int_{\substack{|x|\le N\\ x\in\mathbb{R}^q}} F(x)\, e^{2\pi i x\cdot(a+h)} dV(x)\right)\\ &= \int_{\mathbb{R}^q} G(L+\lambda;x)\, e^{2\pi i a\cdot x}\left(\Delta_x+\lambda+4\pi^2a^2\right) F(x-y)\, dV(x).\end{aligned} \tag{18.36}$$

Clearly, in the case that the parameter λ is given in such a way that there is no lattice point $h \in \Lambda^{-1}$ satisfying $L^\wedge(h) = 4\pi^2\left((h+a)^2 - a^2\right) = \lambda$, the finite lattice point sum in (18.36) is understood to be zero, and we simply arrive at the following statement.

Corollary 18.8. *Suppose that there is no lattice point $h \in \Lambda^{-1}$ satisfying the equation $4\pi^2\left((h+a)^2 - a^2\right) = \lambda$, i.e., $\lambda \notin \mathrm{Spect}_L(\Lambda)$. Then, under the assumption of Theorem 18.7,*

$$\begin{aligned}\lim_{N\to\infty}&\sum_{\substack{|g-y|\le N\\ g\in\Lambda}} e^{2\pi i a\cdot g}F(g-y) \qquad (18.37)\\ &= \int_{\mathbb{R}^q} G(L+\lambda;x)\, e^{2\pi i a\cdot x}\left(\Delta_x+\lambda+4\pi^2a^2\right) F(x-y)\, dV(x).\end{aligned}$$

Theorem 18.7 demonstrates that the convergence of multi-dimensional sums (understood in spherical summation)

$$\sum_{g\in\Lambda} e^{2\pi i a\cdot g}F(g-y) = \lim_{N\to\infty} \sum_{\substack{|g-y|\leq N\\ g\in\Lambda}} e^{2\pi i a\cdot g}F(g-y) \tag{18.38}$$

is closely related to the spectrum of the operator L. In fact, for the class of non-eigenvalues, i.e., for values $\lambda \notin \mathrm{Spect}_L(\Lambda)$, Theorem 18.7 immediately guarantees the convergence of the infinite series.

In order to ensure the convergence of the finite sum in (18.36) for eigenvalues, however, we have to require an additional condition for the remaining volume integrals (i.e., Fourier transforms):

Proposition 18.9. *For a given* $\lambda \in \mathrm{Spect}_L(\Lambda)$, $F \in \mathrm{C}^{(2)}(\mathbb{R}^q)$ *is said to be of class* $\mathrm{Z}_\lambda^{(2)}(\mathbb{R}^q)$, *if the integrals*

$$\int_{\mathbb{R}^q} F(x)\ e^{2\pi i x\cdot(a+h)}dV(x) \tag{18.39}$$

exist for all $h \in \Lambda^{-1}$ *with* $L^\wedge(h) = 4\pi^2\left((h+a)^2 - a^2\right) = \lambda$ *in the (spherical) sense*

$$\int_{\mathbb{R}^q} F(x)\ e^{2\pi i x\cdot(a+h)}dV(x) = \lim_{N\to\infty} \int_{\substack{|x|\leq N\\ x\in\mathbb{R}^q}} F(x)\ e^{2\pi i x\cdot(a+h)}dV(x). \tag{18.40}$$

Observing Proposition 18.9, we are able to formulate the following convergence theorem.

Theorem 18.10. *For a preassigned* $\lambda \in \mathrm{Spect}_L(\Lambda)$, *let* $F \in \mathrm{C}^{(2)}(\mathbb{R}^q)$ *be a member of class* $\mathrm{X}^{(2)}(\mathbb{R}^q) \cap \mathrm{Y}_\lambda^{(2)}(\mathbb{R}^q) \cap \mathrm{Z}_\lambda^{(2)}(\mathbb{R}^q)$.
Then the series

$$\sum_{g\in\Lambda} e^{2\pi i a\cdot g}F(g-y) = \lim_{N\to\infty} \sum_{\substack{|g-y|\leq N\\ g\in\Lambda}} e^{2\pi i a\cdot g}F(g-y) \tag{18.41}$$

is convergent (in the indicated spherical sense).

To be more concrete, we are confronted with the following alternative situation: For given $\lambda \notin \mathrm{Spect}_L(\Lambda)$ and $F \in \mathrm{X}^{(2)}(\mathbb{R}^q) \cap \mathrm{Y}_\lambda^{(2)}(\mathbb{R}^q)$, we have

$$\begin{aligned}&\lim_{N\to\infty} \sum_{\substack{|g-y|\leq N\\ g\in\Lambda}} e^{2\pi i a\cdot g}F(g-y)\\ &\quad = \int_{\mathbb{R}^q} G(L+\lambda;x)\ e^{2\pi i a\cdot x}\left(\Delta_x + \lambda + 4\pi^2 a^2\right)F(x-y)\ dV(x).\end{aligned} \tag{18.42}$$

whereas, for given $\lambda \in \mathrm{Spect}_L(\Lambda)$ and $F \in \mathrm{X}^{(2)}(\mathbb{R}^q) \cap \mathrm{Y}^{(2)}_\lambda(\mathbb{R}^q) \cap \mathrm{Z}^{(2)}_\lambda(\mathbb{R}^q)$, we have

$$\begin{aligned} &\lim_{N\to\infty} \sum_{\substack{|g-y|\leq N \\ g\in\Lambda}} e^{2\pi i a\cdot g} F(g-y) \\ &= \lim_{N\to\infty} \frac{1}{\|\mathcal{F}_\Lambda\|} \; e^{2\pi i a\cdot y} \sum_{\substack{4\pi^2((h+a)^2-a^2)=\lambda \\ h\in\Lambda^{-1}}} e^{2\pi i h\cdot y} \int_{\substack{|x|\leq N \\ x\in\mathbb{R}^q}} F(x)\, e^{2\pi i x\cdot(a+h)} dV(x) \\ &\quad + \int_{\mathbb{R}^q} G(L+\lambda;x)\, e^{2\pi i a\cdot x} \left(\Delta_x + \lambda + 4\pi^2 a^2\right) F(x-y)\, dV(x). \end{aligned} \tag{18.43}$$

As a consequence, our approach canonically leads to the following convention.

Proposition 18.11. *For a given $\lambda \in \mathbb{R}$, a function $F \in \mathrm{C}^{(2)}(\mathbb{R}^q)$ is said to belong to the class $\mathrm{C}^{(2)}_\lambda(\mathbb{R}^q)$, if one of the following conditions is satisfied:*

(i) F is of class $\mathrm{X}^{(2)}(\mathbb{R}^q) \cap \mathrm{Y}^{(2)}_\lambda(\mathbb{R}^q)$, provided that $\lambda \notin \mathrm{Spect}_L(\Lambda)$,

(ii) F is of class $\mathrm{X}^{(2)}(\mathbb{R}^q) \cap \mathrm{Y}^{(2)}_\lambda(\mathbb{R}^q) \cap \mathrm{Z}^{(2)}_\lambda(\mathbb{R}^q)$, provided that $\lambda \in \mathrm{Spect}_L(\Lambda)$.

18.4 Poisson-Type Summation Formula (Heuristic Approach)

We heuristically point out a first consequence leading to a formal variant of the Poisson summation formula over $\mathbb{R}^q$. To this end, under the condition $F \in \mathrm{C}^{(2)}_\lambda(\mathbb{R}^q)$, we start from (18.42) and (18.43), respectively.

By the formal use of the Fourier series of $G(L+\lambda;x)$ (which, however, does not exist in ordinary sense) we obtain for sufficiently large positive values N

$$\begin{aligned} &\int_{\substack{|x|\leq N \\ x\in\mathbb{R}^q}} G(L+\lambda;x)\, e^{2\pi i a\cdot x} \left(\Delta_x + \lambda + 4\pi^2 a^2\right) F(x-y)\, dV(x) \\ &\sim \sum_{\substack{4\pi^2((h+a)^2-a^2)\neq\lambda \\ h\in\Lambda}} \frac{1}{\lambda - 4\pi^2\left((h+a)^2 - a^2\right)} \\ &\qquad \times \int_{\substack{|x|\leq N \\ x\in\mathbb{R}^q}} e^{2\pi i(a+h)\cdot x} \left(\Delta_x + \lambda + 4\pi^2 a^2\right) F(x-y)\, dV(x). \end{aligned} \tag{18.44}$$

Observing that F is of class $\mathrm{X}^{(2)}(\mathbb{R}^q)$, we find by virtue of the second Green's theorem, i.e., integration by parts

$$\begin{aligned} &\int_{\substack{|x|\leq N\\ x\in\mathbb{R}^q}} e^{2\pi i(a+h)\cdot x}\ \left(\Delta_x+\lambda+4\pi^2a^2\right)F(x-y)\ dV(x) \qquad (18.45)\\ &= \ (\lambda-4\pi^2\left((h+a)^2-a^2\right))\int_{\substack{|x|\leq N\\ x\in\mathbb{R}^q}} e^{2\pi i(a+h)\cdot x}\ F(x-y)\ dV(x)\\ &+ \quad o(1)\end{aligned}$$

for $N\to\infty$. Hence, in connection with Theorem 18.10, our heuristic excursion yields the following *Poisson-type summation formula* for a function $F\in \mathrm{C}^{(2)}_{\lambda}(\mathbb{R}^q)$:

$$\begin{aligned} &\lim_{N\to\infty}\sum_{\substack{|g-y|\leq N\\ g\in\Lambda}} e^{2\pi i a\cdot g}F(g-y) \qquad (18.46)\\ &\quad\sim \frac{1}{\|\mathcal{F}_\Lambda\|}\ e^{2\pi i a\cdot y}\sum_{h\in\Lambda^{-1}} e^{2\pi i h\cdot y}\int_{\mathbb{R}^q} F(x)\ e^{2\pi i x\cdot(a+h)}dV(x).\end{aligned}$$

Unfortunately, the formal step (18.44) to interchange the volume integral and the lattice sum has to be justified to guarantee the pointwise equality in (18.46). So, in its present form under the condition that F is of class $\mathrm{X}^{(2)}(\mathbb{R}^q), q\geq 2$, our approach is not constructive and not permissible for lattice point summation and sampling purposes. Instead we have to find a mathematically rigorous way. This will be undertaken by the transition to Λ-Euler–Green functions corresponding to iterated $(L+\lambda)$-operators and additional assumptions on the smoothness of the function F under consideration. It turns out that these modifications finally guarantee a mathematically sound way, on which we can base the requested Poisson-type summation formula and, subsequently, our Shannon-type sampling intentions.

18.5 Euler-Type Summation Formula over Increasing Balls Involving Iterated Euler–Green Functions

We assume that the function F under consideration admits a higher differentiability order, so that iterated Euler–Green functions may be used in an Euler-type summation context. Essential tools for the asymptotic handling of extended Euler-type summation formulas are the integral estimates collected in Theorem 18.2. Our approach naturally implies that the convergence criteria justifying Poisson-type summation formulas over Euclidean spaces $\mathbb{R}^q$ have to be understood in spherical sense, i.e., the Euler-type summation formula involving iterated Euler–Green functions as the point of departure has to be considered over balls with increasing radii, analogously to the procedure of the last section.

To be more concrete, for arbitrary lattices $\Lambda \subset \mathbb{R}^q$, arbitrary vectors $a, y \in \mathbb{R}^q$, and $2m$-times continuously differentiable functions F on closed balls $\overline{\mathbb{B}^q_N(y)}$, Corollary 9.8 implies the summation formula

$$\begin{aligned} &\sum_{\substack{|g-y|\le N \\ g\in\Lambda}} e^{2\pi i a\cdot g} F(g-y) \qquad (18.47) \\ &= \frac{1}{\|\mathcal{F}_\Lambda\|} e^{2\pi i a\cdot y} \sum_{\substack{4\pi^2\left((h+a)^2-a^2\right)=\lambda \\ h\in\Lambda^{-1}}} e^{2\pi i h\cdot y} \int_{\substack{|x|\le N \\ x\in\mathbb{R}^q}} F(x) e^{2\pi i x\cdot(a+h)} dV(x) \\ &\quad + \int_{\substack{|x-y|\le N \\ x\in\mathbb{R}^q}} G((L+\lambda)^m; x)\, e^{2\pi i a\cdot x} \left(\Delta_x + \lambda + 4\pi^2 a^2\right)^m F(x-y)\, dV(x) \\ &\quad + R_m(N), \end{aligned}$$

where the *remainder term* is given by

$$\begin{aligned} &R_m(N) \qquad (18.48) \\ &= \frac{1}{2} \sum_{\substack{|g-y|=N \\ g\in\Lambda}} e^{2\pi i a\cdot g} F(g-y) \\ &\quad - \sum_{k=0}^{m-1} \int_{\substack{|x-y|=N \\ x\in\mathbb{R}^q}} e^{2\pi i a\cdot x} G((L+\lambda)^{k+1}; x) \frac{\partial}{\partial\nu_x} \left(\Delta_x + \lambda + 4\pi^2 a^2\right)^k F(x-y)\, dS(x) \\ &\quad + \sum_{k=0}^{m-1} \int_{\substack{|x-y|=N \\ x\in\mathbb{R}^q}} e^{2\pi i a\cdot x} \left(\Delta_x + \lambda + 4\pi^2 a^2\right)^k F(x-y) \\ &\qquad\qquad \times \left(\frac{\partial}{\partial\nu_x} + 4\pi i a\cdot\nu(x)\right) G((L+\lambda)^{k+1}; x)\, dS(x). \end{aligned}$$

18.6 Spherically-Reflected Convergence Criteria (Higher Differentiability Orders)

As pointed out previously, questions of the convergence as $N \to \infty$ require estimates of the remainder term (18.48), where the integral estimates for Euler–Green functions to iterated $(L+\lambda)$-operators, i.e., Theorem 18.2, play an important part.

For a given $\lambda \in \mathbb{R}$, the function spaces $\mathrm{X}_\lambda^{(2m)}(\mathbb{R}^q)$, $\mathrm{Y}_\lambda^{(2m)}(\mathbb{R}^q)$, respectively, are introduced in extension to Proposition 18.3 and Proposition 18.5 as follows:

Proposition 18.12. *A function F is said to be of class $\mathrm{X}_\lambda^{(2m)}(\mathbb{R}^q)$, if F is of class $\mathrm{C}^{(2m)}(\mathbb{R}^q)$ and satisfies the asymptotic relations*

$$\left(\Delta_x + \lambda + 4\pi^2 a^2\right)^k F(x) \;=\; o\left(|x|^{1-q}\right), \quad |x| \to \infty \tag{18.49}$$

and

$$\left|\nabla_x \left(\Delta_x + \lambda + 4\pi^2 a^2\right)^k F(x)\right| \;=\; o\left(|x|^{1-q}\right), \quad |x| \to \infty, \tag{18.50}$$

for $k = 0, \ldots, m-1$ (note that the case $m = 1$ is independent of $\lambda \in \mathbb{R}$, so that $\mathrm{X}_\lambda^{(2)}(\mathbb{R}^q)$ coincides with $\mathrm{X}^{(2)}(\mathbb{R}^q)$).

Proposition 18.13. *A function F is said to be of class $\mathrm{Y}_\lambda^{(2m)}(\mathbb{R}^q)$, if F is of class $\mathrm{C}^{(2m)}(\mathbb{R}^q)$ and satisfies*

$$\left(\Delta_x + \lambda + 4\pi^2 a^2\right)^m F(x) = O\left(|x|^{-(q+\varepsilon)}\right), \quad \varepsilon > 0, \;\; |x| \to \infty. \tag{18.51}$$

In consistency with our earlier context, the function space $\mathrm{Z}_\lambda^{(2m)}(\mathbb{R}^q)$ is introduced as follows:

Proposition 18.14. *A function $F \in \mathrm{C}^{(2m)}(\mathbb{R}^q)$ is said to be of class $\mathrm{Z}_\lambda^{(2m)}(\mathbb{R}^q)$, if the integrals*

$$\int_{\mathbb{R}^q} F(x)\, \overline{\Phi_h(x)}\, dV(x) \tag{18.52}$$

exist for all $h \in \Lambda^{-1}$ with $L^\wedge(h) = 4\pi^2\left((h+a)^2 - a^2\right) = \lambda$ in the (spherical) sense

$$\int_{\mathbb{R}^q} F(x)\overline{\Phi_h(x)}\, dV(x) = \lim_{N\to\infty} \int_{\substack{|x|\le N \\ x\in\mathbb{R}^q}} F(x)\, \overline{\Phi_h(x)}\, dV(x). \tag{18.53}$$

By virtue of these settings we are prepared to begin our specification of asymptotic criteria from a straightforward discussion of (18.47) and its remainder term (18.48).

Lemma 18.15. *Assume that $F \in \mathrm{C}^{(2m)}(\mathbb{R}^q), m \in \mathbb{N}$, is of class $\mathrm{X}_\lambda^{(2m)}(\mathbb{R}^q)$. Then, letting $N \to \infty$ we have*

$$R_m(N) = o(1). \tag{18.54}$$

Proof. We know from the already-known Gaussian estimate (cf. C.F. Gauss [1826]) that

$$\frac{1}{2} \sum_{\substack{|g-y|=N \\ g\in\Lambda}} e^{2\pi i a\cdot g} F(g-y) = o\left(N^{1-q} \sum_{\substack{|g-y|=N \\ g\in\Lambda}} 1 \right) = o(1), \quad N \to \infty. \tag{18.55}$$

Furthermore it follows that

$$\begin{aligned} & - \sum_{k=0}^{m-1} \int_{\substack{|x-y|=N \\ x\in\mathbb{R}^q}} e^{2\pi i a\cdot x} G((L+\lambda)^{k+1}; x) \frac{\partial}{\partial \nu_x} \left(\Delta_x + \lambda + 4\pi^2 a^2\right)^k F(x-y)\, dS(x) \\ & \quad + \sum_{k=0}^{m-1} \int_{\substack{|x-y|=N \\ x\in\mathbb{R}^q}} e^{2\pi i a\cdot x} \left(\Delta_x + \lambda + 4\pi^2 a^2\right)^k F(x-y) \\ & \qquad \times \left(\frac{\partial}{\partial \nu_x} + 4\pi i a \cdot \nu(x)\right) G((L+\lambda)^{k+1}; x)\, dS(x) \\ & = o\left(N^{1-q} \left(\sum_{k=0}^{m-1} \int_{\substack{|x|=N \\ x\in\mathbb{R}^q}} \left| G((L+\lambda)^{k+1}; x)\right| \right.\right. \\ & \qquad \left.\left. + \left| \frac{\partial}{\partial \nu} G((L+\lambda)^{k+1}; x)\right| \, dS(x)\right)\right) \end{aligned} \tag{18.56}$$

for $N \to \infty$. In connection with Theorem 18.2, we therefore obtain

$$\begin{aligned} & - \sum_{k=0}^{m-1} \int_{\substack{|x-y|=N \\ x\in\mathbb{R}^q}} e^{2\pi i a\cdot x} G((L+\lambda)^{k+1}; x) \frac{\partial}{\partial \nu_x} \left(\Delta_x + \lambda + 4\pi^2 a^2\right)^k F(x-y)\, dS(x) \\ & \quad + \sum_{k=0}^{m-1} \int_{\substack{|x-y|=N \\ x\in\mathbb{R}^q}} e^{2\pi i a\cdot x} \left(\Delta_x + \lambda + 4\pi^2 a^2\right)^k H(x-y) \\ & \qquad \times \left(\frac{\partial}{\partial \nu_x} + 4\pi i a \cdot \nu(x)\right) G((L+\lambda)^{k+1}; x)\, dS(x) \\ & = o(1) \end{aligned} \tag{18.57}$$

for $N \to \infty$.

Collecting all details we arrive at the desired result stated in Lemma 18.15. □

Next we come to the discussion of the volume integral in (18.47) containing the product of $G((L+\lambda)^m; x)\ e^{2\pi i a\cdot x}$ and $\left(\Delta_x + \lambda + 4\pi^2 a^2\right)^m F(x-y)$ as integrand.

Lemma 18.16. *Assume that, for a preassigned $\lambda \in \mathbb{R}$, the function F is of class $\mathrm{Y}_\lambda^{(2m)}(\mathbb{R}^q)$. Then the volume integral*

$$\int_{\mathbb{R}^q} G((L+\lambda)^m; x)\ e^{2\pi i a\cdot x} \left(\Delta_x + \lambda + 4\pi^2 a^2\right)^m F(x-y)\, dV(x) \tag{18.58}$$

is absolutely convergent.

Proof. From Theorem 18.2 we are able to guarantee with sufficiently large positive constants M, N

$$\left| \int_{\substack{M \le |x-y| \le N \\ x \in \mathbb{R}^q}} G((L+\lambda)^m; x) \left(\Delta_x + \lambda + 4\pi^2 a^2\right)^m F(x-y)\, dV(x) \right| \tag{18.59}$$
$$= O\left(\int_M^N \frac{1}{(1+r)^{q+\varepsilon}} \left(\int_{\substack{|x-y|=r \\ x \in \mathbb{R}^q}} |G((L+\lambda)^m; x)|\, dS(x) \right) dr \right)$$
$$= O\left(\int_M^N \frac{r^{q-1}}{(1+r)^{q+\varepsilon}}\, dr \right).$$

Consequently, the absolute convergence of the integral (18.58) is obvious. □

Combining Lemma 18.15 and Lemma 18.16, we obtain as a first consequence.

Theorem 18.17. *Assume that F is of class $\mathrm{X}_\lambda^{(2m)}(\mathbb{R}^q) \cap \mathrm{Y}_\lambda^{(2m)}(\mathbb{R}^q)$. Then, the limit*

$$\lim_{N\to\infty} \left(\sum_{\substack{|g-y| \le N \\ g \in \Lambda}} e^{2\pi i a\cdot g} F(g-y) - \frac{1}{\|\mathcal{F}_\Lambda\|} e^{2\pi i a\cdot y} \sum_{\substack{4\pi^2\left((h+a)^2 - a^2\right) = \lambda \\ h \in \Lambda^{-1}}} e^{2\pi i h\cdot y} \int_{\substack{|x| \le N \\ x \in \mathbb{R}^q}} F(x)\, e^{2\pi i x\cdot(a+h)} dV(x) \right) \tag{18.60}$$

exists, and we have

$$\lim_{N\to\infty} \left(\sum_{\substack{|g-y| \le N \\ g \in \Lambda}} e^{2\pi i a\cdot g} F(g-y) - \frac{1}{\|\mathcal{F}_\Lambda\|} e^{2\pi i a\cdot y} \sum_{\substack{4\pi^2\left((h+a)^2 - a^2\right) = \lambda \\ h \in \Lambda^{-1}}} e^{2\pi i h\cdot y} \int_{\substack{|x| \le N \\ x \in \mathbb{R}^q}} F(x)\, e^{2\pi i x\cdot(a+h)} dV(x) \right)$$
$$= \int_{\mathbb{R}^q} G((L+\lambda)^m; x)\, e^{2\pi i a\cdot x} \left(\Delta_x + \lambda + 4\pi^2 a^2\right)^m F(x-y)\, dV(x). \tag{18.61}$$

In the case that $\lambda \notin \mathrm{Spect}_L(\Lambda)$, i.e., that there is no lattice point $h \in \Lambda^{-1}$ satisfying $L^\wedge(h) = 4\pi^2\left((h+a)^2 - a^2\right) = \lambda$, the finite sum

$$\frac{1}{\|\mathcal{F}_\Lambda\|} e^{2\pi i a\cdot y} \sum_{\substack{4\pi^2((h+a)^2 - a^2) = \lambda \\ h \in \Lambda^{-1}}} e^{2\pi i h\cdot y} \int_{\substack{|x| \le N \\ x \in \mathbb{R}^q}} F(x)\, e^{2\pi i x\cdot(a+h)} dV(x) \tag{18.62}$$

is understood to be zero, so that

$$\lim_{N\to\infty} \sum_{\substack{|g-y|\le N\\ g\in\Lambda}} e^{2\pi i a\cdot g} F(g-y)$$
$$= \int_{\mathbb{R}^q} G((L+\lambda)^m; x)\, e^{2\pi i a\cdot x} \left(\Delta_x + \lambda + 4\pi^2 a^2\right)^m F(x-y)\, dV(x). \tag{18.63}$$

Once again, Theorem 18.17 demonstrates that the convergence of multi-dimensional alternating sums (understood in spherical summation)

$$\sum_{g\in\Lambda} e^{2\pi i a\cdot g} F(g-y) = \lim_{N\to\infty} \sum_{\substack{|g-y|\le N\\ g\in\Lambda}} e^{2\pi i a\cdot g} F(g-y) \tag{18.64}$$

is closely related to the spectrum $\operatorname{Spect}_L(\Lambda)$. As a matter of fact, for the class of non-eigenvalues, i.e., $\lambda \notin \operatorname{Spect}_L(\Lambda)$, Theorem 18.17 immediately guarantees the convergence of the series (18.64) in spherical sense.

In order to guarantee the convergence of the series (18.64) for an eigenvalue $\lambda \in \operatorname{Spect}_L(\Lambda)$, however, the space $\mathrm{Z}_\lambda^{(2m)}(\mathbb{R}^q)$ has to come into play. In fact, for $\lambda \in \operatorname{Spect}_L(\Lambda)$, we are led to formulate the following convergence theorem.

Theorem 18.18. *Let F be a of class $\mathrm{X}_\lambda^{(2m)}(\mathbb{R}^q) \cap \mathrm{Y}_\lambda^{(2m)}(\mathbb{R}^q) \cap \mathrm{Z}_\lambda^{(2m)}(\mathbb{R}^q)$. Then the series*

$$\sum_{g\in\Lambda} e^{2\pi i a\cdot g} F(g-y) = \lim_{N\to\infty} \sum_{\substack{|g-y|\le N\\ g\in\Lambda}} e^{2\pi i a\cdot g} F(g-y) \tag{18.65}$$

is convergent (in the indicated spherical sense).

More specifically, for $\lambda \notin \operatorname{Spect}_L(\Lambda)$ and $F \in \mathrm{X}_\lambda^{(2m)}(\mathbb{R}^q) \cap \mathrm{Y}_\lambda^{(2m)}(\mathbb{R}^q)$, we have

$$\lim_{N\to\infty} \sum_{\substack{|g-y|\le N\\ g\in\Lambda}} e^{2\pi i a\cdot g} F(g-y)$$
$$= \int_{\mathbb{R}^q} G((L+\lambda)^m; x)\, e^{2\pi i a\cdot x} \left(\Delta_x + \lambda + 4\pi^2 a^2\right)^m F(x-y)\, dV(x), \tag{18.66}$$

whereas, for $\lambda \in \operatorname{Spect}_L(\Lambda)$ and $F \in \mathrm{X}^{(2)}(\mathbb{R}^q) \cap \mathrm{Y}_\lambda^{(2)}(\mathbb{R}^q) \cap \mathrm{Z}_\lambda^{(2)}(\mathbb{R}^q)$,

$$\lim_{N\to\infty} \sum_{\substack{|g-y|\le N\\ g\in\Lambda}} e^{2\pi i a\cdot g} F(g-y)$$
$$= \lim_{N\to\infty} \frac{1}{\|\mathcal{F}_\Lambda\|}\, e^{2\pi i a\cdot y} \sum_{\substack{4\pi^2((h+a)^2-a^2)=\lambda\\ h\in\Lambda^{-1}}} e^{2\pi i h\cdot y} \int_{\substack{|x|\le N\\ x\in\mathbb{R}^q}} F(x) e^{2\pi i x\cdot(a+h)} dV(x)$$
$$= \int_{\mathbb{R}^q} G((L+\lambda)^m; x)\, e^{2\pi i a\cdot x} \left(\Delta_x + \lambda + 4\pi^2 a^2\right)^m F(x-y)\, dV(x). \tag{18.67}$$

Analogously to the former setting ($m = 1$) we are led to the following convention.

Proposition 18.19. *For a given $\lambda \in \mathbb{R}$, a function $F \in \mathrm{C}^{(2m)}(\mathbb{R}^q), m \in \mathbb{N}$, is said to belong to the class $\mathrm{C}_\lambda^{(2m)}(\mathbb{R}^q)$, if one of the following conditions is satisfied:*

(i) F is of class $\mathrm{X}_\lambda^{(2m)}(\mathbb{R}^q) \cap \mathrm{Y}_\lambda^{(2m)}(\mathbb{R}^q)$, provided that $\lambda \notin \mathrm{Spect}_L(\Lambda)$,

(ii) F is of class $\mathrm{X}_\lambda^{(2m)}(\mathbb{R}^q) \cap \mathrm{Y}_\lambda^{(2m)}(\mathbb{R}^q) \cap \mathrm{Z}_\lambda^{(2)}(\mathbb{R}^q)$, provided that $\lambda \in \mathrm{Spect}_L(\Lambda)$.

18.7 Poisson-Type Summation Formula (Rigorous Approach)

The convergence criteria, which led to the formulation of Theorem 18.18, enable us to derive a multi-dimensional analogue of the Poisson summation formula. In the context of continuous $(L+\lambda)^m$-Euler–Green functions satisfying the condition $m > \frac{q}{2}$, we are able to do this in a non-heuristic way, i.e., with mathematical rigor. The price, however, is the additional smoothness imposed on the function F under consideration.

Let us explain the details: For the integers $m \in \mathbb{N}$ with $m > \frac{q}{2}$, the iterated Λ–Euler–Green function $G((L+\lambda)^m; \cdot)$ permits an absolutely and uniformly convergent Fourier series in $\mathbb{R}^q$. Thus, Lebesgue's theorem allows us to interchange the order of summation and integration. It follows that

$$\begin{aligned} &\int_{\mathbb{R}^q} G((L+\lambda)^m; x)\, e^{2\pi i a\cdot x} \left(\Delta_x + \lambda + 4\pi^2 a^2\right)^m F(x-y)\, dV(x) \qquad (18.68) \\ &= \frac{1}{\|\mathcal{F}_\Lambda\|} \sum_{\substack{4\pi^2((h+a)^2-a^2)\neq\lambda \\ h\in\Lambda}} \frac{1}{(\lambda - 4\pi^2((h+a)^2-a^2))^m} \\ &\qquad \times \int_{\mathbb{R}^q} e^{2\pi i (a+h)\cdot x} \left(\Delta_x + \lambda + 4\pi^2 a^2\right)^m F(x-y)\, dV(x) \end{aligned}$$

holds true for a function $F \in \mathrm{C}_\lambda^{(2m)}(\mathbb{R}^q)$. Moreover, in connection with (18.49) and (18.50), repeated application of Green's theorem yields the identity

$$\begin{aligned} &\int_{\mathbb{R}^q} e^{2\pi i (a+h)\cdot x} \left(\Delta_x + \lambda + 4\pi^2 a^2\right)^m F(x-y)\, dV(x) \qquad (18.69) \\ &= \int_{\mathbb{R}^q} \left(\Delta_x + \lambda + 4\pi^2 a^2\right)^m e^{2\pi i (a+h)\cdot x}\, F(x-y)\, dV(x) \\ &= (\lambda - 4\pi^2((h+a)^2-a^2))^m \int_{\mathbb{R}^q} e^{2\pi i (a+h)\cdot x}\, F(x-y)\, dV(x). \end{aligned}$$

Inserting (18.69) into (18.68) we therefore find that

$$\begin{aligned}
&\int_{\mathbb{R}^q} G((L+\lambda)^m; x)\, e^{2\pi i a\cdot x} \left(\Delta_x + \lambda + 4\pi^2 a^2\right)^m F(x-y)\, dV(x) \\
&= \frac{1}{\|\mathcal{F}_\Lambda\|}\, e^{2\pi i a\cdot y} \sum_{\substack{h\in\Lambda^{-1} \\ 4\pi^2((h+a)^2-a^2)\neq\lambda}} e^{2\pi i h\cdot y} \int_{\mathbb{R}^q} F(x)\, e^{2\pi i x\cdot(a+h)} dV(x).
\end{aligned} \tag{18.70}$$

In combination with Theorem 18.18, this finally yields to the desired *multi-dimensional Poisson-type summation formula.*

Theorem 18.20. *Let Λ be an arbitrary lattice in the Euclidean space $\mathbb{R}^q$. For a preassigned $\lambda \in \mathbb{R}$, suppose that the function F is of class $\mathrm{C}^{(2m)}_\lambda(\mathbb{R}^q)$, $m > \frac{q}{2}$. Then the limit*

$$\sum_{g\in\Lambda} e^{2\pi i a\cdot g} F(g-y) = \lim_{N\to\infty} \sum_{\substack{|g-y|\leq N \\ g\in\Lambda}} e^{2\pi i a\cdot g} F(g-y) \tag{18.71}$$

is convergent, and we have

$$\begin{aligned}
&\lim_{N\to\infty} \sum_{\substack{|g-y|\leq N \\ g\in\Lambda}} e^{2\pi i a\cdot g} F(g-y) \\
&= \frac{1}{\|\mathcal{F}_\Lambda\|} \sum_{h\in\Lambda^{-1}} e^{2\pi i y\cdot(a+h)} \int_{\mathbb{R}^q} F(x)\, e^{2\pi i x\cdot(a+h)} dV(x).
\end{aligned} \tag{18.72}$$

It should be alluded that the sum on the left side of the identity (18.72) is not necessarily absolutely convergent in $\mathbb{R}^q$, so the process of summation must be taken into account. As already mentioned several times, the convergence of the series on the left side in Theorem 18.20 has to be understood in the spherical sense.

Once more, it should be pointed out that the "wave number" $\lambda \in \mathbb{R}$ reflects the specific character of a function of the class $\mathrm{C}^{(2m)}_\lambda(\mathbb{R}^q)$, $m > \frac{q}{2}$. Our approach shows that it is extremely helpful to adapt the parameter $\lambda \in \mathbb{R}$ specifically to the oscillating properties of the function under consideration. In doing so, our sufficient conditions to establish the validity of the Poisson-type summation formula are particularly suited to the discussion of multi-dimensional alternating series and integrals. Convincing case scenarios within the framework of Hardy–Landau (alternating) identities can be found in the next sections.

We end our considerations involving the Poisson-type summation formula in $\mathbb{R}^q$ with a hybrid variant of the *Parseval identity*, that immediately follws from the identity (18.72).

Corollary 18.21. *Let Λ be an arbitrary lattice in the Euclidean space $\mathbb{R}^q$. For a preassigned $\lambda \in \mathbb{R}$, suppose that the function F is of class $\mathrm{C}_\lambda^{(2m)}(\mathbb{R}^q)$, $m > \frac{q}{2}$. Let $\mathcal{H}$ be a regular region in $\mathbb{R}^q$. Suppose that G is an integrable function on $\overline{\mathcal{H}}$. Then, for an arbitrary $a \in \mathbb{R}^q$,*

$$\int_{\mathcal{H}} \lim_{N\to\infty} \sum_{\substack{|g-y|\le N\\ g\in\Lambda}} e^{2\pi i a\cdot g}\, F(g-y)\, G(y)\, dV(y) \tag{18.73}$$

$$= \frac{1}{\|\mathcal{F}_\Lambda\|} \sum_{h\in\Lambda^{-1}} \int_{\mathcal{H}} G(y)\, e^{2\pi i y\cdot(a+h)}\, dV(y) \int_{\mathbb{R}^q} F(x)\, e^{2\pi i x\cdot(a+h)} dV(x).$$

18.8 Hardy–Landau-Type Identities (Spherical Harmonic Weights)

Next, in agreement with our approach, we apply the developed convergence criteria (stated in Theorem 18.20) to higher-dimensional extensions of the Hardy–Landau identity. As the point of departure we choose a product function consisting of a radial and an angular weight function, i.e., the infinitely often differentiable function $F_R : \mathbb{R}^q \to \mathbb{R}$, $R > 0$ fixed, given by

$$F_R(x-y) = \frac{J_\nu(q; 2\pi|x-y|R)}{|x-y|^\nu}\, H_n(q; x-y),\ y \in \mathbb{R}^q, \tag{18.74}$$

where $H_n(q;\cdot)$ is a homogeneous, harmonic polynomial of degree $n \in \mathbb{N}_0$ in q dimensions, and $J_\nu(q;\cdot)$ is the Bessel function of order $\nu > 0$ and dimension q.

Expanding $J_\nu(q; 2\pi rR)$ in powers of its argument we have

$$\frac{J_\nu(q; 2\pi rR)}{r^\nu} = \frac{(\pi R)^\nu \Gamma\left(\frac{q}{2}\right)}{\Gamma\left(\nu+\frac{q}{2}\right)} \left(1 - \frac{(\pi rR)^2}{1!(\nu+\frac{q}{2})} + - \cdots\right). \tag{18.75}$$

For $n \in \mathbb{N}$ and $r \to \infty$, the standard estimates of the theory of Bessel functions (i.e., Lemma 6.28) yield the relation

$$\begin{aligned} J_\nu(q; 2\pi Rr) &= J_n(q; 2\pi Rr)\ \cos\left((\nu-n)\frac{\pi}{2}\right) \\ &\quad + J_{n+1}(q; 2\pi Rr)\ \sin\left((\nu-n)\frac{\pi}{2}\right) \\ &\quad + O\left(r^{-\frac{q+1}{2}}\right). \end{aligned} \tag{18.76}$$

Under the following assumptions imposed on the values ν, n

$$q \ge 2,\ \nu - n > \frac{q-1}{2},\ \ \nu > 0,\ \ n \ge 0, \tag{18.77}$$

the theory of Bessel functions implies the asymptotic relations

$$\begin{aligned} (\Delta+\lambda)^k F_R(x-y) &= o(|x|^{1-q}), \quad |x|\to\infty, \\ \left|\nabla(\Delta+\lambda)^k F_R(x-y)\right| &= o(|x|^{1-q}), \quad |x|\to\infty \end{aligned} \tag{18.78}$$

for each non-negative integer k and all parameters $\lambda \in \mathbb{R}$.

Moreover, from the differential equation for the Bessel function, we are able to deduce that

$$\left|\left(\Delta_x + 4\pi^2 R^2\right) \frac{J_\nu(q; 2\pi|x-y|R)}{|x-y|^\nu}\right| = O\left(\frac{J_{\nu+1}(q; 2\pi|x-y|R)}{|x-y|^{\nu+1}}\right) \tag{18.79}$$

for $|x|\to\infty$. Thus, in view of the assumptions (18.77), we obtain the estimate

$$(\Delta+\lambda+4\pi^2 a^2)F_R(x) = O\left(|x|^{-(q+\varepsilon)}\right), \quad |x|\to\infty, \tag{18.80}$$

under the special choice

$$\lambda = 4\pi^2(R^2-a^2) \tag{18.81}$$

and

$$\varepsilon = \nu - n - \frac{q-1}{2} > 0. \tag{18.82}$$

Altogether, Theorem 18.17 leads to the following result.

Lemma 18.22. *Under the assumption (18.77),*

$$\begin{aligned} \lim_{N\to\infty} \left(\sum_{\substack{|g-y|\le N \\ g\in\Lambda}} e^{2\pi i a\cdot g} F_R(g-y) \right. & - \left. \frac{1}{\|\mathcal{F}_\Lambda\|} \sum_{\substack{|h+a|=R \\ h\in\Lambda^{-1}}} e^{2\pi i h\cdot y} \int_{\mathbb{R}^q} F_R(x) e^{2\pi i x(a+h)}\, dV(x) \right) \\ & = \frac{1}{\|\mathcal{F}_\Lambda\|} \sum_{\substack{|h+a|\neq R \\ h\in\Lambda^{-1}}} e^{2\pi i h\cdot y} \int_{\mathbb{R}^q} F_R(x) e^{2\pi i x\cdot(a+h)}\, dV(x). \end{aligned}$$

Therefore, for values R with $4\pi^2(R^2-a^2) \notin \mathrm{Spect}_L(\Lambda)$, F_R already allows the identity:

Lemma 18.23. *For all R with* $4\pi^2(R^2-a^2) \notin \mathrm{Spect}_\Delta(\Lambda)$

$$\begin{aligned} & \lim_{N\to\infty} \sum_{\substack{|g-y|\le N \\ g\in\Lambda}} e^{2\pi i a\cdot g} F_R(g-y) \\ = \; & \frac{1}{\|\mathcal{F}_\Lambda\|} \sum_{h\in\Lambda^{-1}} e^{2\pi i h\cdot y} \int_{R^q} F_R(x) e^{2\pi i x\cdot(a+b)}\, dV(x). \end{aligned} \tag{18.83}$$

It remains to investigate the integrals on the right side of (18.83). Introducing standard polar coordinates, we find that

$$
\begin{aligned}
&\frac{1}{\|\mathcal{F}_\Lambda\|}\int_{\mathbb{R}^q} F_R(x) e^{2\pi i x\cdot(a+h)}\, dV(x) \qquad (18.84)\\
&= \frac{i^n}{\|\mathcal{F}_\Lambda\|}\|\mathbb{S}^{q-1}\|\; Y_n\left(q;\frac{h}{|h|}\right)\int_0^\infty J_\nu(q;2\pi r R) J_n(q;2\pi|h+a|r)\; r^{n-\nu+q-1}\, dr.
\end{aligned}
$$

The theory of discontinuous integrals gives detailed information about the convergence of all terms

$$
\int_0^\infty J_\nu(q;2\pi r R) J_n(q;2\pi|h+a|r) r^{n-\nu+q-1} dr. \qquad (18.85)
$$

From (18.76), we are able to deduce that the integrand is asymptotically equal to

$$
\begin{aligned}
& r^{n-\nu+q-1}\; J_n(q;2\pi|h+a|r) J_n(q;2\pi Rr)\cos\left((\nu-n)\frac{\pi}{2}\right) \qquad (18.86)\\
&= r^{n-\nu+q-1}\; J_n(q;2\pi|h+a|r) J_{n+1}(q;2\pi Rr)\sin\left((\nu-n)\frac{\pi}{2}\right)\\
&\quad + O\left(r^{n-\nu-1}\right).
\end{aligned}
$$

Hence, under one of the following conditions imposed on the values ν, n

(i) $4\pi^2(R^2-a^2)\notin \mathrm{Spect}_L(\Lambda)$, *i.e.,* $R\neq|h+a|$ *for all* $h\in\Lambda^{-1}$:

$$
\nu - n > 0, \quad \nu > 0. \qquad (18.87)
$$

(ii) $4\pi^2(R^2-a^2)\in \mathrm{Spect}_L(\Lambda)$, *i.e.,* $R=|h+a|$ *for some* $h\in\Lambda^{-1}$:

$$
\nu - n > 1, \quad \nu > 0, \qquad (18.88)
$$

or

$$
\nu - n > 0, \quad \nu - n \text{ odd integer}, \qquad (18.89)
$$

the integrals (18.85) are convergent (more details on the discontinuous integral theory can be found, e.g., in the monograph by G.N. Watson [1944]).

All in all, under the condition that ν, m obey the specified constraints, we have

$$
\begin{aligned}
&\lim_{N\to\infty}\sum_{\substack{|g-y|\le N\\ g\in\Lambda}} e^{2\pi i a\cdot(g-y)}\; F_R(g-y) \qquad (18.90)\\
&= \frac{1}{\|\mathcal{F}_\Lambda\|}\sum_{h\in\Lambda^{-1}} e^{2\pi i h\cdot y}\int_{\mathbb{R}^2} F_R(x)\; e^{2\pi i x\cdot(a+h)}\; dV(x).
\end{aligned}
$$

More explicitly, the two occurring cases depending on the choice of λ may be characterized as follows:

The case $4\pi^2(R^2 - a^2) \notin \mathrm{Spect}_L(\Lambda)$*:* Under the assumptions (18.77) and (18.87), i.e., $\nu - n > \frac{q-1}{2}$, $\nu > 0, n \in \mathbb{N}_0$, the integrals (18.85) exist. Even more, they are explicitly known from the Hankel transform of discontinuous integrals (Section 14.1)

$$\frac{1}{\|\mathcal{F}_\Lambda\|} \int_{\mathbb{R}^q} F_R(x) e^{2\pi i x\cdot(a+h)} \, dV(x) \tag{18.91}$$
$$= \begin{cases} \frac{2i^n}{\|\mathbb{S}^{q-1}\| \|\mathcal{F}_\Lambda\|} \frac{\pi^{\nu-n}}{R^{2n-\nu+q}\Gamma(\nu-n)} \left(1 - \left(\frac{|h+a|}{R}\right)^2\right)^{\nu-n-1} H_n(q; h+a) & , |h+a| < R, \\ 0 & , |h+a| > R. \end{cases}$$

Therefore, as a first result, we obtain from (18.83):

Theorem 18.24. *Let Λ be an arbitrary lattice in $\mathbb{R}^q$. Then, for all numbers R with $4\pi^2(R^2 - a^2) \notin \mathrm{Spect}_L(\Lambda)$ and for all values ν, n with $\nu - n > \frac{q-1}{2}$ and $\nu > 0$, $n \in \mathbb{N}_0$ (see (18.82)), we have*

$$\frac{2i^n \pi^{\nu-n} e^{2\pi i a\cdot y}}{\|\mathbb{S}^{q-1}\| \|\mathcal{F}_\Lambda\| R^{2n-\nu+q} \Gamma(\nu-n)} \sum_{\substack{|h+a|<R \\ h\in\Lambda^{-1}}} \left(1 - \left(\frac{|h+a|}{R}\right)^2\right)^{\nu-n-1} H_n(q; h+a)$$
$$= \lim_{N\to\infty} \sum_{\substack{|g-y|\le N \\ g\in\Lambda}} e^{2\pi i a\cdot g} \frac{J_\nu(q; 2\pi|g-y|R)}{|g-y|^\nu} H_n(q; g-y). \tag{18.92}$$

The case $4\pi^2(R^2 - a^2) \in \mathrm{Spect}_L(\Lambda)$*:* For all values ν, n with $\nu - n > \frac{q-1}{2}$ and $\nu > 0$, $n \in \mathbb{N}_0$ the identity (18.83) leads to the limit relation

$$\lim_{N\to\infty} \left(\sum_{\substack{|g-y|\le N \\ g\in\Lambda}} e^{2\pi i a\cdot g} \frac{J_\nu(q; 2\pi|g-y|R)}{|g-y|^\nu} H_n(q; g-y) \right. \tag{18.93}$$
$$\left. - \frac{1}{\|\mathcal{F}_\Lambda\|} \sum_{\substack{|h+a|=R \\ h\in\Lambda^{-1}}} \int_{\substack{|x|\le N \\ x\in\mathbb{R}^q}} \left(1 - \left(\frac{|h+a|}{R}\right)^2\right)^{\nu-n-1} H_n(q; h+a) e^{-2\pi i h\cdot x} \, dV(x) \right)$$
$$= \frac{2i^n \pi^{\nu-n}}{\|\mathbb{S}^{q-1}\| \|\mathcal{F}_\Lambda\| R^{2n-\nu+q} \Gamma(\nu-n)} \sum_{\substack{|h|<R \\ h\in\Lambda^{-1}}} \left(1 - \left(\frac{|h+a|}{R}\right)^2\right)^{\nu-n-1} H_n(q; h+a).$$

As a consequence, for values R with $4\pi^2(R^2 + a^2) \in \mathrm{Spect}_L(\Lambda)$, we obtain by combination of (18.87), (18.88), and (18.89) the following result.

Theorem 18.25. *Let Λ be an arbitrary lattice in $\mathbb{R}^q$. Let ν, n satisfy one of the following assumptions:*

$$\nu - n > \max\left(1, \frac{q-1}{2}\right), \quad \nu > 0, \ n \in \mathbb{N}_0, \tag{18.94}$$

or

$$\nu - n > \frac{q-1}{2}, \quad \nu > 0, \ n \in \mathbb{N}_0, \ \ \nu - n \textit{ is an odd integer.} \tag{18.95}$$

Then the following q-dimensional analogue to the Hardy–Landau-type identity holds true:

$$\begin{aligned} &\frac{2 i^n \pi^{\nu-n} e^{2\pi i a\cdot y}}{\|\mathbb{S}^{q-1}\| \|\mathcal{F}_\Lambda\| R^{2n-\nu+q} \Gamma(\nu-n)} \sum_{\substack{|h+a|\le R \\ h\in\Lambda^{-1}}}{}' \left(1 - \left(\frac{|h+a|}{R}\right)^2\right)^{\nu-n-1} H_n(q; h+a) \\ &= \lim_{N\to\infty} \sum_{\substack{|g-y|\le N \\ g\in\Lambda}} e^{2\pi i a\cdot g} \frac{J_\nu(q; 2\pi|g-y|R)}{|g-y|^\nu} H_n(q; g-y). \end{aligned} \tag{18.96}$$

Finally, replacing the inverse lattice $\Lambda^{-1} \subset \mathbb{R}^q$ by the lattice $\Lambda \subset \mathbb{R}^q$ we find:

Corollary 18.26. *Under the assumptions of Theorem 18.25*

$$\begin{aligned} &\frac{2 i^n \pi^{\nu-n} e^{2\pi i a\cdot y}}{\|\mathbb{S}^{q-1}\| R^{2n-\nu+q} \Gamma(\nu-n)} \sum_{\substack{|g+a|\le R \\ g\in\Lambda}}{}' \left(1 - \left(\frac{|g+a|}{R}\right)^2\right)^{\nu-n-1} H_n(q; g+a) \\ &= \lim_{N\to\infty} \frac{1}{\|\mathcal{F}_\Lambda\|} \sum_{\substack{|h-y|\le N \\ h\in\Lambda^{-1}}} e^{2\pi i a\cdot h} \frac{J_\nu(q; 2\pi|h-y|R)}{|h-y|^\nu} H_n(q; h-y). \end{aligned} \tag{18.97}$$

Remark 18.27. *For R especially chosen in such a way that $g' \in \Lambda$ is the only lattice point in Λ satisfying $|g'+a| < R$, we already obtain a sampling-type formula for homogeneous harmonic polynomials*

$$\begin{aligned} &\frac{2 i^n \pi^{\nu-n}}{\|\mathbb{S}^{q-1}\| R^{2n-\nu+q} \Gamma(\nu-n)} \left(1 - \left(\frac{|g'+a|}{R}\right)^2\right)^{\nu-n-1} e^{2\pi i a\cdot y} \ H_n(q; g'+a) \\ &= \lim_{N\to\infty} \frac{1}{\|\mathcal{F}_\Lambda\|} \sum_{\substack{|h-y|\le N \\ h\in\Lambda^{-1}}} e^{2\pi i a\cdot h} \ H_n(q; h-y) \ \frac{J_\nu(q; 2\pi|h-y|R)}{|h-y|^\nu}. \end{aligned} \tag{18.98}$$

We conclude this section with a remarkable collection of some number theoretically important 2D examples (cf. W. Freeden [2011] and the references therein):

Example 18.28. *For $q = 2$ the homogeneous harmonic polynomials of degree $n \in \mathbb{N}_0$ are linear combinations of the functions*

$$(x_1, x_2)^T \mapsto (x_1 + ix_2)^n + (x_1 - ix_2)^n, \tag{18.99}$$

and

$$(x_1, x_2)^T \mapsto i^{-1}\left((x_1 + ix_2)^n - (x_1 - ix_2)^n\right). \tag{18.100}$$

Taking $\Lambda \subset \mathbb{R}^2$ simply as the unit lattice $\mathbb{Z}^2$ and setting as homogeneous harmonic polynomial

$$(x_1, x_2)^T \mapsto H_n(2; x) = H_n(2; x_1, x_2) = (x_1 + ix_2)^n, \ n \in \mathbb{N}_0, \tag{18.101}$$

we are able to deduce from Corollary 18.26 with $\nu = 4k + 1, \ n = 4k, k \in \mathbb{N}_0$, the following extension of the Hardy–Landau identity:

$$\begin{aligned} \sum_{(n_1,n_2)^{\mathrm{T}} \in \mathbb{Z}^2} & \frac{J_{4k+1}(2\pi\sqrt{n_1^2 + n_2^2}\, R)}{\sqrt{n_1^2 + n_2^2}^{\,4k+1}} (n_1 + in_2)^{4k} \\ &= \frac{1}{R^{\,4k+1}} \sum_{\substack{n_1^2 + n_2^2 \le R^2 \\ (n_1,n_2)^{\mathrm{T}} \in \mathbb{Z}^2}}{}' (n_1 + in_2)^{4k}. \end{aligned} \tag{18.102}$$

For $k = 0$, we are led back to the classical Hardy–Landau identity.

Part IX

Multivariate Shannon-Type Sampling Formulas over Euclidean Spaces

Chapter 19

Shannon-Type Sampling Based on Poisson-Type Summation Formulas over Euclidean Spaces

Next we base our sampling obligations on the lattice point identity derived in the last chapter. In fact, it is easily seen that the *Poisson-type summation formula* in $\mathbb{R}^q$ (i.e., Theorem 18.20) allows two equivalent writings

$$\sum_{g\in\Lambda} e^{2\pi i y\cdot(g-a)} \int_{\mathbb{R}^q} F(x)\, e^{-2\pi i x\cdot(g-a)} dV(x) \tag{19.1}$$
$$= \lim_{N\to\infty} \frac{1}{\|\mathcal{F}_\Lambda\|} \sum_{\substack{|h|\le N\\ h\in\Lambda^{-1}}} F(h)\, e^{2\pi i a\cdot(h+y)}, \quad a,y\in\mathbb{R}^q,$$

and

$$\lim_{N\to\infty} \sum_{\substack{|g+y|\le N\\ g\in\Lambda}} e^{-2\pi i a\cdot(g+y)} F(g+y) \tag{19.2}$$
$$= \frac{1}{\|\mathcal{F}_\Lambda\|} \sum_{h\in\Lambda^{-1}} e^{2\pi i h\cdot y} \int_{\mathbb{R}} F(x)\, e^{-2\pi i x\cdot(h+a)} dx, \quad a,y\in\mathbb{R},$$

provided that $F \in \mathrm{C}^{(2m)}_\lambda(\mathbb{R}^q)$, $m > \frac{q}{2}$.

In analogy to the context implied by the Poisson-type summation formula for a regular region (Theorem 15.4), the identities (19.1) and (19.2), respectively, offer two different entry points for Shannon-type sampling, i.e., functional values–based Shannon-type sampling and Fourier transformed values–based Shannon-type sampling characterized by

- the summation of the resulting cardinal series on the left side of (19.1) is extended over functional values of $F \in \mathrm{C}^{(2m)}_\lambda(\mathbb{R}^q)$, $m > \frac{q}{2}$, in y-shifted lattice points of Λ,
- the summation of the cardinal series on the right side of (19.2) is extended over functional values of the Fourier transform of $F \in \mathrm{C}^{(2m)}_\lambda(\mathbb{R}^q)$, $m > \frac{q}{2}$, in a-shifted lattice points of Λ^{-1}.

Once more, it should be noted that the role of functional values and Fourier transformed functional values of F is exchangeable, since the Fourier-transformed values of the Fourier transform canonically lead back to the functional values by virtue of the Fourier inversion formula. Nevertheless, as already seen in the context of Theorem 15.4, we see that the two different entry points (19.1) and (19.2) provided by the same Poisson-type summation formula (Theorem 18.20) subsequently deliver different points of departure to Shannon-type sampling variants. In fact, Section 19.1 and Section 19.3, respectively, will give a deeper insight into this situation.

19.1 Functional Values–Based Shannon-Type Sampling

Let $\mathcal{H}$ be a regular region in $\mathbb{R}^q$. For $y \in \mathbb{R}^q$ and $F \in \mathrm{C}_\lambda^{(2m)}(\mathbb{R}^q)$, $m > \frac{q}{2}$, we obtain from the identity (19.1) by integration over $\mathcal{H}$

$$\begin{aligned}\int_{\mathcal{H}} \sum_{g\in\Lambda} e^{2\pi i y\cdot(g-a)} \int_{\mathbb{R}^q} F(x)\, e^{-2\pi i x\cdot(g-a)} dV(x)\, dV(a) \qquad (19.3)\\ = \lim_{N\to\infty} \sum_{\substack{|h|\le N\\ h\in\Lambda^{-1}}} F(h)\, \frac{1}{\|\mathcal{F}_\Lambda\|} \int_{\mathcal{H}} e^{2\pi i a\cdot(h+y)}\, dV(a).\end{aligned}$$

We are interested in a modification of the left side of (19.3) such that the right side of (19.3) serves as a cardinal series in a Shannon-type sampling process. To this end, we again denote by $\Lambda(\mathcal{H})$ the set of lattice points $g' \in \Lambda \subset \mathbb{R}^q$ such that $\mathcal{F}_\Lambda + \{g'\}$ contains points of $\mathcal{H}$, i.e.,

$$\Lambda(\mathcal{H}) = \{g' \in \Lambda : (\mathcal{F}_\Lambda + \{g'\}) \cap \mathcal{H} \neq \emptyset\}. \qquad (19.4)$$

Using this abbreviation we are able to write

$$\begin{aligned}\sum_{g'\in\Lambda(\mathcal{H})} \int_{\mathcal{H}_{g'}} \sum_{g\in\Lambda} e^{2\pi i y\cdot(g-a)} \int_{\mathbb{R}^q} F(x)\, e^{-2\pi i x\cdot(g-a)} dV(x)\, dV(a)\\ = \lim_{N\to\infty} \sum_{\substack{|h|\le N\\ h\in\Lambda^{-1}}} F(h)\, \frac{1}{\|\mathcal{F}_\Lambda\|}\, K_{\mathcal{H}}(h+y), \qquad (19.5)\end{aligned}$$

where the set $\mathcal{H}_{g'}, g' \in \Lambda(\mathcal{H})$, is given by

$$\mathcal{H}_{g'} = (\mathcal{F}_\Lambda + \{g'\}) \cap \mathcal{H}, \quad g' \in \Lambda(\mathcal{H}). \qquad (19.6)$$

Since the function

$$a \mapsto \sum_{g\in\Lambda} e^{2\pi i y\cdot(g-a)} \int_{\mathbb{R}^q} F(x)\, e^{-2\pi i x\cdot(g-a)} dV(x), \quad a \in \mathbb{R}^q, \qquad (19.7)$$

is Λ–periodic, we are able to reformulate (19.5) as follows:

$$\begin{aligned}
&\sum_{g'\in\Lambda(\mathcal{H})} \int_{\mathcal{H}_{g'}} \sum_{g\in\Lambda} e^{2\pi i y\cdot(g-a)} \int_{\mathbb{R}^q} F(x)\, e^{-2\pi i x\cdot(g-a)} dV(x)\, dV(a) \\
&= \sum_{g'\in\Lambda(\mathcal{H})} \int_{\mathcal{H}^0_{g'}} \sum_{g\in\Lambda} e^{2\pi i y\cdot(g-a)} \int_{\mathbb{R}^q} F(x)\, e^{-2\pi i x\cdot(g-a)} dV(x)\, dV(a) \\
&= \lim_{N\to\infty} \sum_{\substack{|h|\le N\\ h\in\Lambda^{-1}}} F(h)\, \frac{1}{\|\mathcal{F}_\Lambda\|}\, K_{\mathcal{H}}(h+y), \qquad (19.8)
\end{aligned}$$

where $\mathcal{H}^0_{g'}, g'\in\Lambda(\mathcal{H})$, is the shift of $\mathcal{H}_{g'}$ to the fundamental cell, i.e.,

$$\mathcal{H}^0_{g'} = \mathcal{H}_{g'} - \{g'\},\ g'\in\Lambda(\mathcal{H}), \qquad (19.9)$$

forms a subset of $\mathcal{F}_\Lambda$. In doing so, we arrive at the identity

$$\begin{aligned}
&\sum_{g'\in\Lambda(\mathcal{H})} \int_{\mathcal{H}^0_{g'}} \sum_{g\in\Lambda} e^{2\pi i y\cdot(g-a)} \int_{\mathbb{R}^q} F(x)\, e^{-2\pi i x\cdot(g-a)} dV(x)\, dV(a) \\
&= \sum_{g'\in\Lambda(\mathcal{H})} \int_{\mathbb{R}^q} e^{2\pi i y\cdot w}\, \chi_{\mathcal{H}^0_{g'}+\Lambda}(w) \int_{\mathbb{R}^q} F(x)\ e^{-2\pi i x\cdot w}\ dV(x)\, dV(w) \\
&= \lim_{N\to\infty} \sum_{\substack{|h|\le N\\ h\in\Lambda^{-1}}} F(h)\, \frac{1}{\|\mathcal{F}_\Lambda\|}\, K_{\mathcal{H}}(h+y), \qquad (19.10)
\end{aligned}$$

where the characteristic function $x\mapsto \chi_{\mathcal{H}^0_{g'}+\Lambda}(x)$, $x\in\mathbb{R}^q$, is, as usual, given by

$$\chi_{\mathcal{H}^0_{g'}+\Lambda}(x) = \begin{cases} 1, & x\in \bigcup\limits_{g\in\Lambda} \left(\mathcal{H}^0_{g'} + \{g\}\right) \\ 0, & \text{otherwise.} \end{cases} \qquad (19.11)$$

Summarizing our results we therefore obtain the following lattice point Shannon-type sampling formula.

Theorem 19.1. *(Shannon-Type Sampling Theorem). Let $\mathcal{H}$ be a regular region in $\mathbb{R}^q$. Assume that Λ is an arbitrary lattice in $\mathbb{R}^q$. Suppose that F is a member of the class $\mathrm{C}^{(2m)}_\lambda(\mathbb{R}^q)$, $m>\frac{q}{2}$. Then, for all $y\in\mathbb{R}^q$, we have*

$$\begin{aligned}
&\sum_{\substack{(\mathcal{F}_\Lambda+\{g'\})\cap\mathcal{H}\neq\emptyset\\ g'\in\Lambda}} \int_{\mathbb{R}^q} e^{2\pi i y\cdot w}\, \chi_{\mathcal{H}^0_{g'}+\Lambda}(w) \underbrace{\int_{\mathbb{R}^q} F(x)\ e^{-2\pi i x\cdot w}\ dV(x)}_{=F^\wedge_{\mathbb{R}^q}(w)} dV(w) \\
&= \sum_{\substack{(\mathcal{F}_\Lambda+\{g'\})\cap\mathcal{H}\neq\emptyset\\ g'\in\Lambda}} \int_{\bigcup\limits_{g\in\Lambda}(((\mathcal{H}\cap(\mathcal{F}_\Lambda+\{g'\}))-\{g'\})+\{g\})} F^\wedge_{\mathbb{R}^q}(w)\, e^{2\pi i y\cdot w}\ dV(w) \\
&= \lim_{N\to\infty} \sum_{\substack{|h|\le N\\ h\in\Lambda^{-1}}} F(h)\, \frac{1}{\|\mathcal{F}_\Lambda\|}\, K_{\mathcal{H}}(h+y), \qquad (19.12)
\end{aligned}$$

We distinguish the three cases $\overline{\mathcal{H}} \subsetneq \overline{\mathcal{F}_\Lambda}$, $\overline{\mathcal{H}} = \overline{\mathcal{F}_\Lambda}$, and $\overline{\mathcal{H}} \supsetneq \overline{\mathcal{F}_\Lambda}$:

For $\overline{\mathcal{H}} \subsetneq \overline{\mathcal{F}_\Lambda}$ we find that

$$\int_{\bigcup\limits_{g\in\Lambda} (\mathcal{H}+\{g\})} F^{\wedge}_{\mathbb{R}^q}(x)\, e^{2\pi i y\cdot x}\, dV(x) = \lim_{N\to\infty} \sum_{\substack{|h|\leq N\\ h\in\Lambda^{-1}}} F(h)\, \frac{1}{\|\mathcal{F}_\Lambda\|}\, K_{\mathcal{H}}(h+y). \tag{19.13}$$

As an example, by choosing $\overline{\mathcal{H}} = \overline{\mathbb{B}^q_\tau} \subset \overline{\mathcal{F}_\Lambda}$, we obtain

$$\begin{aligned} &\int_{\bigcup\limits_{g\in\Lambda} (\mathbb{B}^q_\tau+\{g\})} F^{\wedge}_{\mathbb{R}^q}(x)\, e^{2\pi i y\cdot x}\, dV(x) \qquad (19.14)\\ &= \lim_{N\to\infty} \sum_{\substack{|h|\leq N\\ h\in\Lambda^{-1}}} F(h)\, \tau^q\, \frac{\|\mathbb{S}^{q-1}\|}{\|\mathcal{F}_\Lambda\|}\, \frac{J_1(q;2\pi|h+y|\tau)}{2\pi|h+y|\tau}. \end{aligned}$$

Thus we are confronted with the following observation: The closer $\overline{\mathcal{H}}$ is to a finite (closed) union of lattice cells of Λ, the smaller is the aliasing error by this variant of Shannon-type sampling.

For $\overline{\mathcal{H}} = \overline{\mathcal{F}_\Lambda}$, we obtain by application of the Fourier inversion formula

$$\begin{aligned} &\int_{\bigcup\limits_{g\in\Lambda} (\mathcal{F}_\Lambda+\{g\})} F^{\wedge}_{\mathbb{R}^q}(x)\, e^{2\pi i y\cdot x}\, dV(x) \qquad (19.15)\\ &= \int_{\mathbb{R}^q} F^{\wedge}_{\mathbb{R}^q}(x)\, e^{2\pi i y\cdot x}\, dV(x)\\ &= F(y)\\ &= \lim_{N\to\infty} \sum_{\substack{|h|\leq N\\ h\in\Lambda^{-1}}} F(h)\, \frac{1}{\|\mathcal{F}_\Lambda\|}\, K_{\mathcal{F}_\Lambda}(h+y). \end{aligned}$$

As a consequence, functions F of the class $\mathrm{C}^{(2m)}_{\lambda}(\mathbb{R}^q)$, $m > \frac{q}{2}$, allows a sampling in the standard way by choosing the fundamental cell $\mathcal{F}_\Lambda$ of the lattice Λ to determine the kernel $K_{\mathcal{F}_\Lambda}$

$$F(y) = \lim_{N\to\infty} \sum_{\substack{|h|\leq N\\ h\in\Lambda^{-1}}} F(h)\, \frac{1}{\|\mathcal{F}_\Lambda\|}\, K_{\mathcal{F}_\Lambda}(h+y), \quad y\in\mathbb{R}^q. \tag{19.16}$$

For $\overline{\mathcal{H}} \supsetneq \overline{\mathcal{F}_\Lambda}$

$$\begin{aligned} &\sum_{\substack{(\mathcal{F}_\Lambda+\{g'\})\cap\mathcal{H}\neq\emptyset\\ g'\in\Lambda}} \int_{\bigcup\limits_{g\in\Lambda} (((\mathcal{H}\cap(\mathcal{F}_\Lambda+\{g'\}))-\{g'\})+\{g\})} F^{\wedge}_{\mathbb{R}^q}(x)\, e^{2\pi i y\cdot x}\, dV(x) \qquad (19.17)\\ &= \lim_{N\to\infty} \sum_{\substack{|h|\leq N\\ h\in\Lambda^{-1}}} F(h)\, \frac{1}{\|\mathcal{F}_\Lambda\|}\, K_{\mathcal{H}}(h+y). \end{aligned}$$

All in all, we are led to the conclusion that, in the case where $\mathcal{H}$ consists of a finite union of lattice cells, aliasing can be avoided. So, the functional-based Shannon-type sampling as discussed in this section does not provide a bundle of results comparable to the bandlimited case involving finite regions. Nevertheless, a way is found in not-necessarily bandlimited sampling to detect the explicit aliasing error and to formulate Shannon-type approximate identities.

19.2 Paley–Wiener Reproducing Kernel Structure

Let us denote by $\mathrm{B}_\lambda^{(2m)}(\mathbb{R}^q)$, $m > \frac{q}{2}$ the closure of the space $\mathrm{C}_\lambda^{(2m)}(\mathbb{R}^q)$, $m > \frac{q}{2}$ with respect to the topology $\|\cdot\|_{\mathrm{L}^2(\mathbb{R}^q)}$:

$$\mathrm{B}_\lambda^{(2m)}(\mathbb{R}^q) \;=\; \overline{\mathrm{C}_\lambda^{(2m)}(\mathbb{R}^q)}^{\|\cdot\|_{\mathrm{L}^2(\mathbb{R}^q)}}, \quad m > \frac{q}{2}. \tag{19.18}$$

Theorem 19.2. *The set* $\mathrm{B}_\lambda^{(2m)}(\mathbb{R}^q)$ *forms a reproducing kernel Hilbert space with the uniquely determined kernel*

$$K_{\mathcal{F}_\Lambda}(y-x) = \int_{\mathcal{F}_\Lambda} e^{2\pi i a\cdot(y-x)}\, dV(a). \tag{19.19}$$

Proof. In fact, by virtue of (19.16), it follows that

$$\begin{aligned}
&\int_{\mathbb{R}^q} F(y)\, K_{\mathcal{F}_\Lambda}(y-z)\, dV(y) && (19.20)\\
&= \lim_{N\to\infty} \sum_{\substack{|h|\le N\\ h\in\Lambda^{-1}}} F(h)\, \frac{1}{\|\mathcal{F}_\Lambda\|} \int_{\mathbb{R}^q} K_{\mathcal{F}_\Lambda}(y+h)\, K_{\mathcal{F}_\Lambda}(y-z)\, dV(y).
\end{aligned}$$

Now we know from (17.16) that

$$K_{\mathcal{F}_\Lambda}(w-z) = \int_{\mathbb{R}^q} K_{\mathcal{F}_\Lambda}(y-z)\, K_{\mathcal{F}_\Lambda}(y-w)\, dV(y). \tag{19.21}$$

Thus we obtain

$$\begin{aligned}
&\int_{\mathbb{R}^q} F(y)\, K_{\mathcal{F}_\Lambda}(y-z)\, dV(y) && (19.22)\\
&= \lim_{N\to\infty} \sum_{\substack{|h|\le N\\ h\in\Lambda^{-1}}} F(h)\, \frac{1}{\|\mathcal{F}_\Lambda\|} \int_{\mathbb{R}^q} K_{\mathcal{F}_\Lambda}(y+h)\, K_{\mathcal{F}_\Lambda}(y-z)\, dV(y)\\
&= \lim_{N\to\infty} \sum_{\substack{|h|\le N\\ h\in\Lambda^{-1}}} F(h)\, \frac{1}{\|\mathcal{F}_\Lambda\|}\; K_{\mathcal{F}_\Lambda}(h+z),
\end{aligned}$$

so that, in connection with (19.16), we are able to verify the reproducing property

$$\int_{\mathbb{R}^q} F(y)\, K_{\mathcal{F}_\Lambda}(y-z)\, dV(y) \;=\; F(z), \quad z \in \mathbb{R}^q. \tag{19.23}$$

□

An immediate consequence is the following lemma.

Lemma 19.3. *$\{K_{\mathcal{F}_\Lambda}(\cdot - h)\}$ constitutes a closed and complete system in $\mathrm{B}_\lambda^{(2m)}(\mathbb{R}^q)$, i.e.,*

$$\overline{\mathrm{span}_{h\in\Lambda^{-1}} K_{\mathcal{F}_\Lambda}(\cdot \;- h)} = \mathrm{B}_\lambda^{(2m)}(\mathbb{R}^q), \tag{19.24}$$

where the completion is understood in the sense of $\|\cdot\|_{\mathrm{L}^2(\mathbb{R}^q)}$.

Proof. $F \in \mathrm{B}_\lambda^{(2m)}(\mathbb{R}^q)$ and the conditions

$$\int_{\mathbb{R}^q} F(x)\, K_{\mathcal{F}_\Lambda}(x-h)\, dV(x) \;=\; 0 \tag{19.25}$$

for all $h \in \Lambda^{-1}$ imply that $F(h) = 0$ for all $h \in \Lambda^{-1}$. Now, we have

$$F(y) = \frac{1}{\|\mathcal{F}_\Lambda\|} \sum_{h\in\Lambda^{-1}} \underbrace{F(h)}_{=0}\, K_{\mathcal{F}_\Lambda}(h+y) \tag{19.26}$$

for all $y \in \mathbb{R}^q$. Consequently, $(F, K_{\mathcal{F}_\Lambda}(h-\cdot))_{\mathrm{L}^2(\mathbb{R}^q)} = F(h) = 0$ is equivalent to $F = 0$. This guarantees the completeness in $\mathrm{B}_\lambda^{(2m)}(\mathbb{R}^q)$. Since closure and completeness are equivalent in $\mathrm{B}_\lambda^{(2m)}(\mathbb{R}^q)$, this also proves the desired result of Lemma 19.3. □

Remark 19.4. *Paley–Wiener splines can be developed in close similarity to the theory presented for regular regions (see Chapter 17).*

19.3 Fourier Transformed Values–Based Shannon-Type Sampling

We come back to the *Poisson summaton formula* in $\mathbb{R}^q$ (cf. Theorem 18.20) in the form (19.2)

$$\lim_{N\to\infty} \sum_{\substack{|g+y|\le N\\ g\in\Lambda}} e^{-2\pi i a\cdot(g+y)} F(g+y) \tag{19.27}$$

$$= \frac{1}{\|\mathcal{F}_\Lambda\|} \sum_{h\in\Lambda^{-1}} e^{2\pi i h\cdot y} \int_{\mathbb{R}} F(x)\, e^{-2\pi i x\cdot(h+a)} dx, \quad a, y \in \mathbb{R}.$$

In what follows, for $a \in \mathbb{R}^q$ and $F \in \mathrm{C}_\lambda^{(2m)}(\mathbb{R}^q)$, $m > \frac{q}{2}$, we use the right side of (19.27) as a cardinal series in a non-bandlimited Shannon-type sampling process. To this end, we notice that the functions

$$y \mapsto \lim_{N\to\infty} \sum_{\substack{|g+y|\leq N \\ g\in\Lambda}} e^{-2\pi i a\cdot(g+y)}\ F(g+y),\ y \in \mathbb{R}^q \tag{19.28}$$

and

$$y \mapsto \frac{1}{\|\mathcal{F}_\Lambda\|} \sum_{h\in\Lambda^{-1}} e^{2\pi i h\cdot y} \int_{\mathbb{R}^q} F(x)\ e^{-2\pi i x\cdot(h+a)}\ dV(x),\ y \in \mathbb{R}^q \tag{19.29}$$

are Λ-periodic. Furthermore, the sum (19.29) may be formally understood as the Fourier series of (19.28), i.e., the Fourier coefficients of the function (19.28) are given by

$$\begin{aligned}
&\int_{\mathcal{F}_\Lambda} \lim_{N\to\infty} \sum_{\substack{|g+y|\leq N \\ g\in\Lambda}} e^{-2\pi i a\cdot(g+y)} F(g+y)\ e^{2\pi i y\cdot h}\ dV(y) \qquad (19.30)\\
&\quad = \lim_{N\to\infty} \int_{\mathcal{F}_\Lambda} \sum_{\substack{|g+y|\leq N \\ g\in\Lambda}} e^{-2\pi i(a+h)\cdot(g+y)}\ F(g+y)\ e^{-2\pi i h\cdot(g+y)} e^{2\pi i h\cdot y}\ dV(y)\\
&\quad = \lim_{N\to\infty} \int_{\mathcal{F}_\Lambda} \sum_{\substack{|g+y|\leq N \\ g\in\Lambda}} e^{-2\pi i(a+h)\cdot(g+y)}\ F(g+y)\ dV(y)\\
&\quad = \lim_{N\to\infty} \int_{\substack{|x|\leq N \\ x\in\mathbb{R}^q}} F(x)\ e^{-2\pi i x\cdot(a+h)}\ dV(x)
\end{aligned}$$

for all $h \in \Lambda$.

Now, by virtue of (19.30), the lattice point identity (19.27) opens the perspective to discuss aspects of Shannon-type sampling by integrating (19.27) over a regular region $\mathcal{H} \subset \mathbb{R}^q$ with respect to the variable y. As an immediate result we obtain the following lattice point formula, that serves as an essential tool for the Shannon-type sampling discussion

$$\begin{aligned}
&\int_{\mathcal{H}} \lim_{N\to\infty} \sum_{\substack{|g+y|\leq N \\ g\in\Lambda}} e^{-2\pi i a\cdot(g+y)} F(g+y)\ dV(y) \qquad (19.31)\\
&= \sum_{\substack{(\mathcal{F}_\Lambda+\{g'\})\cap\mathcal{H}\neq\emptyset \\ g'\in\Lambda}} \int_{\bigcup_{g\in\Lambda}(((\mathcal{H}\cap(\mathcal{F}_\Lambda+\{g'\}))-\{g'\})+\{g\})} F(x)\ e^{-2\pi i a\cdot x}\ dV(x).
\end{aligned}$$

Theorem 19.5. *(Shannon-Type Sampling Theorem). Let $\mathcal{H}$ be a regular region in $\mathbb{R}^q$. Assume that Λ is an arbitrary lattice in Euclidean space $\mathbb{R}^q$.*

Suppose that F is a member of the class $\mathrm{C}_{\lambda}^{(2m)}(\mathbb{R}^q)$, $m > \frac{q}{2}$. Then

$$\sum_{\substack{(\mathcal{F}_\Lambda+\{g'\})\cap\mathcal{H}\neq\emptyset \\ g'\in\Lambda}} F^{\wedge}_{\bigcup\limits_{g\in\Lambda}(((\mathcal{H}\cap(\mathcal{F}_\Lambda+\{g'\}))-\{g'\})+\{g\})}(a) \tag{19.32}$$
$$= \sum_{h\in\Lambda^{-1}+\{a\}} \underbrace{\int_{\mathbb{R}^q} F(x)\, e^{-2\pi i h\cdot x}\, dV(x)}_{=F^{\wedge}_{\mathbb{R}^q}(h)} \frac{1}{\|\mathcal{F}_\Lambda\|} \underbrace{\int_{\mathcal{H}} e^{2\pi i(h-a)\cdot y}\, dV(y)}_{=K_{\mathcal{H}}(h-a)}.$$

The proof of Theorem 19.5 follows by already-known arguments, which will not be repeated here. As a consequence of (19.32) we obtain

$$\sum_{\substack{(\mathcal{F}_\Lambda+\{g'\})\cap\mathcal{H}\neq\emptyset \\ g'\in\Lambda}} F^{\wedge}_{\bigcup\limits_{g\in\Lambda}(((\mathcal{H}\cap(\mathcal{F}_\Lambda+\{g'\}))-\{g'\})+\{g\})}(a) - \frac{\|\mathcal{H}\|}{\|\mathcal{F}_\Lambda\|} \underbrace{\int_{\mathbb{R}^q} F(x)\, e^{-2\pi i a\cdot x}\, dV(x)}_{=F^{\wedge}_{\mathbb{R}^q}(a)}$$
$$= \sum_{\substack{h\in\Lambda^{-1}+\{a\} \\ h\neq a}} \underbrace{\int_{\mathbb{R}^q} F(x)\, e^{-2\pi i h\cdot x}\, dV(x)}_{=F^{\wedge}_{\mathbb{R}^q}(h)} \frac{1}{\|\mathcal{F}_\Lambda\|} \underbrace{\int_{\mathcal{H}} e^{2\pi i(h-a)\cdot y}\, dV(y)}_{=K_{\mathcal{H}}(h-a)}. \tag{19.33}$$

First we distinguish the two examples $\overline{\mathcal{H}} \subsetneq \overline{\mathcal{F}_\Lambda}$ and $\overline{\mathcal{H}} \supsetneq \overline{\mathcal{F}_\Lambda}$:

For $\overline{\mathcal{H}} \subsetneq \overline{\mathcal{F}_\Lambda}$ we obtain

$$F^{\wedge}_{\bigcup\limits_{g\in\Lambda}(\mathcal{H}+\{g\})}(a) - \frac{\|\mathcal{H}\|}{\|\mathcal{F}_\Lambda\|} \underbrace{\int_{\mathbb{R}^q} F(x)\, e^{-2\pi i a\cdot x}\, dV(x)}_{=F^{\wedge}_{\mathbb{R}^q}(a)}$$
$$= \sum_{\substack{h\in\Lambda^{-1}+\{a\} \\ h\neq a}} \underbrace{\int_{\mathbb{R}^q} F(x)\, e^{-2\pi i h\cdot x}\, dV(x)}_{=F^{\wedge}_{\mathbb{R}^q}(h)} \frac{1}{\|\mathcal{F}_\Lambda\|} \underbrace{\int_{\mathcal{H}} e^{2\pi i(h-a)\cdot y}\, dV(y)}_{=K_{\mathcal{H}}(h-a)}, \tag{19.34}$$

while for $\overline{\mathcal{H}} \supsetneq \overline{\mathcal{F}_\Lambda}$

$$\sum_{\substack{(\mathcal{F}_\Lambda+\{g'\})\cap\mathcal{H}\neq\emptyset \\ g'\in\Lambda}} F^{\wedge}_{\bigcup\limits_{g\in\Lambda}(((\mathcal{H}\cap(\mathcal{F}_\Lambda+\{g'\}))-\{g'\})+\{g\})}(a) - \frac{\|\mathcal{H}\|}{\|\mathcal{F}_\Lambda\|} \underbrace{\int_{\mathbb{R}^q} F(x)\, e^{-2\pi i a\cdot x}\, dV(x)}_{=F^{\wedge}_{\mathbb{R}^q}(a)}$$
$$= \sum_{\substack{h\in\Lambda^{-1}+\{a\} \\ h\neq a}} \underbrace{\int_{\mathbb{R}^q} F(x)\, e^{-2\pi i h\cdot x}\, dV(x)}_{=F^{\wedge}_{\mathbb{R}^q}(h)} \frac{1}{\|\mathcal{F}_\Lambda\|} \underbrace{\int_{\mathcal{H}} e^{2\pi i(h-a)\cdot y}\, dV(y)}_{=K_{\mathcal{H}}(h-a)}. \tag{19.35}$$

In particular, with $\mathcal{H} = \mathbb{B}^q_\tau \subset \mathcal{F}_\Lambda$, we find that

$$F^\wedge_{\bigcup_{g\in\Lambda}(\mathbb{B}^q_\tau+\{g\})}(a) - \frac{\|\mathbb{B}^q_\tau\|}{\|\mathcal{F}_\Lambda\|} \underbrace{\int_{\mathbb{R}^q} F(x)\, e^{-2\pi i a\cdot x}\, dV(x)}_{=F^\wedge_{\mathbb{R}^q}(a)} \tag{19.36}$$

$$= \sum_{\substack{h\in\Lambda^{-1}+\{a\}\\ h\neq a}} \underbrace{\int_{\mathbb{R}^q} F(x)\, e^{-2\pi i h\cdot x}\, dV(x)}_{=F^\wedge_{\mathbb{R}^q}(h)}\ \tau^q\ \frac{\|\mathbb{S}^{q-1}\|}{\|\mathcal{F}_\Lambda\|}\ \frac{J_1(q; 2\pi|h-a|\tau)}{2\pi|h-a|\tau}.$$

The particular case $\overline{\mathcal{H}} = \overline{\mathcal{F}_\Lambda}$ is trivial. In fact, it follows for every $a \in \mathbb{R}^q$ that

$$\sum_{\substack{h\in\Lambda^{-1}+\{a\}\\ h\neq a}} \underbrace{\int_{\mathbb{R}^q} F(x)\, e^{-2\pi i h\cdot x}\, dV(x)}_{=F^\wedge_{\mathbb{R}^q}(h)}\ \frac{1}{\|\mathcal{F}_\Lambda\|} \underbrace{\int_{\mathcal{F}_\Lambda} e^{2\pi i(h-a)\cdot y}\, dV(y)}_{=K_{\mathcal{F}_\Lambda}(h-a)} = 0. \tag{19.37}$$

19.4 Shannon-Type Sampling Involving Dilated Fundamental Cells

For the case $\overline{\mathcal{H}} = \sigma\,\overline{\mathcal{F}_\Lambda}$, $\sigma > 1$, we obtain

$$\sum_{\substack{(\mathcal{F}_\Lambda+\{g'\})\cap\sigma\mathcal{F}_\Lambda\neq\emptyset\\ g'\in\Lambda}} F^\wedge_{\bigcup_{g\in\Lambda}(((\sigma\mathcal{F}_\Lambda\cap(\mathcal{F}_\Lambda+\{g'\}))-\{g'\})+\{g\})}(a) - \sigma^q \underbrace{\int_{\mathbb{R}^q} F(x)\, e^{-2\pi i a\cdot x}\, dV(x)}_{=F^\wedge_{\mathbb{R}^q}(a)}$$

$$= \sum_{\substack{h\in\Lambda^{-1}+\{a\}\\ h\neq a}} \underbrace{\int_{\mathbb{R}^q} F(x)\, e^{-2\pi i h\cdot x}\, dV(x)}_{=F^\wedge_{\mathbb{R}^q}(h)}\ \frac{1}{\|\mathcal{F}_\Lambda\|} \underbrace{\int_{\sigma\mathcal{F}_\Lambda} e^{2\pi i(h-a)\cdot y}\, dV(y)}_{=K_{\sigma\mathcal{F}_\Lambda}(h-a)}, \tag{19.38}$$

while, for $\overline{\mathcal{H}} = \sigma\,\overline{\mathcal{F}_\Lambda}$, $\sigma < 1$, we have

$$F^\wedge_{\bigcup_{g\in\Lambda}(\sigma\mathcal{F}_\Lambda+\{g\})}(a) - \sigma^q \underbrace{\int_{\mathbb{R}^q} F(x)\, e^{-2\pi i a\cdot x}\, dV(x)}_{=F^\wedge_{\mathbb{R}^q}(a)} \tag{19.39}$$

$$= \sum_{\substack{h\in\Lambda^{-1}+\{a\}\\ h\neq a}} \underbrace{\int_{\mathbb{R}^q} F(x)\, e^{-2\pi i h\cdot x}\, dV(x)}_{=F^\wedge_{\mathbb{R}^q}(h)}\ \frac{1}{\|\mathcal{F}_\Lambda\|} \underbrace{\int_{\sigma\mathcal{F}_\Lambda} e^{2\pi i(h-a)\cdot y}\, dV(y)}_{=K_{\sigma\mathcal{F}_\Lambda}(h-a)}.$$

In other words, as $\sigma \to 1$, the error tends to zero, i.e.,

$$\sum_{\substack{h\in\Lambda^{-1}+\{a\}\\ h\neq a}} \underbrace{\int_{\mathbb{R}^q} F(x)\, e^{-2\pi i h\cdot x}\, dV(x)}_{=F^\wedge_{\mathbb{R}^q}(h)}\ \frac{1}{\|\mathcal{F}_\Lambda\|} \underbrace{\int_{\sigma\mathcal{F}_\Lambda} e^{2\pi i(h-a)\cdot y}\, dV(y)}_{=K_{\sigma\mathcal{F}_\Lambda}(h-a)} \to 0 \tag{19.40}$$

so that

$$F^{\wedge}_{\mathbb{R}^q}(a) \approx \sum_{h \in \Lambda^{-1}+\{a\}} F^{\wedge}_{\mathbb{R}^q}(h) \, \frac{1}{\|\mathcal{F}_\Lambda\|} \, K_{\sigma\mathcal{F}_\Lambda}(h-a) \tag{19.41}$$

for $\sigma < 1$, but sufficiently close to 1. Moreover, we are able to deduce from (19.40) that

$$\sum_{\substack{h \in \Lambda^{-1}+\{a\} \\ h \neq a, h \neq h'}} F^{\wedge}_{\mathbb{R}^q}(h) \; K_{\sigma\mathcal{F}_\Lambda}(h-a) \approx - \, F^{\wedge}_{\mathbb{R}^q}(h') \; K_{\sigma\mathcal{F}_\Lambda}(h'-a) \tag{19.42}$$

for $h' \in \Lambda^{-1} + \{a\}, h' \neq a$, and $\sigma < 1$, but sufficiently close to 1 (note that an analogous argumentation also applies to the case $\sigma > 1$, but sufficiently close to 1). In other words, an approximate procedure is found to obtain a value $F^{\wedge}_{\mathbb{R}^q}(h')$ from values $F^{\wedge}_{\mathbb{R}^q}(h)$ with $h \in \Lambda^{-1} + (a), h' \neq h, h \neq a$.

All in all, we are allowed to conclude that, for all values $\sigma > 0$, non-bandlimited Shannon-type sampling does not provide results free of aliasing. Nevertheless, a way is found to detect the explicit aliasing error and to formulate Shannon-type approximate identities.

19.5 Bivariate Locally-Supported Sampling Functions

Some bivariate examples, where the function F represents a bandlimited Fourier transform, should be mentioned in more detail:

(1) Constant Weight. The function

$$x \mapsto \frac{1}{R} \int_{\substack{|x| \leq R \\ x \in \mathbb{R}^2}} e^{-2\pi i x \cdot w} dV(w) = \frac{J_1(2\pi|x|R)}{|x|}, \quad x \in \mathbb{R}^2, R > 0, \tag{19.43}$$

is of class $\mathrm{C}^{(\infty)}_{4\pi^2(R^2-a^2)}(\mathbb{R}^2)$. Thus we obtain for $\sigma < 1$, but sufficiently close to 1 from (19.41)

$$\begin{aligned} &\int_{\mathbb{R}^2} \int_{\substack{|x| \leq R \\ x \in \mathbb{R}^2}} e^{-2\pi i x \cdot w} dV(w) \; e^{2\pi i x \cdot a} \; dV(x) \\ &\approx \frac{1}{\|\mathcal{F}_\Lambda\|} \sum_{h \in \Lambda^{-1}+\{a\}} \int_{\mathbb{R}^2} \int_{\substack{|x| \leq R \\ x \in \mathbb{R}^2}} e^{-2\pi i x \cdot w} dV(w) \; e^{2\pi i h \cdot x} \; dV(x) \;\; K_{\sigma\mathcal{F}_\Lambda}(h-a). \end{aligned} \tag{19.44}$$

The Fourier inversion formula (Theorem 11.1) yields

$$\alpha_{\mathbb{B}^2_R}(a) \approx \frac{1}{\|\mathcal{F}_\Lambda\|} \sum_{h \in \Lambda^{-1}+\{a\}} \alpha_{\mathbb{B}^2_R}(h) \;\; K_{\sigma\mathcal{F}_\Lambda}(h-a) \tag{19.45}$$

with the solid angle $\alpha_{\mathbb{B}^2_R}(a)$ at $a \in \mathbb{R}^2$ subtended by the circle $\partial\mathbb{B}^2_R$.

(2) Spherical Harmonics Weight. From the Hankel transform of discontinuous integrals (Section 14.1) we know that

$$\begin{aligned}
&\int_{\mathbb{R}^2} \frac{J_\nu(2;2\pi|x|R)}{|x|^\nu} H_n(2;x)\, e^{2\pi i x\cdot a}\, dV(x) \qquad (19.46)\\
&= \alpha_{\mathbb{B}^2_R}(a)\, \frac{2i^n}{\|\mathbb{S}^1\|} \frac{\pi^{\nu-n}}{R^{2n-\nu+2}\Gamma(\nu-n)} \left(1-\left(\frac{|a|}{R}\right)^2\right)^{\nu-n-1} H_n(2;a).
\end{aligned}$$

Thus we obtain for $\sigma < 1$, but sufficiently close to 1

$$\begin{aligned}
&\alpha_{\mathbb{B}^2_R}(a) \left(1-\left(\frac{|a|}{R}\right)^2\right)^{\nu-n-1} H_n(2;a) \qquad (19.47)\\
&\approx \frac{1}{\|\mathcal{F}_\Lambda\|} \sum_{h\in\Lambda^{-1}+\{a\}} \alpha_{\mathbb{B}^2_R}(h) \left(1-\left(\frac{|h|}{R}\right)^2\right)^{\nu-n-1} H_n(2;h)\, K_{\sigma\mathcal{F}_\Lambda}(h-a)
\end{aligned}$$

for $\sigma < 1$, but sufficiently close to 1.

(3) General Weight. Let $\mathcal{G} \subset \mathbb{R}^2$ be a regular region (as specified earlier in Chapter 11). Suppose that F is of class $\mathrm{C}^{(0)}(\overline{\mathcal{G}})$.

Then, the function $F^\wedge_\mathcal{G}$ given by

$$F^\wedge_\mathcal{G}(y) = \int_{\substack{x\in\mathcal{G}\\ x\in\mathbb{R}^2}} F(x)\, e^{-2\pi i x\cdot y}\, dV(x) \qquad (19.48)$$

is of class $\mathrm{C}^{(\infty)}_{4\pi^2(R^2-a^2)}(\mathbb{R}^2)$ under the conditions of Corollary 11.10, as well as Corollary 11.12 and of class $\mathrm{C}^{(\infty)}_{4\pi^2(-a^2)}(\mathbb{R}^2)$ under the conditions of Corollary 11.45.

It follows that

$$(F^\wedge_\mathcal{G})^\wedge_{\mathbb{R}^q}(a) \approx \sum_{h\in\Lambda^{-1}+\{a\}} (F^\wedge_\mathcal{G})^\wedge_{\mathbb{R}^q}(h)\, \frac{1}{\|\mathcal{F}_\Lambda\|}\, K_{\sigma\mathcal{F}_\Lambda}(h-a) \qquad (19.49)$$

for $\sigma < 1$, but sufficiently close to 1. Hence, the Fourier inversion formula (Theorem 11.1) yields for $\sigma < 1$, but sufficiently close to 1

$$\alpha_\mathcal{G}(a)\, F(a) \approx \frac{1}{\|\mathcal{F}_\Lambda\|} \sum_{\substack{h\in\Lambda^{-1}+\{a\}\\ h\in\overline{\mathcal{G}}}} \alpha_\mathcal{G}(h)\, F(h)\;\; K_{\sigma\mathcal{F}_\Lambda}(h-a) \qquad (19.50)$$

with the solid angle $\alpha_\mathcal{G}(a)$ at $a \in \mathbb{R}^2$ subtended by $\partial\mathcal{G}$.

More explicitly, for all values $\sigma \leq 1$, we have

$$F^{\wedge}_{\bigcup\limits_{g\in\Lambda}(\sigma\mathcal{F}_\Lambda+\{g\})}(a) \; - \; \sigma^2 \; \alpha_{\mathcal{G}}(a) \; F(a) \tag{19.51}$$
$$= \sum_{\substack{h\in\Lambda^{-1} \\ h\neq 0, a+h\in\overline{\mathcal{G}}}} \alpha_{\mathcal{G}}(a+h)F(a+h) \; \frac{1}{\|\mathcal{F}_\Lambda\|} \; K_{\sigma\mathcal{F}_\Lambda}(h)$$
$$= \sum_{\substack{h\in\Lambda^{-1}+\{a\} \\ h\neq a, h\in\overline{\mathcal{G}}}} \alpha_{\mathcal{G}}(h) \; F(h) \; \frac{1}{\|\mathcal{F}_\Lambda\|} \; K_{\sigma\mathcal{F}_\Lambda}(h-a)$$

for F being of class $\mathrm{C}^{(0)}(\overline{\mathcal{G}})$ (note a similar result holds true for $\sigma > 1$).

19.6 From Gaussian to Ordinary Non-Bandlimited Shannon-Type Sampling

Let $\mathcal{H}, \mathcal{K}$ be regular regions in $\mathbb{R}^q$. It is clear that the function

$$a \mapsto e^{-\pi\tau a^2} F^{\wedge}_{\mathcal{K}}(a) = e^{-\pi\tau a^2} \int_{\mathcal{K}} F(x) e^{2\pi i a\cdot x} \; dV_{(q)}(x), \quad a \in \mathbb{R}^q, \tag{19.52}$$

is of class $\mathrm{C}^{(\infty)}_{-4\pi^2 a^2}(\mathbb{R}^q)$, provided that F is of class $\mathrm{C}^{(0)}(\overline{\mathcal{K}})$. Furthermore, because of the exponential decay, we are able to derive from Theorem 18.20 the identity

$$\lim_{N\to\infty} \sum_{\substack{|g-y|\leq N \\ g\in\Lambda}} e^{2\pi i a\cdot(g-y)} \; e^{-\pi\tau(g-y)^2} F^{\wedge}_{\mathcal{K}}(g-y) \tag{19.53}$$
$$= \frac{1}{\|\mathcal{F}_\Lambda\|} \sum_{h\in\Lambda^{-1}} e^{-2\pi i h\cdot y} \int_{\mathbb{R}^q} e^{-\pi\tau x^2} F^{\wedge}_{\mathcal{K}}(x) \; e^{-2\pi i x\cdot(h-a)} dV(x), \quad a, y \in \mathbb{R}^q,$$

so that

$$\sum_{\substack{(\mathcal{F}_\Lambda+\{g'\})\cap\mathcal{H}\neq\emptyset \\ g'\in\Lambda}} \int_{\bigcup\limits_{g\in\Lambda}(((\mathcal{H}\cap(\mathcal{F}_\Lambda+\{g'\}))-\{g'\})+\{g\})} e^{-\pi\tau x^2} F^{\wedge}_{\mathcal{K}}(x) \; e^{2\pi i a\cdot x} \; dV(x)$$
$$= \sum_{h\in\Lambda^{-1}} \underbrace{\int_{\mathbb{R}^q} e^{-\pi\tau x^2} F^{\wedge}_{\mathcal{K}}(x) \; e^{2\pi i(h-a)\cdot x} \; dV(x)}_{=\,\mathcal{G}^{\mathcal{K}}_\tau(F)(h-a)} \; \frac{1}{\|\mathcal{F}_\Lambda\|} \underbrace{\int_{\mathcal{H}} e^{-2\pi i h\cdot y} \; dV(y)}_{=\,K_{\mathcal{H}}(-h)}$$
$$= \sum_{h+a\in\Lambda^{-1}} \underbrace{\int_{\mathbb{R}^q} e^{-\pi\tau x^2} F^{\wedge}_{\mathcal{K}}(x) \; e^{2\pi i h\cdot x} \; dV(x)}_{=\,\mathcal{G}^{\mathcal{K}}_\tau(F)(h)} \; \frac{1}{\|\mathcal{F}_\Lambda\|} \underbrace{\int_{\mathcal{H}} e^{-2\pi i(h+a)\cdot y} \; dV(y)}_{=\,K_{\mathcal{H}}(-(h+a))}. \tag{19.54}$$

Thus we arrive at the following *Shannon-type sampling result about the $\mathcal{K}$-bandlimited Gauss transform $\mathcal{G}_\tau^{\mathcal{K}}(F)$ of a function F.*

Corollary 19.6. *Let Λ be an arbitrary lattice in $\mathbb{R}^q$. Let $\mathcal{H}, \mathcal{K}$ be regular regions in $\mathbb{R}^q$. Suppose that F is of class $F \in \mathrm{C}^{(0)}(\overline{\mathcal{K}})$. Then, for $\tau > 0$ and $a \in \mathbb{R}^q$,*

$$\sum_{\substack{(\mathcal{F}_\Lambda+\{g'\})\cap\mathcal{H}\neq\emptyset \\ g'\in\Lambda}} \int_{\bigcup_{g\in\Lambda}(((\mathcal{H}\cap(\mathcal{F}_\Lambda+\{g'\}))-\{g'\})+\{g\})} e^{-\pi\tau x^2} F_{\mathcal{K}}^{\wedge}(x)\, e^{2\pi i a\cdot x}\, dV(x)$$
$$= \sum_{h\in\Lambda^{-1}} \mathcal{G}_\tau^{\mathcal{K}}(F)(h-a)\, \frac{1}{\|\mathcal{F}_\Lambda\|}\, K_{\mathcal{H}}(-h). \qquad (19.55)$$

In particular, for $\overline{\mathcal{H}} \subset \overline{\mathcal{F}_\Lambda}$, we have

$$\int_{\bigcup_{g\in\Lambda}(\mathcal{H}+\{g\})} e^{-\pi\tau x^2} F_{\mathcal{K}}^{\wedge}(x)\, e^{2\pi i a\cdot x}\, dV(x)$$
$$= \sum_{h\in\Lambda^{-1}} \mathcal{G}_\tau^{\mathcal{K}}(F)(h-a)\, \frac{1}{\|\mathcal{F}_\Lambda\|}\, K_{\mathcal{H}}(-h). \qquad (19.56)$$

Clearly, for $\overline{\mathcal{H}} = \overline{\mathcal{F}}_\Lambda$, both sides of (19.56) are equal to $G_\tau^{\mathcal{K}}(F)(a)$, however, at the price of the fact that

$$\sum_{\substack{h\in\Lambda^{-1} \\ h\neq 0}} \mathcal{G}_\tau^{\mathcal{K}}(F)(h-a)\, \frac{1}{\|\mathcal{F}_\Lambda\|}\, K_{\mathcal{H}}(-h) = 0. \qquad (19.57)$$

The relation between the Gauss transform and the Weierstrass transform (see (14.7)) yields the following sampling result:

Corollary 19.7. *Let Λ be an arbitrary lattice in $\mathbb{R}^q$. Let $\mathcal{H}, \mathcal{K}$ be regular regions in $\mathbb{R}^q$. Suppose that F is of class $F \in \mathrm{C}^{(0)}(\overline{\mathcal{K}})$. Then, for $\tau > 0$ and $a \in \mathbb{R}^q$,*

$$\sum_{\substack{(\mathcal{F}_\Lambda+\{g'\})\cap\mathcal{H}\neq\emptyset \\ g'\in\Lambda}} \int_{\bigcup_{g\in\Lambda}(((\mathcal{H}\cap(\mathcal{F}_\Lambda+\{g'\}))-\{g'\})+\{g\})} e^{-\pi\tau x^2} F_{\mathcal{K}}^{\wedge}(x)\, e^{2\pi i a\cdot x}\, dV(x)$$
$$= \sum_{h\in\Lambda^{-1}} \mathcal{W}_\tau(F_{\mathcal{K}}^{\wedge})(-(h-a))\, \frac{1}{\|\mathcal{F}_\Lambda\|}\, K_{\mathcal{H}}(-h)$$
$$= \sum_{h+a\in\Lambda^{-1}} \mathcal{W}_\tau(F_{\mathcal{K}}^{\wedge})(-h)\, \frac{1}{\|\mathcal{F}_\Lambda\|}\, K_{\mathcal{H}}(-(h+a)). \qquad (19.58)$$

In particular, for $\overline{\mathcal{H}} \subset \overline{\mathcal{F}_\Lambda}$, *we have*

$$\int_{\bigcup\limits_{g\in\Lambda}(\mathcal{H}+\{g\})} e^{-\pi\tau x^2} F_{\mathcal{K}}^{\wedge}(x)\, e^{2\pi i a\cdot x}\, dV(x) \tag{19.59}$$

$$= \sum_{h\in\Lambda^{-1}} \mathcal{W}_\tau(F_{\mathcal{K}}^{\wedge})(-(h-a))\, \frac{1}{\|\mathcal{F}_\Lambda\|}\, K_{\mathcal{H}}(-h). \tag{19.60}$$

Applying the limit $\lim_{\substack{\tau\to 0\\ \tau>0}}$ to the identity (19.58) in Corollary 19.7, i.e.,

$$\lim_{\substack{\tau\to 0\\ \tau>0}} \sum_{\substack{(\mathcal{F}_\Lambda+\{g'\})\cap\mathcal{H}\neq\emptyset\\ g'\in\Lambda}} \int_{\bigcup\limits_{g\in\Lambda}(((\mathcal{H}\cap(\mathcal{F}_\Lambda+\{g'\}))-\{g'\})+\{g\})} e^{-\pi\tau x^2} F_{\mathcal{K}}^{\wedge}(x)\, e^{2\pi i a\cdot x}\, dV(x)$$

$$= \lim_{\substack{\tau\to 0\\ \tau>0}} \sum_{h\in\Lambda^{-1}} \mathcal{W}_\tau(F_{\mathcal{K}}^{\wedge})(-(h-a))\, \frac{1}{\|\mathcal{F}_\Lambda\|}\, K_{\mathcal{H}}(-h)$$

we are led to the following result.

Theorem 19.8. *Let* Λ *be an arbitrary lattice in* $\mathbb{R}^q$. *Let* $\mathcal{H},\mathcal{K}$ *be regular regions in* $\mathbb{R}^q$. *Suppose that* F *is of class* $\mathrm{C}^{(0)}(\overline{\mathcal{K}})$. *Then, for* $a\in\mathbb{R}^q$,

$$\lim_{\substack{\tau\to 0\\ \tau>0}} \sum_{\substack{(\mathcal{F}_\Lambda+\{g'\})\cap\mathcal{H}\neq\emptyset\\ g'\in\Lambda}} \int_{\bigcup\limits_{g\in\Lambda}(((\mathcal{H}\cap(\mathcal{F}_\Lambda+\{g'\}))-\{g'\})+\{g\})} e^{-\pi\tau x^2} F_{\mathcal{K}}^{\wedge}(x)\, e^{2\pi i a\cdot x}\, dV(x)$$

$$= \sum_{\substack{h\in\Lambda^{-1}\\ a-h\in\overline{\mathcal{K}}}} \alpha_{\mathcal{K}}(a-h)\, F(a-h)\, \frac{1}{\|\mathcal{F}_\Lambda\|}\, K_{\mathcal{H}}(-h). \tag{19.61}$$

In particular, for $\overline{\mathcal{H}} \subset \overline{\mathcal{F}_\Lambda}$, *we have*

$$\lim_{\substack{\tau\to 0\\ \tau>0}} \int_{\bigcup\limits_{g\in\Lambda}(\mathcal{H}+\{g\})} e^{-\pi\tau x^2} F_{\mathcal{K}}^{\wedge}(x)\, e^{-2\pi i a\cdot x}\, dV(x) \tag{19.62}$$

$$= \sum_{\substack{h\in\Lambda^{-1}\\ a-h\in\overline{\mathcal{K}}}} \alpha_{\mathcal{K}}(a-h)\, F(a-h)\, \frac{1}{\|\mathcal{F}_\Lambda\|}\, K_{\mathcal{H}}(-h).$$

Now, for $\overline{\mathcal{H}} \subsetneq \overline{\mathcal{F}_\Lambda}$, we have

$$\int_{\bigcup\limits_{g\in\Lambda}(\mathcal{H}+\{g\})} e^{-\pi\tau x^2} F_{\mathcal{K}}^{\wedge}(x)\, e^{2\pi i a\cdot x}\, dV(x) \tag{19.63}$$

$$= \underbrace{\int_{\mathbb{R}^q} e^{-\pi\tau x^2} F_{\mathcal{K}}^{\wedge}(x)\, e^{2\pi i a\cdot x}\, dV(x)}_{=\,\mathcal{G}_\tau^{\mathcal{K}}(F)(a)}$$

$$- \int_{\mathbb{R}^q\setminus(\bigcup\limits_{g\in\Lambda}(\mathcal{H}+\{g\}))} e^{-\pi\tau x^2} F_{\mathcal{K}}^{\wedge}(x)\, e^{2\pi i a\cdot x}\, dV(x).$$

Summarzing our results, we therefore obtain the following theorem.

Theorem 19.9. *Let Λ be an arbitrary lattice in $\mathbb{R}^q$. Let $\mathcal{K}$ be a regular region in $\mathbb{R}^q$. Let $\mathcal{H}$ be a regular region in $\mathbb{R}^q$ with $\overline{\mathcal{H}} \subset \overline{\mathcal{F}_\Lambda}$. Suppose that F is of class $\mathrm{C}^{(0)}(\overline{\mathcal{K}})$. Then, for $a \in \mathbb{R}^q$,*

$$\alpha_{\mathcal{K}}(a)\ F(a) - \int_{\mathbb{R}^q \setminus (\bigcup\limits_{g\in\Lambda} (\mathcal{H}+\{g\}))} F_{\mathcal{K}}^{\wedge}(x)\ e^{2\pi i a\cdot x}\ dV(x)$$
$$= \sum_{\substack{h\in\Lambda^{-1} \\ a-h\in\overline{\mathcal{K}}}} \alpha_{\mathcal{K}}(a-h)\ F(a-h)\ \frac{1}{\|\mathcal{F}_\Lambda\|}\ K_{\mathcal{H}}(-h).$$

In other words, the value $\alpha_{\mathcal{K}}(a)\ F(a)$ evaluated by the finite sum on the right side of (19.64)

$$\sum_{\substack{h\in\Lambda^{-1} \\ a-h\in\overline{\mathcal{K}}}} \alpha_{\mathcal{K}}(a-h)\ F(a-h)\ \frac{1}{\|\mathcal{F}_\Lambda\|}\ K_{\mathcal{H}}(-h) \tag{19.64}$$

leads to an error given in the form

$$\int_{\mathbb{R}^q \setminus (\bigcup\limits_{g\in\Lambda} (\mathcal{H}+\{g\}))} F_{\mathcal{K}}^{\wedge}(x)\ e^{2\pi i a\cdot x}\ dV(x). \tag{19.65}$$

Obviously, the error becomes zero, if $\mathcal{H}$ "fills up" the fundamental cell $\mathcal{F}_\Lambda$.

Part X

Conclusion

Chapter 20

Trends, Progress, and Perspectives

In the classical context, the 1D-WKS theorem is directly applicable to time-dependent signals and is normally formulated in that framework.

20.1 Trendsetting Extensions of Shannon Sampling

Already the time-dependent form of Shannon sampling is of great practical importance and the basis of many developments. It has many appealing applications in sciences and engineering (for additional information and references concerning the applicability see, e.g., J.A. Jerri [1976, 1977], F. Stenger [1981], J.R. Higgins [1985], P.L. Butzer et al. [1988], I. Daubechies [2013], A.I. Zayed [1993], J.R. Higgins, R.L. Stens [2000], J.J. Benedetto, P.J.S.G Ferreira [2001], J.J. Benedetto, A.I. Zayed [2003], G.E. Pfander [2015]). Multivariate, in the sense of iterated 1D applications of sampling theory permeate many branches of engineering, such as signal analysis, radar as well as laser technology, and many others. These aspects naturally make sampling theory an offspring of 1D-interpolation but do not reflect specifically multi-dimensional constructive approximation.

The list of extensions and generalizations of the classical Shannon sampling theory given during the last decades is long. Our remarks only intend to present a collection of key aspects: *Sampling* provides the bridge between continuous and discrete versions of a signal. In its customary 1D formulation, known in communication and electrical engineering, the *Shannon sampling theorem* is related to time-dependent signals and 1D-lattices, so that a condition between a bandwidth and sample rate has to be established (see, e.g., E.T. Whittaker [1915], E.T. Whittaker [1923], E.T. Whittaker [1929], H. Nyquist [1928], C.E. Shannon [1949a,b]). In classical Shannon theory it has been recognized that the space of bandlimited functions is the same as the Paley–Wiener space of entire functions whose restriction to the real line is of exponential growth. This space is a reproducing kernel Hilbert space with the sinc-function as the reproducing kernel (see, e.g., M.Z. Nashed [2010]). Astonishingly, the reproducing kernel property did not enter early derivations of sampling approaches to the cardinal series (for various perspectives and surveys on earlier sampling expansions, see, e.g., P.L. Butzer [1983], P.L. Butzer et al. [1988], P.L. Butzer and R.L. Stens [1992], J.R. Higgins [1985, 1996], J.R. Higgins et al. [2000], J.A. Jerri [1969, 1976, 1977], A.I. Zayed [1993], and others). Three decades ago, M.Z. Nashed, G.G. Walter [1991] introduced a new approach to general sampling theorems for functions in reproducing kernel Hilbert spaces. In fact, M.Z. Nashed, G.G. Walter [1991] showed how many of the sampling results in the earlier literature are special cases of their approach. In another paper, M.Z. Nashed, G.G. Walter [1995] showed how to construct a reproducing kernel Hilbert space from a function space that admits sampling expansions. In view of this affinity between reproducing kernel Hilbert spaces and sampling expansions, it is not surprising that in the past 25 years, reproducing kernel Hilbert and Banach spaces have played major roles in signal analysis and applications of inverse problems and imaging (for some perspectives on the role of various function spaces in sampling expansions see M.Z. Nashed [2010], M.Z. Nashed, Q. Sun [2010], M.Z. Nashed, Q. Sun [2013], and also the introductory remarks in W. Freeden, M.Z. Nashed [2015] and W. Freeden et al. [2018]). In fact, it can be stated that seminal ideas such as the use of reproducing kernel Hilbert space framework are often triggered by amazing insights.

20.2 Methodological Progress in Sampling

Many of the basic ideas of sampling are drastically generalized in new directions and applied in diverse fields of sampling, far beyond what anyone could have envisioned in the early days of sampling theory. So, there have been many advances in sampling theory and its applications to signal and image processing in past three decades. Many authors developed sampling theorems based on the theory of regular and singular boundary value problems and also by use of transforms other than the Fourier transform, including such techniques as the Sturm–Liouville, Legendre, Laguerre, Jacobi transform (see, e.g., M.Z. Nashed [2010], M.Z. Nashed, Q. Sun [2010], A.I. Zayed [1993]) originated by Kramer's approach (see H.P. Kramer [1959]). Another field of research has been in non-uniform sampling for non-bandlimited signals. More recently, methods of functional analysis and harmonic analysis have played a pivotal role to provide major advances in sampling theory and its foundational aspects. In particular, new directions have been proposed involving various function spaces that admit sampling expansions such as Sobolev spaces, Wiener amalgam spaces, shift invariant spaces, translation-invariant spaces, convolution sampling, non-linear sampling and other fundamental issues in sampling theory (for various perspectives the reader is referred to, e.g., A. Aldroubi, K. Gröchenig [2001], A. Aldroubi, K. Gröchenig [2001],

A. Aldroubi [2002], A. Aldroubi et al. [2004a], A. Aldroubi et al. [2004b], N.D. Atreas [2011], N.D. Atreas [2012], J. Artonig et al. [2012], C. Bardaro et al. [2007], C. Bardaro et al. [2010], J.J. Benedetto, P.J.S.G Ferreira [2001], J.J. Benedetto, A.I. Zayed [2003], J.J. Benedetto, M.J. Begué [2015], A. Beurling, L. Carleson [1989], A. Bhandari, A. I. Zayed [2017], H. Boche et al. [2015], P.L. Butzer [1983], P.L. Butzer et al. [1988], P.L. Butzer and R.L. Stens [1992], P.L. Butzer et al. [2005], W. Cheney, W.Light [2007], O. Christensen [2003], A. De Maio et al. [2019], P. Devaraj, S. Yugesh [2017], R.J. Duffin, J.J. Eachus [1942], Y.C. Eldar [2015], Y.C. Eldar, G.Kutyniok [2012], H.G. Feichtinger, K. Gröchenig [1994], H.G. Feichtinger et al. [2008], H.R. Fernandez-Morales, A,G. Garcia, Perez-Villalon, G. [2013], S. Foucart, H. Rauhut [2013], K. Gröchenig [1992], K. Gröchenig, J. Stöckler [2013], K Hamm, J. Ledford [2018], G.H. Hardy, J.E. Littlewood, G. Polya [1952], J.R. Higgins [1985], J.R. Higgins [1996], J.R. Higgins et al. [2000], A.E. Ingham [1936], M. Ismail et al. [1995], J.A. Jerri [1977], M.I. Kadec [1964], S.H. Kulkarni et al. [2009], A. Kumar, S. Sampath [2018], H.J. Landau [1967], D. Larson et al. [2008], R.J. Marks II [1991], R.J. Marks II [1993], F.A. Marvasti [2001], D. Middleton [1960], D. Middleton [1965], D.S. Mitrinovic [1970], M.Z. Nashed, X. Li [2018], M.Z. Nashed, G.G. Walter [1991], M.Z. Nashed, O. Scherzer [2002], M.Z. Nashed et al. [2013], A.V. Oppenheim, R.W. Schäfer [1989], R.E.A.C. Paley, N. Wiener [1934], B.S. Pavlov [1979], G.E. Pfander [2013], G.E. Pfander [2015], R. Radha, A. Saswata [2017], T. Schlumprecht, N. Sivakumar [2009], K. Seip [1987], A.A. Selvan, R. Radha [2015], A.A. Selvan, R. Radha [2017], X. Shen, A.I. Zayed [2013], W. Sun, X. Zhou [2002a], W. Sun, X. Zhou [2002b], W. Sun, X. Zhou [2002c], Q. Sun [2006], M. Unser [2000], P.P. Vaidyanathan [2001], M. Vetterli et al. [2002], G. Vinti, L. Zampogni [2009], P. Walk, P. Jung [2020], G.G. Walter, X. Shen [2000], J. Xian, S. Li [2007], A.I. Zayed [1993], A.I. Zayed [1994a], A.I. Zayed [1994b], A.I. Zayed, G. Schmeisser [2014] and the references therein). The papers R.A. Haddad, T.W. Parsons [1991], A.V. Oppenheim, R.W. Schäfer [1989], A. Papoulis [1977] and many references therein deal with foundational sampling aspects, particularly in engineering. Nevertheless, even today, 1D-sampling has a number of problems, for example, bandlimiting operation generates Gibb's oscillations, the sinc-function has a very slow decay at infinity which makes computation in the signal domain rather difficult, aso.

Multi-dimensional sampling can be found in, e.g., A. Aldroubi, K. Gröchenig [2001], A. Aldroubi et al. [2004a], P.L. Butzer et al. [1993], H. Behmard, A. Faridani [2001], D. Costarelli, G. Vinti [2013], F. Cluny et al. [2015], T. Kalker [1999], R.J. Marks II [1991], E. Parzen [1956] and many other contributions. Usually, it is restricted to the fundamental cell of $\mathbb{Z}^q, q \geq 2$, and based on iterated one-dimensional framework.

The close relationship of multivariate Shannon sampling to lattice point identities of analytic and geometric number theory seems to be a new field of interest. W. Freeden, M.Z. Nashed [2015] provides the point of departure for a palette of so-called *Shannon-type sampling formulas* and explicit over- and undersampling characterizations for a diversity of geometries different from the fundamental cell $\mathcal{F}_\Lambda$ of the chosen lattice Λ whose practical usefulness, applicability, economy, and efficiency are evident (for example, in geoexploration discussing Fredholm integral equations of the first kind).

20.3 Bridging Role of Sampling in Recovery Problems

Since the middle of the last century, the rudiments of sampling are covered in almost every engineering textbook on signal analysis, but sampling theory could be found only in rare exceptions in the mathematical literature. In the last three decades, however, with the surge of new techniques in analysis, sampling theory has started to take a new prominent role within the traditional branches of mathematics, thereby leading to discoveries in other areas of mathematics, such as inverse problems, signal processing, and image analysis. Researchers in one of these areas were often unfamiliar with the techniques and relevance of the other areas. In fact, the connections among all the areas were rather tenuous. Today the situation, however, has changed drastically, not least because of the bridging links of the sampling methodology.

Following M.Z. Nashed [2010], our recent understanding of these areas is as follows:

- *Signal Analysis/Processing* (SA) deals with digital representations of signals and their analog reconstructions from digital representations.
- *Image Analysis/Processing* (IA) is concerned with image refinement and recovery, and particularly includes geophysical as well as medical imaging.
- *Inverse Problems* (IP's) deal with determining for a given input-output system an input that produces an observed output, or of determining an input that produces a desired output (or comes as close to it as possible), often in the presence of noise.

The common thread among inverse problems, signal analysis, and imaging is seen in a unifying way as recovering of an object (function, signal, picture) usually from certain known discrete information about the object, what is usually meant of continuous nature. In this respect, sampling methods play pivotal roles in the recovery of an object from discrete information. This is the reason why we are canonically led to conclude that sampling methodologies of providing continuous objects from discrete information takes a significant role as a unifying bridge in the solution processes of recovery problems. Subsequently, the areas of inverse problems, signal processing, and image analysis are in a positive changing process themselves.

Today, the applicability and application of sampling techniques arise in many branches, including computer vision, natural language processing, machine learning, statistics, imaging (such as tomography), remote sensing, non-destructive testing, astronomy, geosystems mathematics, especially involving inverse problems of geosciences and exploration, and many other fields. The overall rationale in most practical methods involving sampling is to construct a "continuous solution" from discrete information, so that it is acceptable physically as a meaningful "recovery approximation".

20.4 SampTA Conference Series

Following almost literally W. Freeden, M.Z. Nashed [2015], we mention that Sampling Theory (in the sense of representation of an analog signal in terms of its samples) has attracted considerable interest in the past six decades. Major advances in the theory and applications of sampling expansion in different function spaces have been made by mathematicians and engineers. There are several journals on signal processing and computational harmonic analysis that publish papers on sampling theory and its applications. The field also sports its own journal *Sampling Theory in Signal and Image Processing* and has its own "professional society".

We quote from Wikipedia:

"SampTA (Sampling Theory and Applications) is a biennial interdisciplinary conference for mathematicians, engineers, and applied scientists. The main purpose of SampTA is to exchange recent advances in sampling theory and to explore new trends and directions in related areas of applications. The SampTA conference series began as a small workshop in 1995 in Riga, Latvia, but the meetings grew into full-fledged conferences attracting an even mix of mathematicians and engineers as the interest in sampling theory blossomed."

Bibliography

Adhikari, S.D., Petermann, Y.–F.S., Lattice Points in Ellipsoids. *Acta Arith.*, (59):329-338, 1991.

Ait–Haddou, R., Biard, L., Slawinski, M.A., Minkowski Isoperimetric–Holograph Curves. *Comput. Aided Geomatric Design*, (17):835-861, 2000.

Aldroubi A., Non-uniform Weighted Average Sampling and Reconstruction in Shift-Invariant and Wavelet Spaces. *Appl. Comput. Harmon. Anal.*, (13):151-161, 2002.

Aldroubi A., Gröchenig, K., Beurling-Landau-Type Theorems for Nonuniform Sampling in Shift-Invariant Spline Spaces *J. Fourier Anal. Appl.*, (6):93-103, 2000.

Aldroubi A., Gröchenig, K., Non-uniform Sampling and Reconstruction in Shift Invariant Spaces. *SIAM Rev.*, (43):585-620, 2001.

Aldroubi A., Sun, Q., Tang, W.-S., Non-uniform Average Sampling and Reconstruction in Multiply Generated Shift-Invariant Spaces. *Constr. Approx.*, (20):173-189, 2004a.

Aldroubi A., Sun, Q., Tang, W.-S., Convolution, Average Sampling and a Calderon Resolution of the Identity for Shift-Invariant Spaces. *J. Fourier Anal. Appl.*, (22): 215-244, 2004b.

Aldroubi A., Unser, M., Sampling Procedure in Function Spaces and Asymptotic Equivalence with Shannon's Sampling Theory, *Numer. Funct. Anal. Optimization*, (15):1-21, 1994.

Apostol, T.M., An Elementary View of Euler's Summation Formula. *The American Mathematical Monthly*, (106):409-418, 1999.

Aronszajn, N., Theory of Reproducing Kernels. *Trans. Am. Math. Soc.*, (68):337-404, 1950.

Aronszajn, N., Creese T.M., Lipkin L.J., Polyharmonic Functions. Clarendon Press, Oxford, 1983.

Artonig, J., Eldar, Y.C., Draggtti, l., Ben-Haim, H., In: Compressed Sensing: Theory and Applications. J., Eldar, G. Kutyniok (Eds.). Cambridge University Press, Cambridge, 148-209, 2012.

Atreas, N.D., Perturbed Sampling Formulas and Local Reconstruction in Shift-Invariant Spaces. *J. Math. Anal. Appl.*, (377):841-852, 2011.

Atreas, N.D., On a Class of Non-uniform Average Sampling Expansions and Partial Reconstruction in Subspaces of $L^2(\mathbb{R})$. *Adv. Comput. Math.*, (36):21-38, 2012.

Bardaro, C., Butzer, P.L., Stens, R.L., Vinti, G., Kantorovich-Type Generalized Sampling Series in the Setting of Orlicz Spaces. *Sampling Theory in Signal and Image Processing*, (6):29-52, 2007.

Bardaro, C., Butzer, P.L., Stens, R.L., Vinti, G., Prediction by Samples from the Past with Error Estimates Covering Discontinuous Signals. *IEEE Trans. Inform. Theory*, (56):614-633, 2010.

Behmard, H., Faridani, A., Sampling of Bandlimited Functions on Unions of Shifted Lattices. *J. Fourier Anal. Appl.*, (8):43-58, 2001.

Benedetto, J.J., Begué, M.J., Fourier Operators in Applied Harmonic Analysis. In: Sampling Theory, a Renaissance, G.E. Pfander (Ed.), Birkhäuser, Springer International Publishing AG Switzerland, 185-216, 2015.

Benedetto, J.J., Ferreira P.J.S.G. (Eds.), Modern Sampling Theory: Mathematics and Applications. Birkhäuser, Boston, 2001.

Benedetto, J.J., Zayed, A.I. (Eds.), Sampling, Wavelets, and Tomography. Birkhäuser, Boston, 2003.

Beurling, A. and Carleson, L., The Collected Works of Arne Beurling: Complex Analysis, Vol. 2, Boston (MA), Birkhäuser, 1989.

Bhandari, A. and Zayed, A.I., Shift-Invariant and Sampling Spaces Associated with the Special Affine Fourier Transform. *Appl. Comput. Harmon. Anal.*, DOI: 10.1016/j.acha.2017.07.002, 2017.

Boche, H., Calderbank, R., Kutyniok, G., Vybiral, J. (Eds.), Compressed Sensing and Its Applications. Birkhäuser, Boston, 2015.

de Boor, C., Höllig, K., Riemenschneider, S., Bivariate Cardinal Interpolation by Splines on a Three-Direction Mesh. *Ill. J. Math.*, (29):533-566, 1985.

Butzer, P.L., A Survey of the Whittaker Shannon Sampling Theorem and some of its Extensions. *J. Math. Res. Exposition,* (3):185-212, 1983.

Butzer, P. L., Engels, W., Ries, S., Stens, R.L., The Shannon Sampling Series and the Reconstruction of Signals in Terms of Linear, Quadratic and Cubic Splines, *SIAM J. Appl. Math.*, (46):299-323, 1986.

Butzer, P. L., Splettstößer, W., Stens, R.L., The Sampling Theorem and Linear Prediction in Signal Analysis. *Jber. Deutsch. Math. Vereinigung (DMV)* (90):1-60, 1988.

Butzer, P. L., Stens, R.L., Sampling Theory for Not-Necessarily Bandlimited Functions: a Historical Overview. *SIAM Rev.*, (34):40-53, 1992.

Butzer, P.L., Fischer, A., Stens, R.L., Generalized Sampling Approximation of Multivariate Signals: General Theory. *Atti Sem. Mat. Fis. Univ. Modena*, (41):17-37, 1993.

Butzer, P.L., Higgins, J.R., Stens, R.L., Classical and Approximate Sampling Theorems: Studies in the $L^p(\mathbb{R})$ and the Uniform Norm. *J. Approx. Theory*, (137):250-263, 2005.

Butzer, P.L. Schmeiser, G., Stens, R.L., The Classical and Approximate Sampling Theorems and Their Equivalence for Entire Functions of Exponential Type. *J. Approx. Theory*, (179):94-111, 2014.

Cassels, J.W.S., An Introduction to the Geometry of Numbers. Springer, Berlin, Heidelberg, New York, 1968.

Cauchy, A. L., Mémoire sur diverses formules d'analyze. *Comptes Rendus des Séances de l'Académie des Sciences*, (12):283-298, 1841.

Chamizo, F., Iwaniec, H., On the Sphere Problem. *Rev. Mat. Iberoamericana*, (11):417-429, 1995.

Chen, J.-R., The Lattice Points in a Circle. *Sci. Sinica*, (12):633-649, 1963.

Cheney, W., Light, W., Course in Approximation Theory. American Mathematical Society, Providence, 2007.

Chui, C.K., Jetter, K., Ward, J.D., Cardinal Interpolation by Multivariate Splines. *Mathematics of Computation*, (48):711-724, 1987.

Courant, R., Hilbert, D., Methoden der Mathematischen Physik I,II. Springer, Berlin, 1924.

Cluny, F., Costarelli, D., Vinti, G., Approximation by Nonlinear Multivariate Sampling Kontorovich Type Operators and Applications to Image Processing. *Numer. Funct. Anal. Optimization*, (34):819-844, 2013.

Cluny, F,. Costarelli, D., Minotti, A, Vinti, G., Enhancement of Thermographic Images as Tool for Structural Analysis in Earthquake Engineering. *NDT & E International*, (70):60-72, 2015.

Christensen, O., An Introduction to Frames and Riesz Bases. Vol. 7, Birkhäuser, 2003.

Daubechies, I., Time-Frequency Localization Operators: A Geometric Phase Space Approach. *IEEE Transactions on Informat. Theory*, (34):961-1005, 1988.

Davis, P.J., Interpolation and Approximation. Blaisdell Publishing Company, Waltham, MA, 1963.

De Maio, A., Eldar, Y. C., Haimovich, A.M. (Eds.), Compressed Sensing in Radar Signal Processing. Cambridge University Press, Cambridge, 2019.

Devaraj, P., Yugesh, S., A Local Weighted Average Sampling and Reconstruction Theorem over Shift-Invariant Subspaces. *Result. Math.*, (71):319-332, 2017.

Devaraj, P., Massopust, P., Yugesh, S., Local Average Sampling and Reconstruction with Fundamental Splines of Fractional Order. *Numer. Funct. Anal. Optimization*, accepted for publication, 2019.

Duffin, R.J., Eachus, J.J., Some Notes on an Expansion Theorem of Paley and Wiener. *Bull. Amer. Math. Soc.*, (48):850-855, 1942.

Eldar, Y.C., Sampling Theory Beyond Bandlimited Systems. Cambridge University Press, Cambridge, 2015.

Eldar, Y.C., Kutynick, G., (Eds.), Compressed Sensing: Theory and Applications. Cambridge University Press, Cambridge, 2012.

Erdös, P., Gruber, M., Hammer, J., Lattice Points, Pitman Monographs and Surveys in Pure and Applied Mathematics. 39, Longman Scientific-Technical, John Wiley Inc., New York, 1989.

Euler, L., Methodus universalis serierum convergentium summas quam proxime inveniendi, Commentarii academie scientiarum Petropolitanae. (8):3-9, *Opera Omnia* (XIV): 101-107, 1736.

Feichtinger H.G., Gröchenig, Theory and Practice of Irregular Sampling. J.J. Benedetto, W. Frazier (Eds.), Wavelets: Mathematics and Applications, *Stud. Adv. Math.*, 305-363. CRC, Boca Raton, FL, 1994.

Feichtinger H.G., Molter, U., Romero,J.L., Perturbation Techniques in Irregular Spline-Type Spaces. *Int. J. Wavelets Multiresolut. Inf. Process.*, (6):249-277, 2008.

Fernandez-Morales H.R., Garcia, A.G., Perez-Villalon G., Generalized Sampling of Multivariate Signals on Euclidean Spaces in Shift-Invariant Subspaces with Multiple Stable Generators. In: Multiscale Signal Analysis and Modeling, X. Shen, A. I. Zayed (Eds.), Springer, 51-80, 2013.

Fomenko, O.M., On the Problem of Gauss. *Acta. Arith.*, (6):277-284, 1961.

Foucart, S., Rauchut, H., A Mathematical Introduction to Compressive Sensing, Birkhäuser, Basel, 2013.

Freeden, W., Eine Verallgemeinerung der Hardy–Landauschen Identität. Ph.D.-Thesis, RWTH Aachen, 1975.

Freeden, W., Über eine Verallgemeinerung der Hardy–Landauschen Identität. *Manuscr. Math.*, (24):205-216, 1978a.

Freeden, W., Über gewichtete Gitterpunktsummen in kreisförmigen Bereichen, *Mitt. Math. Seminar Giessen*, (132):1-22, 1978b.

Freeden, W., Über eine Klasse von Integralformeln der Mathematischen Geodäsie. *Veröff. Geod. Inst. RWTH Aachen*, Report No. 27, 1979.

Freeden, W., On Integral Formulas of the (Unit) Sphere and Their Application to Numerical Computation of Integrals. *Computing*, (25):131-146, 1980.

Freeden, W., On Spherical Spline Interpolation and Approximation. *Math. Meth. Appl. Sci.*, (3):551-575, 1981.

Freeden, W., Multidimensional Euler Summation Formulas and Numerical Cubature. *ISNM*, (57):77-88, 1982.

Freeden, W., Interpolation by Multidimensional Periodic Splines. *J. Approx. Theory*, (55):104-117, 1988.

Freeden, W., Metaharmonic Lattice Point Theory. CRC Press, Chapman & Hall, Boca Raton, 2011.

Freeden, W., Fleck,. J., Numerical Integration by Means of Adapted Euler Summation Formulas. *Numer. Math.*, (51):37-64, 1987.

Freeden, W., Gerhards, C.: Geomathematically Oriented Potential Theory. Chapman and Hall, CRC Press, Boca Raton, London, New York, 2013.

Freeden, W., Gerhards, C., Romberg Extrapolation for Euler Summation - Based Cubature on Regular Regions. *GEM Int. J. Geomath.*, (8):169-182, 2017.

Freeden, W., Gutting, M., On the Completeness and Closure of Vector and Tensor Spherical Harmonics. *Integral Transforms and Special Functions*, (19):713-734, 2008.

Freeden, W., Gutting, M., Special Functions of Mathematical (Geo-)Physics. Birkhäuser, Basel, 2013.

Freeden, W., Gutting, M., Integration and Cubature Methods. CRC Press, Taylor & Francis Group, Boca Raton, 2017.

Freeden, W., Nashed, M.Z., Multivariate Hardy-Type Lattice Point Summation and Shannon-Type Sampling. *GEM Int. J. Geomath.*, (6):163-249, 2015.

Freeden, W., Nashed, M.Z., From Gaussian Circle Problen to Multivariate Shannon Sampling. *Contemporary Mathematics and Its Applications*, Frontiers in Orthogonal Polynomials and q-Series, M.Z. Nashed, X. Li (Eds.), World Scientific, 213-238, 2017a.

Freeden, W., Nashed, M.Z., Operator-Theoretic and Regularization Approaches to Ill-Posed Problems. *GEM Int. J. Geomath.*, (9):1-115, 2018a.

Freeden, W., Nashed, M.Z., Inverse Gravimetry: Background Material and Multiscale Mollifier Approaches. *GEM Int. J. Geomath.*, (9):199-264, 2018b.

Freeden, W., Nashed, M.Z, Schreiner, M., Spherical Sampling. *Geosystems Mathematics*, Birkhäuser, Springer International Publishing, Basel, 2018.

Freeden, W., Schreiner, M., Spherical Functions of Mathematical Geosciences: A Scalar, Vectorial, and Tensorial Setup. Springer, Berlin, Heidelberg, 2009.

Freeden, W., Gervens, T., Schreiner, M., Constructive Approximation on the Sphere (With Applications to Geomathematics). Oxford Science Publications, Clarendon Press, Oxford, 1998.

Freeden, W., Sonar, T., Witte, B., Gauss as Scientific Mediator between Mathematics and Geodesy from the Past to the Present. Handbook of Mathematical Geodesy, W. Freeden, M.Z. Nashed (Eds.), *Geosystems Mathematics*, Springer International Publishing, 1-163, 2018.

Freeden, W., Witte, B., A Combined (Spline-)Interpolation and Smoothing Method for the Determination of the Gravitational Potential from Heterogeneous Data. *Bull. Géod.*, (56):53-62, (1982).

Fricker, F., Einführung in die Gitterpunktlehre. Birkhäuser, Basel, 1982.

Funk, H., Beiträge zur Theorie der Kugelfunktionen. *Math. Ann.*, (77):136-152, 1916.

Gauss, C.F., Disquisitions arithmetica. Gerhard Fleischer, Lipsiae (Leipzig), 1801.

Gauss, C.F., De nexu inter multitudinem classicum, in quas formae binariae secondi gradus distribuuntur, earumque determinantem. *Werke*, (2):269-291, 1826.

Gröchenig, K., Reconstruction Algorithms in Irregular Sampling. *Math. Comp.*, (59):181-194, 1992.

Gröchenig, K., Stöckler, J., Gabor Frames and Totally Positive Functions. *Duke Math. J.*, (162):1003-1031, 2013.

Grosswald, E., Representations of Integers as Sums of Squares. Springer–Verlag, New York, 1985.

Haddad R.A., Parsons, T.W., Digital Signal Processing: Theory, Applications and Hardware. Computer Science Press, 1991.

Hamm, K., Ledford, J., On the Structure and Interpolation Properties of Quasi Shift-Invariant Spaces. *J. Funct. Anal.*, (274):1959-1992, 2018.

Hardy, G.H., On the Expression of a Number as the Sum of Two Squares. *Quat. Journ. Math. (Oxford)*, (46):263-283, 1915.

Hardy, G.H., Notes on Special Systems of Orthogonal Functions, IV: The Whittaker's Cardinal Series. *Proc. Cambr. Phil. Soc.*, (37):331-348, 1941.

Hardy, G.H., Landau, E., The Lattice Points of a Circle. *Proc. Royal Society, A*, (105):244-258, 1924.

Hardy, G.H., Littlewood, J.E., Polya, G., Inequalities. Cambridge Mathematical Library. Cambridge University Press, Cambridge, 1988. Reprint of the 1952 edition.

Hardy, G.H., Wright, E.M., Einführung in die Zahlentheorie. R. Oldenbourg, München, 1958.

Heath-Brown, D. R., Lattice Points in the Sphere. In: Proc. Number Theory Conf. Zakopane, K. Györy (Ed.), Vol. 2, 883-892, 1999.

Higgins, J.R., Five Short Stories about the Cardinal Series. *Bull. Am. Math. Soc.*, (12):54-89, 1985.

Higgins, J.R., Sampling Theory in Fourier and Signal Analysis Volume 1: Foundations. Oxford University Press, Oxford, 1996.

Higgins, J.R., Schmeisser, G., Voss, J.J., The Sampling Theorem and Several Equivalent Results in Analysis. *J. Comput. Anal. Appl.*, (2):333-371, 2000.

Higgins, J.R., Stens, R.L., Sampling Theory in Fourier and Signal Analysis: Vol. 2: Advanced Topics. Oxford Science Publications, Oxford, 2000.

Hilbert, D., Grundzüge einer allgemeinen Theorie der linearen Integralgleichungen. Teubner, Leipzig, 1912.

Hlawka, E., Zur Geometrie der Zahlen. *Math. Z.*, (49):285-312, 1943.

Hlawka, E., Grundbegriffe der Geometrie der Zahlen. *Jber. Deutsch. Mathem. Vereinigung*, (57):37-55, 1954.

Hlawka, E., Schoißengeier, J., Taschner, R., Geometric and Analytic Number Theory. Springer, Berlin, Heidelberg, New York, 1991.

Hua, Loo–Keng, Die Abschätzung von Expotentialsummen und ihre Anwendung in der Zahlentheorie. *Enzyklopädie der Mathematischen Wissenschaften*, Bd. 12, Heft 13, Teil 1, 105-107, 1959.

Huxley, M.N., Exponential Sums and Lattice Points III. *Proc. London Math. Soc.*, (87):591-609, 2003.

Ingham, A.E., Some Trigonometrical Inequalities with Applications to the Theory of Series. *Math. Z.*, (41):367-379, 1936.

Ismail, M., Nashed, M.Z., Zayed, A., Ghaleb, A., Mathematical Analysis, Wavelets and Signal Processing. *Contemporary Mathematics*, Volume 190, American Mathematical Society. Providence, RI, 1995.

Ivanow, V.K., A Generalization of the Voronoi–Hardy Identity. *Sibirsk. Math. Z.*, (3):195-212, 1962.

Ivanow, V.K., Higher-Dimensional Generalization of the Euler Summation Formula (Russian). *Izv. Vyss. Ucebn. Zaved. Mathematika*, 6 (37):72-80, 1963.

Ivic, A., Krätzel, E., Kühleitner, Nowak, W.G., Lattice Points in Large Regions and Related Arithmetic Functions: Recent Developments in a Very Classical Topic. arXiv:0410522v1, 1-39, 2004.

Jerri, J.A., On the Application of Some Interpolating Functions in Physics. *J. Res. Nat. Bur. Standards, Sec. B*, (73B):241-245, 1969.

Jerri, J.A., Sampling Expansion for Laguerre L^2-Transforms. *J. Res. Nat. Bur. Standards, Sect. B*, (80):415-418, 1976.

Jerri, J.A., The Shannon Sampling Theorem - Its Various Extensions and Applications: A Tutorial Review. *Proc IEEE*, (65):1565-1596, 1977.

Kadec M.I,. The Exact Value of the Paley–Wiener Constant. *Sov. Math., Dokl.*, (5):559-561, 1964.

Kalker, T., On Multi-Dimensional Sampling. In: Digital Signal Processing Handbook, Madisetti, V.K., Williams D.B. (Eds.), CRC Press, Taylor & Francis Group, Boca Raton, 1999.

Kellogg, O.D., Foundations of Potential Theory. Frederick Ungar Publishing Company, New York, 1929.

Kolesnik, G., On the Method of Exponent Pairs. *Acta Arith.*, (45):115-143, 1985.

Kotel'nikov, V.A., On the Carrying Capacity of the Ether and Wire in Telecommunications. Material for the First All-Union Conference on Questions of Communication. *Izd. Red. Upr. Svyazi RKKA, Moscow*, 1933 (in Russian).

Kramer, H.P., A Generalized Sampling Theorem. *J. Mat. Phys.*, (38):68-72, 1959.

Krätzel, E., Lattice Points. Kluwer Dordrecht, Boston, London, 1988.

Krätzel, E., Analytische Funktionen in der Zahlentheorie. Teubner Stuttgart, Leipzig, Wiesbaden, 2000.

Kulkarni S.H., Radha, R., Sivananthan S. Nonuniform Sampling Problem. *J. Appl. Funct. Anal.*, (4):58-74, 2009.

Kumar, A., Sampath S., Average Sampling and Reconstruction in Shift-Invariant Spaces and Variable Bandwidth Spaces. *Appl. Anal.*, DOI: 10.1080/00036811.2018.1508652, 2018.

Landau, E., Über die Gitterpunkte in einem Kreis I. *Gött. Nachr.*, 148-160, 1915.

Landau, E., Über die Gitterpunkte in einem Kreis IV. *Gött. Nachr.*, 58-65, 1924.

Landau, E., Vorlesungen über Zahlentheorie, Chelsea Publishing Compagny, New York, 1969 (*Reprint from the Orignal Version Published by S. Hirzel, Leipzig*, 1927).

Landau, H.J., Necessary Density Conditions for Sampling and Interpolation of Certain Entire Functions. *Acta Math.*, (117):37-52, 1967.

Larson, D., Massopust, P., Nashed, M.Z., Nguyen, M.C., Papadakis, M., Zayed, A. (Eds.), Frames and Operator Theory in Analysis and Signal Proccessing, *Contempory Mathematics*, Vol. 451, American Mathematical Society, Providence, RI, 2008.

Lüke, H.D., The Origins of the Sampling Theorem. *IEEE Communications Magazine*, (37):106-108, 1999.

Magnus, W., Fragen der Eindeutigkeit und des Verhaltens im Unendlichen für Lösungen von $\Delta U + k^2 U = 0$. *Abh. Math. Sem. Univ. Hamburg*, (16):77-88, 1949.

Magnus, W., Oberhettinger, and F., Soni, R.P., Formulas and Theorems for the Special Functions of Mathematical Physics. In: Die Grundlehren der mathematischen Wissenschaften in Einzeldarstellungen, Bd. 52, Springer, Berlin, 1966.

Marks II, R.J., Introduction to Shannon Sampling and Interpolation Theory. Springer, Berlin, 1991.

Marks II, R.J., Advanced Topics in Shannon Sampling and Interpolation Theory. Springer, Berlin, Heidelberg, 1993.

Marvasti, F.A. (Ed.), Nonuniform Sampling: Theory and Practice. Information Technology Transmission, Processing, and Storage, Plenum Publ. Corp., New York, 2001.

Michel, V., Lectures on Constructive Approximation. Birkhäuser, Basel, 2013.

Michlin, S.G., Mathematical Physics, an Advanced Course. North Holland, Amsterdam, London, 1970.

Middleton, D., Introduction to Statistical Communication Theory. IEEE Press, 1996, originally published by McGraw-Hill, New York, 1960.

Middleton, D., Topics in Communication Theory. McGraw-Hill, New York, 1965.

Minkowski, H., Geometrie der Zahlen. Teubner, Leipzig, 1896.

Mitrinovic D.S. Analytic Inequalities. Springer-Verlag, New York-Berlin, 1970. In cooperation with P. M. Vasic. Die Grundlehren der Mathematischen Wissenschaften, Bd. 165, 1970.

Mordell, L.J., Poisson's Summation Formula in Several Variables and Some Applications to the Theory of Numbers. *Cambr. Phil. Soc.*, (25):412-420, 1928.

Mordell, L.J., Some Applications of Fourier Series in the Analytic Theory of Numbers. *Cambr. Phil. Soc.*, (24):585-595, 1928.

Mordell, L.J., Poisson's Summation Formula and the Riemann Zeta Function. *J. London Math. Soc.*, (4):285-296, 1929.

Mordell, L.J., On Some Arithmetical Results in the Geometry of Numbers. *Compositio Math.*, (1):248-253, 1935.

Mordell, L.J., Diophantine Equations. Academic Press, 1969.

Müller, C., Über die ganzen Lösungen der Wellengleichung (nach einem Vortrag von G. Herglotz). *Math. Ann.*, (124):235–264, 1952.

Müller, C., Eine Verallgemeinerung der Eulerschen Summenformel und ihre Anwendung auf Fragen der analystischen Zahlentheorie. *Abh. Math. Sem. Univ. Hamburg*, (19):41-61, 1954a.

Müller, C., Eine Formel der analytischen Zahlentheorie. *Abh. Math. Sem. Univ. Hamburg*, (19):62-65, 1954b.

Müller, C., Eine Erweiterung der Hardyschen Identität. *Abh. Math. Sem. Univ. Hamburg*, (19):66-76, 1954c.

Müller, C., Die Grundprobleme der Geometrie der Zahlen. *Mathem.-Phys. Semesterberichte*, (5):63-70, 1956.

Müller, C., Spherical Harmonics. Lecture Notes in Mathematics, 17, Springer, Berlin, 1966.

Müller, C., Analysis of Spherical Symmetries in Euclidean Spaces. Springer, New York, Berlin, Heidelberg, 1998.

Müller, C., Dressler, A., Über eine gewichtete Mittelung der Gitterpunkte in der Ebene, *J. Reine Angew. Mathematik*, (252):82-87, 1972.

Nashed, M.Z., Operator-Theoretic and Computational Approaches to Ill-Posed Problems with Applications to Antenna Theory. *IEEE Trans. Antennas Propagation*, (29):220-231, 1981.

Nashed, M.Z., Inverse Problems, Moment Problems and Signal Processing: Un Menage a Trois, Mathematics in Science and Technology. Siddiqi, A.H., Singh, R.C. Manchanda, P. (Eds.), World Scientific, New Jersey, 1-19, 2010.

Nashed, M.Z., Scherzer, O., Inverse Problems, Image Analysis and Medical Imaging. *Contemporary Mathematics*, Volume 313, American Mathematical Society, Providence, R.I., 2002.

Nashed, M.Z., Sun, Q., Sampling and Reconstruction of Signals in a Reproducing Kernel Subspace of $L^p(\mathbb{R}^d)$. *J. Function. Anal.*, (258):2422-2452, 2010.

Nashed, M.Z., Sun, Q., Function Spaces for Sampling Expansions. In: *Multiscale Signal Analysis and Modeling*, X. Shen, A.I. Zayed (Eds.), 81-104, 2013.

Nashed, M.Z., Sun, Q., Xian, J., Convolution Sampling and Reconstruction of Signals in a Reproducing Kernel Subspace. *Proc. Amer. Math. Soc.*, (141):1995-2007, 2013.

Nashed, M.Z., Walter, G.G., General Sampling Theorems for Functions in Reproducing Kernel Hilbert Spaces. *Math. Contr. Signals Syst.*, (4):363-390, 1991.

Nashed, M.Z., Walter, G.G., Reproducing Kernel Hilbert Space from Sampling Expansions. *Contemporary Math.*, (190):221-226, 1995.

Nashed, M. Z. and Li, X. (Eds.), Frontiers in Orthogonal Polynomials and q-Series. World Scientific, Singapore, 2018.

Natanson, P.I., Theorie der Funktionen einer reellen Veränderlichen, Akademie-Verlag, Berlin, 1961.

Nyquist, H., Certain Factors Affecting Telegraph Speed. *Bell Syst. Tech. J.*, (3):324-346, 1924.

Nyquist, H., Certain Topics in Telegraph Transmission Theory. *Trans. AIEE.*, (47):617-644, 1928.

Ogura, K., On a Certain Transcendental Integral Function in the Theory of Interpolation. *Tôhoku. Math. J.*, (17):64-72, 1920.

Oppenheim, A.V., Schäfer, R.W., Digital Signal Processing. Prentice-Hall, New Jersey, 1989.

Paley, R.E.A.C., Wiener, N. Fourier Transforms in the Complex Domain, Volume 19. Providence(RI): *Amer. Math. Soc. Colloq. Publ.*, 1934.

Parzen E., A Simple Proof and Some Extensions of the Sampling Theorem. Technical Report No. 7, Department of Statistics, Stanford University, 1-10, 1956.

Papoulis A., Signal Analysis. McGraw-Hill, New York, 1971.

Papoulis A., Generalized Sampling Expansion. *IEEE Trans. Circ. Syst.*, (24):652-654, 1977.

Pavlov, B.S., Basicity of an Exponential System and Muckenhoupts Condition. Dokl. Akad. Nauk SSSR, (247):37-40, 1979.

Pfander, G.E., Sampling of Operators, *J. Fourier Analysis and Applications*, 19 (3):612-650, 2013.

Pfander, G.E. (Ed.), Sampling Theory, a Renaissance-Compressed Sensing and Other Developments, Birkhäuser, Boston, 2015.

Pinsky, M.A., Introduction to Fourier Analysis and Wavelets. Brooks/Cole, Thomson Learning, Pacific Grove, CA., 2002.

Raabe, H., Untersuchungen an der wechselseitigen Mehrfachübertragung (Multiplexübertragung). *Elektrische Nachrichtentechnik*, (16):213–228, 1939.

Radha, R., Saswata, A., A Sampling Theorem for the Twisted Shift-Invariant Space. *Adv. Pure Appl. Math.*, (8):293-305, 2017.

Schlumprecht T., Siakumar, N., On the Sampling and Recovery of Band-Limited Functions via Scattered Translates of the Gaussian. *J. Approx. Theory*, (159):128-153, 2009.

Seip, K., An Irregular Sampling Theorem for Functions Band-Limited in a Generalized Sense. *SIAM J. Appl. Math.*, (47):1112-1116, 1987.

Selvan, A.A., Radhan, R., Sampling and Reconstruction in Shift-Invariant Spaces on $\mathbb{R}^d$. *Ann. Math. Pure Appl.*, (194):1683-1706, 2015.

Selvan, A.A., Radhan, R., An Optimal Result for Sampling Sensity in Shift-Invariant Spaces Generated by Meyer Scaling Function. *J. Math. Anal. Appl.*, (451):197-208, 2017.

Shannon, C.E., Communication in the Presence of Noise. *Proc. Institute of Radio Engineers*, (37):10-21, 1949a.

Shannon, C.E., The Mathematical Theory of Communication. University of Illinois Press, Urbana, Ill., 1949b.

Shen, X., Zayed, A.I., (Eds.), Multiscale Signal Analysis and Modeling. Springer, 2013.

Sierpit'nski, W., O pewnem zagadnieniu z rachunku funckcyj asmptotycznych (Über ein Problem des Kalküls der asymptotischen Funktion (polnisch)). *Prace Math.-Fiz.*, (17):77-118, 1906.

Stein, E.M., Weiss, G., Introduction to Fourier Analysis on Euclidean Spaces, Princeton University Press, Princeton, NJ, 1971.

Stenger, F., Numerical Methods Based on Whittaker Cardinal or sinc Functions. *SIAM Rev.*, (23):165-224, 1981.

Sun, Q., Non-Uniform Sampling and Reconstruction for Signals with Finite Rate of Innovations. *SIAM J. Math. Anal.*, (38):1389-1422, 2006.

Sun, W., Zhou, X., Average Sampling in Spline Subspaces. *Appl. Math. Letters*, (15):233-237, 2002a.

Sun, W., Zhou, X., Reconstruction of Band-Limited Functions from Local Averages. *Constr. Approx.*, (18):205-222, 2002b.

Sun, W., Zhou, X., Reconstruction of Band-Limited Signals from Local Averages. *IEEE Trans. Inform. Theory*, (48):2955-2963, 2002c.

Theis, M., Über eine Interpolationsformel von de la Vallée Poussin. *Math. Z.*, (3):93-113, 1919.

Unser, M., Sampling–50 Years after Shannon. *Proc. IEEE*, (88):569-587, 2000.

Vaidyanathan, P.P., Generalizations of the Sampling Theorem. *IEEE Trans. Circuits and Systems*, (48):1094-1109, 2001.

de la Vallée Poussin, Ch.–J., Sur la convergence des formules d'interpolation entre ordonnées équidistantes. *Bull. Cl. Ski. Akad. Roy. Belg.*, (4):319-410, 1908.

Vetterli, M., Marziliano, P. Blu, T., Sampling Signals with Finite Rate of Innovation. *IEEE Trans. Signal Process.*, (50):1417-1428, 2002.

Vinti, G., Zampogni, L., Approximation by Means of Nonlinear Kantorovich Sampling Type Operators in Orlicz Spaces. *J. Approx. Theory*, (161):511-528, 2009.

Voronoi, G., Sur un probleme du calcul des fonctions asymptotiques, *JRAM* (126):241-282, 1903.

Walfisz, A., Über Gitterpunkte in vierdimensionalen Ellipsoiden, *Math. Z.*, (72):259-278, 1960.

Walk, P., Jung, P., Compressed Sensing and Applications to Communication and Digital Signal Processing. Springer, Heidelberg, 2020 (to be published).

Walter, G.G., Shen, X., Wavelets and Other Orthogonal Systems. 2nd ed., CRC Press, Taylor & Francis Group, Boca Raton, 2000.

Watson, G.N., A Treatise on the Theory of Bessel Functions, 2nd ed., Cambridge University Press, Cambridge, 1944.

Weyl, H., Über die Gleichverteilung von Zahlen mod. Eins. *Math. Ann.*, (77):313-352, 1916.

Winogradow, I.M., Verbesserungen der asymptotischen Formeln für die Anzahl der ganzzahligen Punkte in dreidimensionalen Bereichen (Russisch). *Izvestija Akad. Nauk SSSR, Ser. Mat.*, (19):3-9, 1955.

Whittaker, E.T., On the Functions Which are Represented by the Expansions of the Interpolation Theory. *Proc. Royal Soc. Edinburgh, Sec. A*, (35):181-194, 1915.

Whittaker, E.T., On a New Method of Graduation. *Proc. Edin. Math. Soc.*, (41):63-75, 1923.

Whittaker, E.T., The Fourier Theory of Cardinal Functions. *Proc. Math. Cox. Edinburgh*, (1):169-176, 1929.

Xian, J., Li, S., Sampling Set Conditions in Weighted Multiply Generated Shift-Invariant Spaces and their Applications. *Appl. Comput. Harmon. Anal.*, (23):171-180, 2007.

Zayed, A. I., Advances in Shannon's Sampling Theory. CRC Press, Boca, Raton, London, New York, 1993.

Zayed, A. I., Sampling Theorem for Functions Band-Limited to a Disc. *Complex Var. Elliptic Equ.*, (26):245-254, 1994.

Zayed, A. I., A Sampling Theorem for Signals Band-Limited to a General Domain in Several Dimensions. *J. Math. Anal. Appl.*, (187):196-211, 1994.

Zayed, A. I., Schmeisser G. (Eds.), New Perspectives on Approximation and Sampling. Applied and Harmonic Analysis, Birkhäuser, Basel 2014.

Index